Feature Papers in Bone Biomaterials

Feature Papers in Bone Biomaterials

Guest Editors

Feng Chen
Zifei Zhou

MDPI

Basel • Beijing • Wuhan • Barcelona • Belgrade • Novi Sad • Cluj • Manchester

Guest Editors
Feng Chen
Shanghai Stomatological Hospital & School of Stomatology
Fudan University
Shanghai
China

Zifei Zhou
Shanghai Tenth People's Hospital
School of Medicine
Tongji University
Shanghai
China

Editorial Office
MDPI AG
Grosspeteranlage 5
4052 Basel, Switzerland

This is a reprint of the Special Issue, published open access by the journal *Journal of Functional Biomaterials* (ISSN 2079-4983), freely accessible at: https://www.mdpi.com/journal/jfb/special_issues/FP_bone_biomater.

For citation purposes, cite each article independently as indicated on the article page online and as indicated below:

Lastname, A.A.; Lastname, B.B. Article Title. *Journal Name* **Year**, *Volume Number*, Page Range.

ISBN 978-3-7258-5793-7 (Hbk)
ISBN 978-3-7258-5794-4 (PDF)
https://doi.org/10.3390/books978-3-7258-5794-4

Contents

About the Editors

Feng Chen

Feng Chen is a professor and principal investigator (PI) of Shanghai Stomatological Hospital & School of Stomatology, Fudan University. He obtained his Ph.D. from Shanghai Jiao Tong University in 2014. Dr. Chen started his academic career at the Shanghai Institute of Ceramics, Chinese Academy of Science in 2008, and moved to Shanghai Tenth People's Hospital in 2018. In 2024, Dr. Chen joined the Shanghai Stomatology Hospital affiliated with Fudan University and served as the director of the central laboratory. The aim of Chen's research is to explore novel biomimetic materials and investigate their applications in the biomedical field. His group has developed numerous techniques for the synthesis of biomimetic materials with controllable structures and functions. He has obtained more than authorized patents and published more than 150 papers, which have been cited more than 10000 times (H-index of 56).

Zifei Zhou

Zifei Zhou graduated with a doctoral degree in Surgery (Orthopedics) from Tongji University in 2018. Zifei Zhou is an associate researcher at the Institute for Advanced Study at Tongji University and an attending physician at the Department of Orthopedics of Shanghai 10th People's Hospital. Zifei Zhou worked as a postdoctoral fellow in the Laboratory of Biomaterials and Regenerative Medicine, Mayo Clinic. He has more than 15 years of experience as a surgeon and researcher in treatment of bone diseases. His research field is the application of autologous biomaterials in bone repair, nerve repair and bone tumor diagnosis and treatment. To date, Zifei Zhou has participated in 13 scientific projects, being the principal investigator in 7 projects. He has contributed to 43 scientific publications in international journals with referees.

Journal of
Functional Biomaterials

MDPI

Editorial

Feature Papers in Bone Biomaterials

Zifei Zhou [1,*] and Feng Chen [2,*]

1 Department of Orthopedics, Shanghai Tenth People's Hospital, School of Medicine, Tongji University, Shanghai 200072, China
2 Shanghai Key Laboratory of Craniomaxillofacial Development and Diseases, Shanghai Stomatological Hospital & School of Stomatology, Fudan University, Shanghai 201102, China
* Correspondence: zhouzf@tongji.edu.cn (Z.Z.); chen_feng@fudan.edu.cn (F.C.)

1. Introduction

Bone biomaterials have garnered significant attention in the field of regenerative medicine due to their application in a wide range of clinical scenarios. These materials are often designed to mimic the properties or contents of natural bone. The primary focus for researchers lies in enhancing their biocompatibility, mechanical strength, and osteoconductivity, essential attributes for successful bone integration and healing [1]. The biomaterials used for bone regeneration can be categorized into several different types, such as metal, ceramic, polymer, and composite biomaterials. Advancements in the structure of biomaterials have also further expanded their potential. Nanostructured materials exhibit unique properties that enhance cellular responses and promote bone regeneration at the molecular level. This Special Issue contains 10 papers: 8 articles which explore several newly designed bone biomaterials and 2 reviews which explain the importance of calcium phosphate and LIPUS stimulation.

2. Overview of Published Articles

The use of photobiomodulation as a biological intervention therapy is promising, as it may improve bone matrix protein deposition, periosteum development, and trabecular matrix production. It has been found that photobiomodulation does not increase the formation of new bone in critical-size defects when combined with deproteinized bovine bone mineral and collagen membrane. However, absorption of deproteinized bovine bone mineral could be accelerated using photobiomodulation; in such cases, new bone formation has been observed around the collagen membrane, surpassing the original boundaries of the defect [2].

Functional pressure from muscles and ligaments near the hyoid can cause biological apatite crystallites, periosteal insertions, and fibrocartilage entheses to adopt a preferred orientation, in addition to increasing bone mineral density. This indicates that functional pressure at these locations could have a significant impact on the morphological characteristics of the entheses, as well as in bone quality [3].

It has been widely proven that both bone marrow mesenchymal stem cells (BMSCs) and adipose-derived stem cells (ASCs) show great potential in bone repair. Furthermore, the recapitulated osteogenesis ability of BMSC or ASC spheroids makes them an advanced tool in bone regeneration. ASC spheroids seeded in a 3D-printed scaffold showed a good spreading morphology and secreted vascular endothelial growth factor. In 3D-printed scaffolds with ASC spheroids attached, new bone tissue formation was demonstrated through histological in vivo results [4].

Biodegradable metallic biomaterials (BMBs) are promising candidates for orthopedic implants. Among the BMBs available, Mg and its alloys demonstrate multiple biological effects, including osteogenic and angiogenic properties. Compared to pure Mg, the Mg-Sc-Sr alloy induced a higher osteoblastic differentiation of BMSCs and angiogenic differentiation of HUVECs. The Mg-Sc-Sr alloy is an attractive candidate for future bone implants [5].

Various calcium phosphate (CaP)-based biomaterials have been used in bone repair. Hou et al. reviewed the influence of CaPs' species, sizes, and morphologies on its physicochemical properties and biological behavior during the promotion of bone regeneration. This review will be beneficial in understanding the characteristics of CaPs and in creating new high-performance biomaterials for bone repair [6].

Biophysical stimulations play a role as non-invasive and controllable therapy in clinical practice. Moreover, the application of biophysical stimulations could work as "biophysical switch" in the treatment of diseases, with or without biomaterials. Among these biophysical stimulations, low-intensity pulsed ultrasound (LIPUS) has shown many advantages when combined with musculoskeletal-related biomaterials. Jia et al. present a review regarding the synergistic action of LIPUS and biomaterials in the repair of bone, cartilage, or nerves and in the treatment of osteonecrosis and osteolysis. Their crucial acoustic excitation parameters and current limitations and future perspectives are discussed [7].

Wood-derived materials are another candidate for bone scaffolds, as they exhibit similar density and mechanical properties to those of bone tissue. Andze et al. produced a dry-state wood material using a three-step method, which included partial delignification, extraction, and densification. The swelling property of this wooden material was reduced after a chemical treatment. Moreover, the wood sample showed good cytocompatibility [8].

Immunoregulatory topography could modulate the body's response to the surface of biomaterials. RAW264.7 macrophages showed different responses to nanopillar and nanopit substrates. The morphology of the macrophages was large and spread well due to the topography of the nanopillars, which had a diameter of 450 nanometers and an inter-pillar space of 300 nanometers. A significant increase in cell elongation was observed in the macrophages in response to nanopits with a depth of 150 nm and 800 nm edge–edge spaces. The phenotypic markers and integrin β1 expression of M2 macrophages were elevated and inflammatory cytokines we less expressed after these cells exhibited spreading or elongation [9].

Bone tissue engineering can greatly benefit from three-dimensional printing technology. Such technology can enable users to craft personalized designs, create prototypes rapidly, and fabricate intricate scaffolds [10]. Varisized scaffolds were fabricated using three-dimensional printed scaffolds formed of polylactic acid (PLA) and different amounts of gelatin and chitosan (CH) hydrogel. The behaviors of human bone mesenchymal stromal cells (hBMSCs) were unaffected by the hydrogel's content or the core–shell structure's geometry. All scaffolds have the potential to promote cell adhesion, proliferation, and osteogenic differentiation. Future studies could be aided by the strut thickness, hole height, hole width, and CH contents of the PLA-CH three-dimensional scaffolds described in this article [11].

Mn can be introduced onto the surface of titanium platforms by employing the plasma immersion ion implantation and deposition technique, leading to the surface having heightened zeta potentials and nano-hardness. Under mono-culture conditions, a Mn scaffold could stimulate the M1 phenotypes of Raw264.7 macrophages while displaying little promotion of cellular osteogenic differentiation in mouse BMSCs. However, a significant transformation to the M2 phenotypes of these macrophages and significant osteogenic differentiation of the mouse BMSCs were observed when they were co-cultured together. Consequently, Mn-modified biomaterials demonstrated the ability to trigger immunomodulation and stimulate osteogenic differentiation [12].

3. Conclusions and Future Perspectives

Bone biomaterials are an interdisciplinary field within bone tissue engineering [13]. Continuous basic and preclinical research is required to address the difficulties in bone repair. A thorough understanding of the interaction mechanisms between bone biomaterials and bone tissue is essential for the development of personalized bone biomaterials. We should encourage the study of the interaction of bone biomaterials with bone and its surrounding tissues, and especially the influence multiple cells in the local microenvironment have on bone biomaterials. Advances in these areas will enable the design and manufacture of bionic bone biomaterials that are more suitable for bone repair processes.

When designing bone biomaterials, the balance between the degradation of biomaterials and the bone regeneration process is another factor to be considered. If the degradation rate of the biomaterial is consistent with the regeneration process of the bone tissue, then it can be ensured that the biomaterial scaffold will support the formation of new bone tissue. Achieving this optimal balance prevents the premature loss of the structure of the biomaterial while it is still providing continuous mechanical support to bone tissue.

Multiple physical stimuli, such as near-infrared, LIPUS, or electrical signals, when combined with biomaterials, can also further accelerate the bone regeneration process. The combination of physical stimulation and biomaterials often promotes cell proliferation, osteogenic differentiation, and angiogenesis. Therefore, physical energy can be used as an "energy switch" to control the bone tissue regeneration process or the rate of degradation of the scaffold by turning them on or off.

In addition, we should encourage the use of large-animal bone defect models in future studies. The important role of these models in preclinical research is reflected by the fact that these models can provide a local microenvironment closer to that of human bone defects. Large animals can bridge the gap between small-animal research and clinical translational applications. The innovative development of bone biomaterials is expected to bring new hope to patients with bone nonunion or bone defects, who need bone biomaterials.

Funding: This work was supported by grants from the National Key R&D Program of China (2022YFE0123500), National Natural Science Foundation of China (81901892 and 31771081), and Shanghai Tenth People's Hospital (2021SYPDRC010).

Conflicts of Interest: The authors declare no conflicts of interest.

References

1. Shao, R.; Dong, Y.; Zhang, S.; Wu, X.; Huang, X.; Sun, B.; Zeng, B.; Xu, F.; Liang, W. State of the art of bone biomaterials and their interactions with stem cells: Current state and future directions. *Biotechnol. J.* **2022**, *17*, e2100074. [CrossRef] [PubMed]
2. de Freitas, N.R.; Guerrini, L.B.; Esper, L.A.; Sbrana, M.C.; dos Santos, C.C.V.; de Almeida, A.L.P.F. Photobiomodulation and Inorganic Bovine Bone in Guided Bone Regeneration: Histomorphometric Analysis in Rats. *J. Funct. Biomater.* **2023**, *14*, 281. [CrossRef] [PubMed]
3. Kasahara, M.; Someya, T.; Kitamura, K.; Watanabe, G.; Matsunaga, S.; Abe, S.; Hattori, M. Analysis of Bone Mineral Density and Bone Quality of Cortical Bone in the Human Hyoid Body and Histological Observation of the Entheses. *J. Funct. Biomater.* **2024**, *15*, 56. [CrossRef] [PubMed]
4. Kronemberger, G.S.; Palhares, T.N.; Rossi, A.M.; Verçosa, B.R.F.; Sartoretto, S.C.; Resende, R.; Uzeda, M.J.; Alves, A.T.N.N.; Alves, G.G.; Calasans-Maia, M.D.; et al. A Synergic Strategy: Adipose-Derived Stem Cell Spheroids Seeded on 3D-Printed PLA/CHA Scaffolds Implanted in a Bone Critical-Size Defect Model. *J. Funct. Biomater.* **2023**, *14*, 555. [CrossRef] [PubMed]
5. Aboutalebianaraki, N.; Neal, C.J.; Seal, S.; Razavi, M. Biodegradable Mg-Sc-Sr Alloy Improves Osteogenesis and Angiogenesis to Accelerate Bone Defect Restoration. *J. Funct. Biomater.* **2022**, *13*, 261. [CrossRef] [PubMed]
6. Hou, X.; Zhang, L.; Zhou, Z.; Luo, X.; Wang, T.; Zhao, X.; Lu, B.; Chen, F.; Zheng, L. Calcium Phosphate-Based Biomaterials for Bone Repair. *J. Funct. Biomater.* **2022**, *13*, 187. [CrossRef] [PubMed]
7. Jia, W.; Zhou, Z.; Zhan, W. Musculoskeletal Biomaterials: Stimulated and Synergized with Low Intensity Pulsed Ultrasound. *J. Funct. Biomater.* **2023**, *14*, 504. [CrossRef] [PubMed]
8. Andze, L.; Nefjodovs, V.; Andzs, M.; Skute, M.; Zoldners, J.; Kapickis, M.; Dubnika, A.; Locs, J.; Vetra, J. Chemically Pretreated Densification of Juniper Wood for Potential Use in Osteosynthesis Bone Implants. *J. Funct. Biomater.* **2024**, *15*, 287. [CrossRef] [PubMed]
9. Li, K.; Lv, L.; Shao, D.; Xie, Y.; Cao, Y.; Zheng, X. Engineering Nanopatterned Structures to Orchestrate Macrophage Phenotype by Cell Shape. *J. Funct. Biomater.* **2022**, *13*, 31. [CrossRef] [PubMed]
10. Yang, J.; Fatima, K.; Zhou, X.; He, C. Meticulously engineered three-dimensional-printed scaffold with microarchitecture and controlled peptide release for enhanced bone regeneration. *Biomater. Transl.* **2024**, *5*, 69–83. [PubMed]
11. Re, F.; Sartore, L.; Pasini, C.; Ferroni, M.; Borsani, E.; Pandini, S.; Bianchetti, A.; Almici, C.; Giugno, L.; Bresciani, R.; et al. In Vitro Biocompatibility Assessment of Bioengineered PLA-Hydrogel Core-Shell Scaffolds with Mesenchymal Stromal Cells for Bone Regeneration. *J. Funct. Biomater.* **2024**, *15*, 217. [CrossRef] [PubMed]

12. Ye, K.; Zhang, X.; Shangguan, L.; Liu, X.; Nie, X.; Qiao, Y. Manganese-Implanted Titanium Modulates the Crosstalk betweer Bone Marrow Mesenchymal Stem Cells and Macrophages to Improve Osteogenesis. *J. Funct. Biomater.* **2023**, *14*, 456. [CrossRef [PubMed]
13. Liu, H.; Su, J. Organoid extracellular vesicle-based therapeutic strategies for bone therapy. *Biomater. Transl.* **2023**, *4*, 199–21 [PubMed]

Journal of
Functional
Biomaterials

MDPI

Article

Engineering Nanopatterned Structures to Orchestrate Macrophage Phenotype by Cell Shape

Kai Li [1,†], Lin Lv [1,†], Dandan Shao [1,2], Youtao Xie [1], Yunzhen Cao [1] and Xuebin Zheng [1,*]

[1] Key Laboratory of Inorganic Coating Materials CAS, Shanghai Institute of Ceramics, Chinese Academy of Sciences, 1295 Dingxi Road, Shanghai 200050, China; likai@mail.sic.ac.cn (K.L.); lvlin@mail.sic.ac.cn (L.L.); shaodandan@student.sic.ac.cn (D.S.); xieyoutao@mail.sic.ac.cn (Y.X.); yzhcao@mail.sic.ac.cn (Y.C.)

[2] Center of Materials Science and Optoelectronics Engineering, University of Chinese Academy of Sciences, 19 Yuquan Road, Beijing 100049, China

* Correspondence: xbzheng@mail.sic.ac.cn

† These authors contributed equally to this work.

Abstract: Physical features on the biomaterial surface are known to affect macrophage cell shape and phenotype, providing opportunities for the design of novel "immune-instructive" topographies to modulate foreign body response. The work presented here employed nanopatterned polydimethylsiloxane substrates with well-characterized nanopillars and nanopits to assess RAW264.7 macrophage response to feature size. Macrophages responded to the small nanopillars (SNPLs) substrates (450 nm in diameter with average 300 nm edge-edge spacing), resulting in larger and well-spread cell morphology. Increasing interpillar distance to 800 nm in the large nanopillars (LNPLs) led to macrophages exhibiting morphologies similar to being cultured on the flat control. Macrophages responded to the nanopits (NPTs with 150 nm deep and average 800 nm edge-edge spacing) by a significant increase in cell elongation. Elongation and well-spread cell shape led to expression of anti-inflammatory/pro-healing (M2) phenotypic markers and downregulated expression of inflammatory cytokines. SNPLs and NPTs with high availability of integrin binding region of fibronectin facilitated integrin β1 expression and thus stored focal adhesion formation. Increased integrin β1 expression in macrophages on the SNPLs and NTPs was required for activation of the PI3K/Akt pathway, which promoted macrophage cell spreading and negatively regulated NF-κB activation as evidenced by similar globular cell shape and higher level of NF-κB expression after PI3K blockade. These observations suggested that alterations in macrophage cell shape from surface nanotopographies may provide vital cues to orchestrate macrophage phenotype.

Keywords: nanopillar; nanopit; macrophage cell shape; phenotype; integrin β1

1. Introduction

Advances in nanotechnology are increasing the likelihood of programmable nano-bio interfaces, particularly in the novel nanotopographical design to foster cellular modulation, genome editing and biomolecular delivery [1,2]. Engineered nano-bio interfaces fabricated by novel nanopatterning technology have improved the fundamental understanding of cellular functions at the nanoscale [3–5]. Nanopatterned structures are now providing great advantages in manipulating cellular functions and processes in a more controlled manner, which assists the translation into practical applications [6].

In the field of tissue engineering, three basic nanopatterned geometries, nanogrooves, nanopillars and nanopits, are frequently used as a means to investigate stem cell attachment and to control cell differentiation and phenotype in vitro [7]. A study by Yim et al. reported that human mesenchymal stem cells (hMSCs) grown on nanogrooves were preferentially differentiated into a neural lineage as evidenced by the existence of nestin and synaptophysin markers and by the regulation of MAP2 [8]. Dalby et al. reported that hMSCs grown on slightly irregular nanopits produced considerable quantities of bone-specific extracellular matrix proteins osteopontin and osteocalcin. In contrast, cells grown on completely ordered or random substrates did not result in the expression of those two proteins [9]. Fu et al. reported that enhanced osteogenic differentiation of hMSCs was observed on patterned nanopillars with shorter heights (0.97 um) [10].

Although tissue engineering strategies highly focus on stem cells, immune cells are equally essential for tissue repair and regeneration [11]. Of them, macrophages are early responders to tissue injury and are extremely plastic cells involved in host defense, immune regulation and tissue repair [12]. To achieve such diverse functions, macrophages polarize to different phenotypes, ranging from M1 to M2 (pro-inflammatory to anti-inflammatory/pro-healing, respectively), in response to various microenvironment stimuli [13]. The effect of physical stimuli, such as mechanical stiffness and topography, on macrophage behaviors has long been known [14]. Various types of nanotopographies have been shown to have an effect on macrophage cell shape and phenotype, especially when the topography is highly ordered [5]. In a study by Frances et al., elongated cell morphology alone was found to induce bone marrow-derived macrophages (BMMs) polarization into the anti-inflammatory M2 phenotype with increased expression of arginase-I (ARG-I) and mannose receptor CD206 [15]. A surface with nanopatterned grooves drove BMMs polarization to M2 phenotype, and the level of anti-inflammatory cytokine interleukin-10 (IL-10) production in macrophages peaked on the substrate with groove widths of 400–500 nm.

While correlations between nanogrooves, macrophage cell shape and phenotype have been established, only few studies used well-defined nanopillars or nanopits for the evaluation of the specific effect of surface parameters (e.g., diameter, height and spacing) in macrophage polarization. A study by Padmanabhan et al. suggested that patterned nanopillars of diameter larger than 200 nm were required to induce morphological changes in BMMs, leading to more elongated and larger cell morphology [4]. Conflicting results have demonstrated the decrease in macrophage cell spreading on nanopillars [16,17]. The overall effect of these nanotopographies on macrophage cell shape and phenotype is unclear. It has however been shown that nanopit and nanopillar substrates could be employed to investigate the role of filopodia formation, focal adhesion dynamics and cytoskeletal organization in a controlled manner [18]. A systematic examination of nanotopography-cell interaction between nanopatterned pillars or pits and macrophages may provide data on cell shape modulation as the underlying mechanisms driving macrophage polarization status changes.

Here, nanopatterned pillars and pits in varying sizes were fabricated on polydimethylsiloxane (PDMS) substrates by replica molding. The effects of nanopattern features on macrophage morphology, phenotype and gene expression profile were evaluated. In order to study the biochemical mechanism underlying nanopattern-cell interactions, nanopattern-induced changes in fibronectin adsorption, integrin expression, focal adhesion formation and expression of downstream effector kinases were analyzed.

2. Materials and Methods

2.1. Fabrication and Characterization of the Nanopatterned Surfaces

Silicon wafers with highly ordered nanopits (390 and 780 nm pit diameter) in a hexagonal conformation pattern (In-situ Technology Co., Ltd., Shanghai, China) were used as templates. Nanopatterned PDMS substrates were prepared via mixing the curing agent (Sylgard 184, Dow Corning, Midland, MI, USA) with silicone elastomer in an 1:10 ratio by weight, pouring the prepolymer over the patterned silicon wafers (10 mm × 10 mm, 20 mm × 20 mm) and curing 2 h at room temperature and 3 h at 60 °C. Once cured, the PDMS was peeled from the wafer. By replica molding, the large nanopillars (LNPLs) and small nanopillars (SNPLs) substrates were prepared. The SNPLs substrate was further used to fabricate the patterned nanopit (NPT) substrate. A flat PDMS substrate served as control. The surfaces of nanopatterned substrates were observed by field emission scanning electron microscope (FE-SEM, Magellan 400, FEI Company, Hillsboro, OR, USA).

Surface topography was analyzed by atomic force microscope (AFM, Bruker Dimension Icon, Karslruhe, Germany). The imaging was conducted in tapping mode with a cantilever of spring constant of 2.8 N m^{-1} with a resonance frequency of 75 kHz. The images were taken at a scan size of 3 μm × 3 μm. Surface wettability was measured by contact angle meter (SL200B, KINO Industry Co., Ltd., Boston, MA, USA).

2.2. *Effects of the Nanopatterned Structures on Fibronectin Distribution and Conformation*

Each sample was immersed in 20 μg/mL of fibronectin (FN, Sigma, St. Louis, MO, USA) solution for 10 min at 37 °C. After this, each sample was rinsed by phosphate-buffered saline (PBS) to remove non-adsorbed FN. The distribution of adsorbed FN on the surfaces was observed by AFM in a tapping mode. The amount of adsorbed FN was quantitatively measured using the Micro BCA Protein Assay Kit (Beyotime Biotechnology, Nantong, China). RGD availability of FN on the surfaces was assessed by examining the binding of the anti-FN antibody (HFN7.1) with enzyme-like immunosorbent assay (ELISA) as described elsewhere [14]. Considering the difference among the samples in the amount of adsorbed FN, RGD availability of FN was calculated by normalizing the intensity of HFN7.1 to the amount of adsorbed FN.

2.3. *Cell Culture*

The murine-derived macrophage RAW264.7 cell line was obtained from the Cell Bank of Type Culture Collection of Chinese Academy of Sciences (Shanghai, China). The cells were incubated in α-minimum essential medium (α-MEM, HyClone, Logan, UT, USA) supplemented with 1% (*v*/*v*) penicillin/streptomycin (Wisent, Nanjing, China) and 10% (*v*/*v*) fetal bovine serum (FBS, Wisent) in an incubator at 37 °C with 5% CO_2. The medium was refreshed every two days, and the cells were passaged upon reaching 80% confluence.

2.4. *Effects of the Nanopatterned Structures on Macrophage Cell Shape and Phenotype*

2.4.1. Cell Morphology

RAW264.7 cells were cultured on the samples in 48-well plates at a density of 2×10^4 cells/well. After incubation for 24 h, the cells were fixed overnight in 2% (*v*/*v*) glutaraldehyde, dehydrated with a graded ethanol series and air-dried. FE-SEM was used for cell morphology observation. Average cell area of cells on each sample was calculated using 50 measurements from three random areas of a SEM image.

2.4.2. Cytoskeleton and Focal Adhesion Observation

RAW264.7 cells were seeded onto each sample in 48-well plates at a density of 2×10^4 cells/well. After incubated for 3 days, the cells were treated by 4% (*v*/*v*) paraformaldehyde fixation and 0.5% (*v*/*v*) Triton X-100 permeabilization. After incubation in blocking buffer (5% bovine serum albumin in 0.1 M PBS), the cells were treated by primary antibody against vinculin (1:150, Cell Signaling Technology, Danvers, MA, USA) diluted in blocking buffer overnight at 4 °C. Afterward, the cells were incubated with fluorescence-conjugated secondary antibody against vinculin (1:250, Cell Signaling Technology) for 1 h. Meanwhile, the cells were treated by DAPI (Molecular Probes) and rhodamine phalloidin (Molecular Probes) for nucleus and actin staining, respectively. Confocal microscope system (TCS SP5 II, Leica, Nussloch, Germany) was used for immunofluorescence observation. Elongation factor was calculated as the length of the cell's long axis divided by the length of the short axis across the nucleus as described by McWhorter et al. [15]. Approximately one hundred cells were evaluated for the quantitative analysis of cell morphology changes.

2.4.3. Flow Cytometry

RAW264.7 cells were cultured on the samples in 6-well plates at a density of 5×10^5 cells/well and collected after 3 days. After centrifugation and re-suspension, the cells were stained with anti-mouse CD11c-FITC and anti-mouse CD206-PE (Thermo Scientific, Waltham, MA, USA) at 4 °C in darkness for 30 min. The flow cytometry was performed using a BD FACSAria flow cytometry system (BD Biosciences, Franklin Lakes, NJ, USA). FlowJo software was employed for data analysis.

2.4.4. Quantitative Real-Time Polymerase Chain Reaction (qRT-PCR) Assay

RAW264.7 cells were incubated on the samples in 6-well plates at a density of 5×10^5 cells/well for 3 days. Total RNA was isolated using the Trizol reagent (Life Technologies, Carlsbad, CA, USA) and reverse-transcribed by a reverse transcription kit according to the manufacturer's instructions (TaKaRa Biotechnology, Beijing, China). qRT-PCR was performed using Applied Biosystems 7500 RT-PCR system (Applied Biosystems, Waltham, MA, USA) with a PCR kit (Applied Biosystems) to measure the expression of the integrin receptors (*Itga5, Itgb1*), the surface markers (*Cd11c, Ccr7, Cd206, Cd163*), the pro-inflammatory cytokines (*Tnfa, Il6, Il1b*) and the anti-inflammatory cytokines (*Il10, Il1ra*). All mRNA values were normalized against the housekeeping gene (GAPDH). All primer sequences are presented in Table 1.

Table 1. Sequences of all primers used in this study.

Target Gene	Direction	5′-3′ Primer Sequence
Itga5	F	5′-CTTCTCCGTGGAGTTTTACCG-3′
	R	5′-GCTGTCAAATTGAATGGTGGTG-3′
Itgb1	F	5′-CGTGGTTGCCGGAATTGTTC-3′
	R	5′-ACCAGCTTTACGTCCATAGTTTG-3′
Cd11c	F	5′ -CTGGATAGCCTTTCTTCTGCTG- 3′
	R	5′ -GCACACTGTGTCCGAACTCA- 3′
Ccr7	F	5′-TGTACGAGTCGGTGTGCTTC-3′
	R	5′-GGTAGGTATCCGTCATGGTCTTG-3′
Cd163	F	5′-ATGGGTGGACACAGAATGGTT-3′
	R	5′-CAGGAGCGTTAGTGACAGCAG-3′
Cd206	F	5′-CTCTGTTCAGCTATTGGACGC-3′
	R	5′-CGGAATTTCTGGGATTCAGCTTC-3′
Il6	F	5′-ACTCACCTCTTCAGAACGAATTG-3′
	R	5′-CCATCTTTGGAAGGTTCAGGTTG-3′
Tnfa	F	5′-CCTCTCTCTAATCAGCCCTCTG-3′
	R	5′-GAGGACCTGGGAGTAGATGAG-3′
Il1b	F	5′-TGGAGAGTGTGGATCCCAAG-3′
	R	5′-GGTGCTGATGTACCAGTTGG-3′
Il10	F	5′-GACTTTAAGGGTTACCTGGGTTG-3′
	R	5′-TCACATGCGCCTTGATGTCTG-3′
Il1ra	F	5′-CATTGAGCCTCATGCTCTGTT-3′
	R	5′-CGCTGTCTGAGCGGATGAA-3′
GAPDH	F	5′-TGACCACAGTCCATGCCATC-3′
	R	5′-GACGGACACATTGGGGGTAG-3′

2.5. *Western Blot*

RAW264.7 cells were incubated on the samples in 6-well plates at a density of 5×10^5 cells/well for 3 days. The cells were collected and lysed with Ripa buffer (Beyotime Biotechnology, China), which was supplemented with 1% (v/v) protease inhibitor. After denaturation and separation, the proteins were treated by primary rabbit Anti-phospho-PI3-kinase p85-α/γ (pTyr467/199, Sigma), phospho-Akt Ser473 (D9E, CST), NF-κB p65 (D14E12, CST) and GAPDH (CST), respectively. Afterward, anti-rabbit IgG secondary antibodies conjugated to horseradish peroxidase (Proteintech Group, Inc., Rosemont, IL, USA) were used. An ECL system (Amersham Pharmacia Biotech, Piscataway, NJ, USA) was employed to reveal the protein bands.

Small interfering RNA (siRNA) was utilized to inhibit integrin β1 expression. RAW264.7 cells were incubated on the samples in 6-well plates at a density of 5×10^5 cells/well and then infected with siRNA (GenePharma, Shanghai, China) to specifically knockdown β1. Inhibition of PI3K was performed by adding PI3K inhibitor LY294002 (L9908, Sigma) at a concentration of 10 and 20 μM. RAW264.7 cells in the absence of the inhibitor served as control. After inhibition, western blot assay of p-PI3K, p-Akt and NF-κB was performed as described above.

2.6. Statistical Analysis

The data were shown as mean ± standard deviation (SD) of three independent experiments (sample size n = 3 performed in triplicate). One-way ANOVA followed by SNK post hoc test was employed for analysis of statistical difference. Significance was indicated by * (p value below 0.05).

3. Results and Discussion

3.1. Surface Characterization

PDMS substrates with different surface topographies were fabricated by a replica molding method with three different patterns. Generally, the topographical features as observed by SEM and AFM (Figure 1) consisted of a hexagonal placement of cylindrical nanopillars or nanopits. The large nanopillars (LNPLs) were 400 nm high pillars with a diameter of 700 nm and an average edge-edge distance of 800 nm; the small nanopillars (SNPLs) were 150 nm in height and 450 nm in diameter with average 300 nm edge-edge spacing; the nanopits (NPTs) were 700-nm-diameter pits (150 nm deep, average 800 nm edge-edge spacing). A flat PDMS substrate served as control. The water contact angle of the flat, LNPLs, SNPLs and NPTs substrates was measured as 118.28 ± 0.22°, 120.29 ± 1.94°, 112.71 ± 1.48° and 135.33 ± 2.30°, respectively.

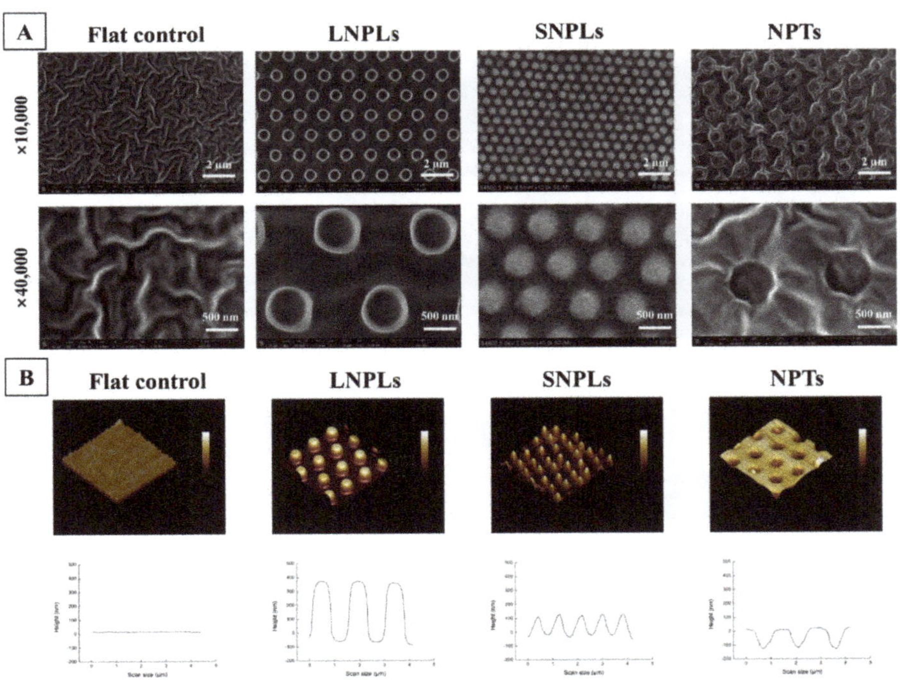

Figure 1. Surface characterization of the nanopatterned and flat substrates. (**A**) SEM images of the substrate surfaces. (**B**) AFM images of the substrate surfaces: three dimensional images (upper row) and height profiles (bottom row).

3.2. Effects of Nanopatterned Topographies on Macrophage Morphology

To monitor changes in cell morphology caused by nanopatterned topographies, RAW 264.7 macrophage cells were cultured on nanopatterned or flat PDMS substrates for 24 h. In general, cells exhibited heterogenous morphologies; nevertheless, certain morphologies were more predominant on specific nanopatterned substrates, which were presented in SEM images (Figure 2).

Macrophages on the flat control displayed a native globular shape with almost no protrusions. For LNPLs, the morphological responses of macrophages due to the topographical features were slightly but not considerably influenced by the nanopillars, and thus were close to those found on the flat control. A few cells with an elongated shape were observed as well, which had long and thin extensions following the nanopillar array. Such extensions (i.e., confined by the spacing between the nanopillars) were not found on the flat surface. The magnified image (8000 times) for LNPLs showed that the cell body was mostly suspended on top of the nanopillars, whereas the cell gripped the nanopillars with filopodia to induce some pillar bending. On SNPL substrate, most cells displayed a well-spread polygonal shape with the formation of large lamellipodium. Moreover, filopodia can be seen emerging from the lamellipodium and locate the end to the nanopillars, as shown in the 8000 times SEM image. In contrast, cells on NPT substrate were more frequently spindle-shaped. The filopodial formation was observed between the nanopit features at the inter-pit region in the magnified image (8000 times).

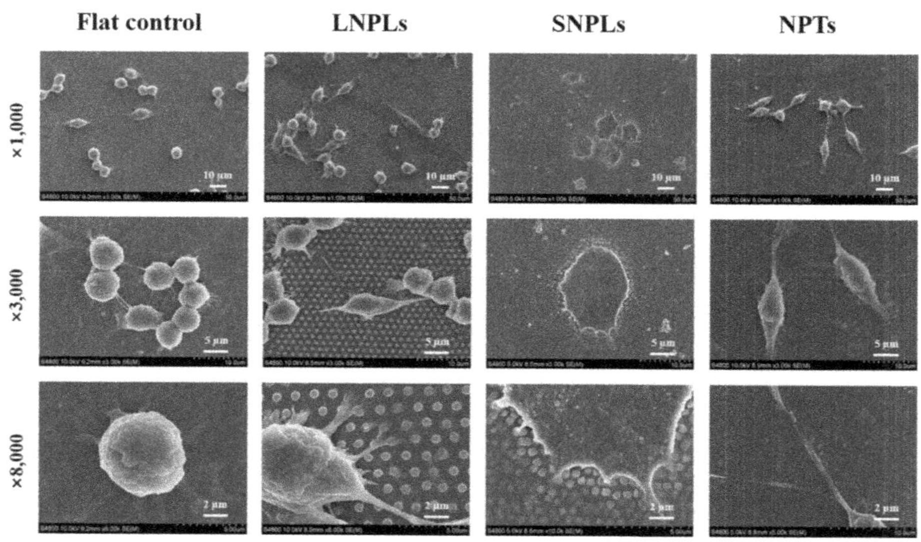

Figure 2. SEM images of RAW264.7 cells on the nanopatterned and flat substrates after 1 day of incubation.

Fluorescence microscopy images (Figure 3A) were in agreement with SEM observations of macrophage adaptation to the different nanotopographies. In addition, the absence of organized actin stress fibers was observed for the macrophages, which was consistent with previously reported observations. Instead, the cells exhibited diffuse actin staining with some clustering, particularly around their periphery. A quantitative analysis of cell morphology changes (Figure 3B) was performed by calculating the projected spreading area of the RAW264.7 cells on each sample and their elongation factor. The spreading cell area remained constant for the macrophage grown on the flat control, LNPL and NPT substrates. SNPLs promoted a statistically significant increase in spreading cell area. Concerning the elongation factor, the value is high when cell shape is elongated and decreases when the cell becomes more polygonal. Cells on the NPTs substrate showed a considerably higher degree of elongation compared with cells on the other three substrates. Macrophages on the SNPLs had the lowest degree of elongation.

The dimensions of the nanopillars, such as diameter, height and spacing, were shown to influence the cell's morphology and spreading. When these dimensions are such that cell membrane cannot accommodate surface recess, the cell is localized on the top of the nanopillars. Generally, this is achieved when the nanopillars are densely packed and their height is more than 40 nm. These observations were in agreement with the SEM results in our study. When the height and density of the nanopillars are sufficient to isolate the cells from the underlying planar substrate, the effect of the nanopillar edge-edge spacing and diameter becomes the determining factors on integrin clustering regulation and subsequent focal adhesion formation [19]. To facilitate integrin activation and clustering in the cells suspended on a nanopillar array, the feature diameter must exceed 70 nm, which was verified by multiple works using nanopillar arrays >400 nm in height [20]. In this work, considering the feature diameter of LNPLs and SNPLs both exceeding 70 nm, it is reasonably to believe that edge-edge spacing plays a vital role in determining cellular adhesion morphology. Compared to SNPLs with edge-edge distance of 300 nm, increasing the distance to 800 nm in LNPLs markedly reduced macrophage cellular adhesion. This indicated that a higher interpillar distance might inhibit integrin clustering and focal adhesion formation at the bridging site between two adjacent nanopillars.

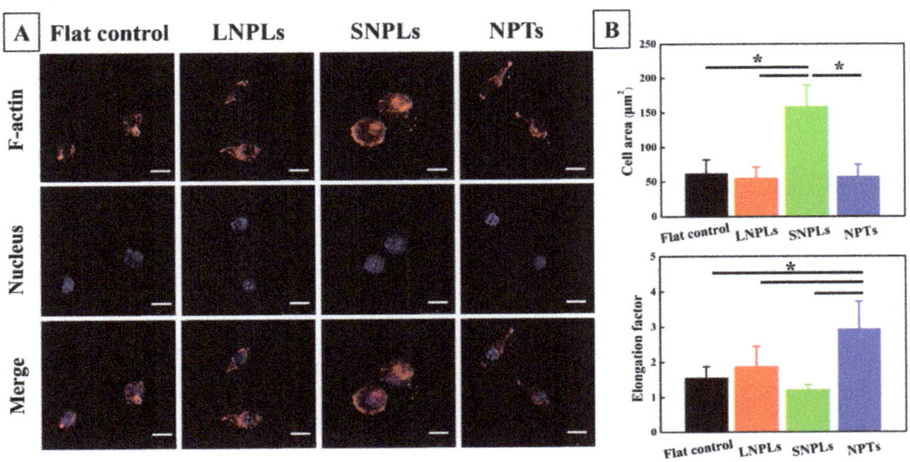

Figure 3. (**A**) Typical immunofluorescence images of RAW264.7 cells on the nanopatterned and flat substrates after 3 days of incubation. Scale bar = 15 μm. (**B**) Quantitative analysis of cell morphology changes. Asterisks indicate significant difference (* $p < 0.05$).

Similarly, pitted topographies have been observed to exert different effects on cell adhesion in vitro, depending on pit depth, spacing and diameter. Increasing pit diameter over 70 nm has been shown to perturb integrin clustering [21,22]. This was consistent with the results of cellular adhesion on nanpillar arrays with feature diameter of <70 nm. When pit depth exceeds 100 nm, the sites of focal adhesions seem to be confined to the inter-pit area [23,24]. This might explain the spindle-shaped macrophage cell morphology on NPTs and the filopodial formation observed at the inter-pit region.

3.3. Correlation of Cell Shape and Macrophage Polarization Status

To investigate whether cell shape may have a direct effect on macrophage polarization status, we employed flow cytometry to calculate the percentage of M1 and M2 cells on different substrates by measuring the level of the surface markers CD11c and CD206, respectively. As shown in Figure 4A, a decreased percentage of M1-type macrophages expressing CD11c was found for the SNPLs (44.6%) and NPTs (49.8%) compared with the flat control (66.2%) and LNPLs (61.0%). In addition, the percentage of CD206-positive cells increased from 31.2% for the flat control to 37.8%, 60.9% and 65.0% for the LNPLs, SNPLs and NPTs, respectively.

To confirm the phenotypic polarization of RAW264.7 macrophages, we measured the gene expression level of M1/M2 phenotypic markers by qRT-PCR (Figure 4B). There was no significant difference between the flat control and LNPL substrates in the expression of *Ccr7* and *Cd11c* (M1 surface markers) as well as *Tnfa*, *Il6* and *Il1b* (pro-inflammatory genes). However, the cells on the LNPLs led to significant up-regulation of *Cd163* and *Cd206* (M2 surface markers) as well as *Il10* and *Il1ra* (anti-inflammatory genes) expression when compared with the flat control. Cells grown on the SNPLs and NPTs showed further increases in *Cd206* and *Il1ra* expression. Compared to the flat control and LNPLs, considerably decreased expression of *Ccr7*, *Il6* and *Il1b* were observed for the SNPLs and NPTs. SNPLs and NPTs exerted a greater effect on the macrophage polarization status through inducing the M2 phenotype. Together, these data suggested that macrophage cell shape was associated with the macrophage polarization status; besides, well-spread cell shape induced by SNPLs and elongated shape induced by NPTs correlated with M2 polarization.

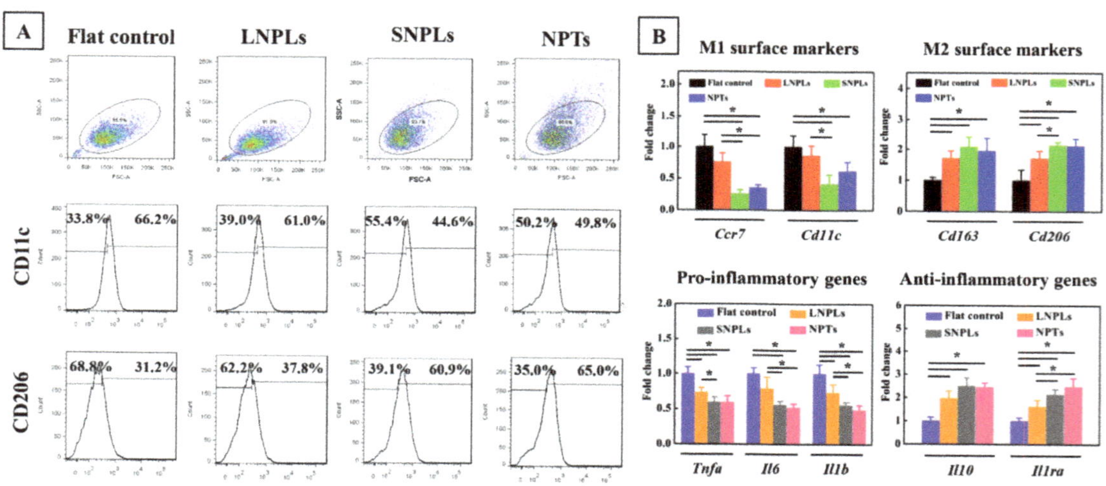

Figure 4. Assessment of macrophage polarization status on the nanopatterned and flat surfaces according to (**A**) flow cytometry assay of M1/M2 surface markers (CD11c, CD206) and (**B**) fold change of gene expression of M1/M2 surface markers (*Ccr7* and *Cd11c* for M1, *Cd163* and *Cd206* for M2), pro-inflammatory cytokines (*Tnfa, Il6, Il1b*) and anti-inflammatory cytokines (*Il10, Il1ra*). Asterisks indicate significant difference (* $p < 0.05$).

Growing evidence has revealed that morphology is heavily associated with phenotype in certain macrophages; however, the phenomenon may not be universal. We and others have observed that while M2-polarized RAW264.7 exhibited well-spread cell shape, other macrophages did not display such dramatic changes [25–27]. Likely contributing to conflicting reports were several factors, including the differences between murine and human macrophages used in different reports, and the sensitivity of macrophages to different substrates including polymers, ceramics and metals [28–31].

3.4. Biochemical Cues Underlying Cell-Nanopattern Interactions Govern Macrophage Polarization Status

The initial adsorption of extracellular matrix (ECM) proteins, such as FN, which forms a temporal matrix on biomaterial surface, becomes a vital link between the biomaterial and host response. Therefore, it is imperative to know the manner by which the adsorbed proteins interact with the nanopatterned surface when determining the subsequent behavior of macrophages. The distribution of adsorbed FN was investigated by AFM in a tapping mode (Figure 5A). Globular aggregates appeared on the flat surface. In contrast, interconnected nanofibrils were observed on the nanopillar array surfaces. Imbedded FN molecules were found in the nanopits. The adsorbed FN proteins have been reported to have different configurations, such as filamentous or globular and extended or close-packed forms, depending on the local environment and the substrate [32]. Thus, the native conformation of globular FN in solution was maintained following adsorption onto the flat control, while FN proteins were adsorbed in an extended conformation on the nanopillar array surfaces. The distinct distribution of FN at the material interface might have consequences at the molecular level for availability of the integrin binding domain of FN ($FNIII_{9-10}$) [33].

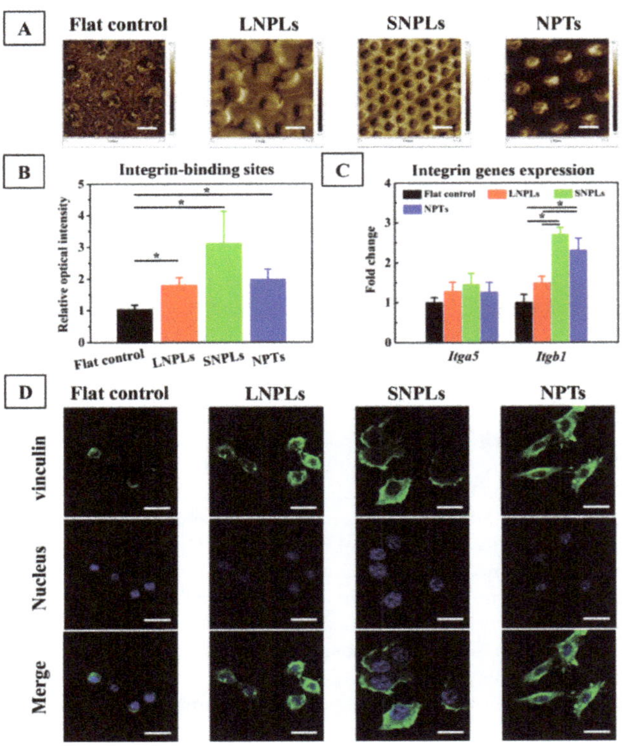

Figure 5. (**A**) FN distribution on the nanopatterned and flat surfaces as observed by AFM. Scale bar = 5 μm. (**B**) Relative intensity of integrin-binding sites of adsorbed FN on the nanopatterned and flat surfaces measured by ELISA. (**C**) Fold change of gene expression of integrin receptors α5 (*Itga5*) and β1 (*Itgb1*)) by RAW264.7 cells on day 3. (**D**) Vinculin (green), cell nucleus (blue) and merged staining images of RAW264.7 cells on the nanopatterned and flat substrates after 3 days of incubation. Scale bar = 20 μm. Asterisks indicate significant difference (* $p < 0.05$).

The RGD amino acid sequence, located in the FN repeat III_{9-10}, is of great importance in modulating cell adhesion, enhancing α5β1 integrin binding to FN [34]. The RGD availability of FN on the sample surfaces was evaluated by examining the binding of the anti-FN antibody (HFN7.1). As shown in Figure 5B, the intensity of HFN7.1 for the nanopatterned substrates was greatly increased compared to the flat control. Moreover, the intensity was increased on SNPLs in comparison to LNPLs and NPTs. However, the difference did not reach statistical significance. Generally, $\alpha 5\beta 1$ integrin is considered to be a major receptor for macrophage adhesion to FN [35]. Figure 5C showed that the expression level of integrin $\beta 1$ (*Itgb1*) was significantly higher on the nanopatterned substrates compared to the flat control. This was consistent with the result of the RGD availability of FN. However, among the nanopatterned substrates, the SNPLs and NPTs showed higher expression of *Itgb1* than the LNPLs. Subsequent to binding to ECM proteins, integrins cluster and develop into complexes called focal adhesion. Feature dimension on the nanopatterned substrates might play a role in integrin clustering and result in different *Itgb1* expression. Massia and Hubbell found that clustering of at least 6 RGD-ligand-occupied integrin heterodimers per micron (corresponding to a spacing of 440 nm) was sufficient to support cell spreading [36]. Previous work with defined arrays of bound RGD fragments demonstrated that substratum-integrin interactions were disrupted when the integrin spacing was in the range 70–300 nm [3,37,38]. Hence, it can be inferred that LNPLs with nanopillar spacing of 800 nm disrupted the lateral spacing of integrin clustering, which restricted focal adhesion formation between two adjacent nanopillars and the macrophages had to extend their filopodia to find more adsorbed proteins. By contrast, SNPLs with nanopillar spacing of 300 nm and NPTs with inter-pit region of 800 nm could facilitate integrin clustering, thus storing focal adhesion formation.

Stable focal adhesion requires integrin clustering and stability of the binding between the two (integrin cluster size), and it also involves the stabilization of the cytoplasmic side with reinforcement of adhesion proteins, including vinculin, paxillin, talin and focal adhesion kinase [39]. In this work, we employed vinculin as a marker of focal adhesion considering that it is recruited at the adhesion site via integrin $\beta 1$ receptor. The localization of vinculin was probed via immune-fluorescence microscopy (Figure 5D). Relatively smaller focal adhesions were observed on the flat control surface. For LNPLs, focal adhesions were localized regularly along the cell periphery, while cells on SNPLs formed adhesion contact not only at cell periphery but also within the lamellipodium leading edge. Focal adhesion on NPTs were distributed predominantly to the cell periphery; meanwhile, they were observed to be involved in cell protrusions.

NF-κB plays a vital role in the innate immune response [40,41]. As shown in Figure 6A, enhanced levels of NF-κB protein were measured on the LNPLs and flat substrates when compared with the SNPLs and NPTs. This was consistent with the results of macrophage polarization status on the substrates. NF-κB subunits have sites for phosphorylation and other post-translational modification which are vital for crosstalk with other signaling pathways. The crosstalk can occur through various kinases, including glycogen synthase kinase and Akt triggered by phosphoinositide 3-kinase (PI3K), which regulate NF-κB transcriptional activity [42]. In this work, the phosphorylation levels of PI3K and Akt on the SNPLs and NPTs substrates were higher than those of the LNPLs and flat substrates. Similar results were observed by Zhao et al. that PI3K/Akt pathway negatively regulated NF-κB pathway and the expression of inflammatory genes [43]. As shown in Figure 6B, the phosphorylation of PI3K/Akt was inhibited after $\beta 1$-specific knockdown, which indicated that $\beta 1$ integrin was implicated in PI3K/Akt pathway activation.

To further investigate the effect of PI3K/Akt pathway on the transactivation potential of NF-κB, NF-κB transactivation was monitored in the presence of PI3K inhibitor LY294002. As shown in Figure 6C, blockade of PI3K increased NF-κB activation. Moreover, higher concentration of LY294002 resulted in higher level of NF-κB activation. These confirmed that PI3K/Akt pathway negatively regulated NF-κB activation. However, it is necessary to point out that the effect of PI3K/Akt pathway on NF-κB activity often highly relies on the micro-environment or the cell type and that even an opposite effect can occur in distinct cell types. For instance, the effect of Akt on NF-κB was reported, which was activating in epithelial cells, but can be inhibitory in macrophages [44,45]. PI3K has also been recognized to regulate actin assembly and cytoskeleton organization facilitating cell spreading [27,46,47]. After PI3K inhibition, the macrophages on all the substrates showed globular morphology (Figure 6D). This confirmed the pivotal role of PI3K in promoting cell spreading and thus explained the more spreading cell morphology on SNPL and NPT substrates.

Because macrophages bridge innate immunity and adaptive immunity, a new paradigm was established to harness macrophage responses by immunomodulating biomaterials for endogenous repair. During macrophage-material interaction in vivo, the material surface acts as a temporary niche for macrophages to reside in, and the properties of the material surface weigh in on macrophage phenotypes. A broad range of material surface properties, such as pore size, shape and geometry, stiffness, topography and surface modification (e.g., hydrophilicity, integrin engagement), have been proven to modulate macrophage behavior and tune implantation outcomes. Although a lot of these studies observed correlations between material designs and macrophage phenotypes, a refocus on directing macrophage behavior to achieve suitable immune responses needs to be emphasized.

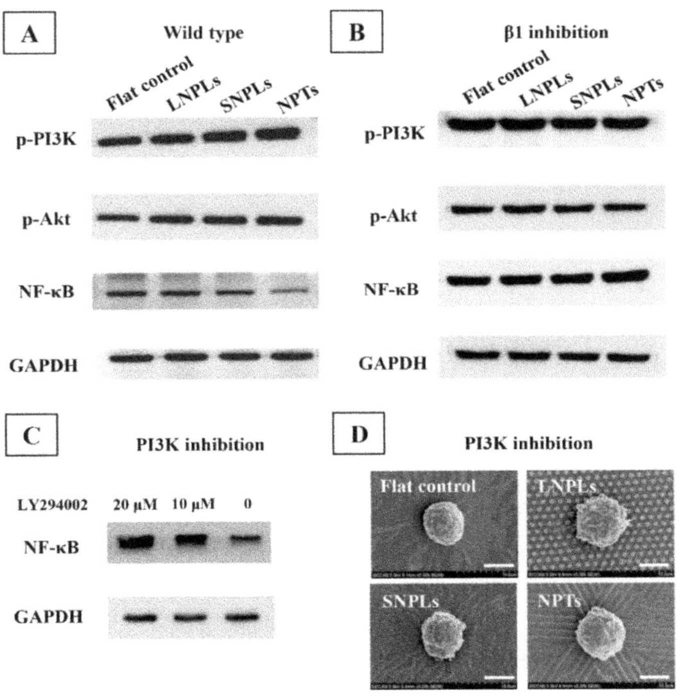

Figure 6. Western blot analysis of p-PI3K, p-Akt and NF-κB in (**A**) wild-type and (**B**) β1 knockdown RAW264.7 macrophages. (**C**) Western blot analysis of NF-κB activation in the presence of PI3K inhibitor LY294002. (**D**) SEM images of RAW264.7 cells after 1 day of incubation in the presence of 20 μM LY294002.

4. Conclusions

Utilizing a nanopatterning approach to modulate macrophage cell shape directly, we demonstrated here that elongation and well-spread cell shape led to expression of M2 phenotypic markers and downregulated expression of inflammatory cytokines. Changes in macrophage phenotype polarization from surface nanotopographies depended on a mechanism involving changes in cell adhesion and shape. In contrast to SNPLs, a higher interpillar distance in LNPLs (800 nm) inhibited integrin clustering, *Itgb1* expression, and focal adhesion formation, resulting in reduced macrophage cellular adhesion. The NPTs substrates with 150 nm pit depth confined the sties of focal adhesions to the inter-pit area (average 800 nm edge-edge spacing), resulting in elongation cell morphology. SNPLs with smaller interpillar distance and NTPs with higher inter-pit area facilitated the exposure of availability of integrin binding region of FN, expression of *Itgb1*, and formation of focal adhesion. Higher expression of *Itgb1* in macrophages on the SNPLs and NTPs activated the intracellular PI3K/Akt pathway, which promoted macrophage cell spreading and negatively regulated NF-κB activation as evidenced by similar globular cell shape and higher level of NF-κB activation after PI3K blockade. Our findings highlighted the potential of using nanopatterned polymers to modulate macrophage functions and thus the foreign body response.

Author Contributions: Conceptualization, K.L. and Y.C.; methodology, K.L. and L.L.; software, L.L.; validation, K.L., L.L. and Y.X.; formal analysis, L.L.; investigation, D.S.; resources, Y.X. and Y.C.; data curation, K.L. and L.L.; writing—original draft preparation, L.L.; writing—review and editing, K.L.; visualization, K.L. and L.L.; supervision, X.Z.; project administration, Y.X.; funding acquisition, K.L. All authors have read and agreed to the published version of the manuscript.

Funding: This research was funded by the National Natural Science Foundation of China (Grant No. 51971236), the Shanghai Rising-Star Program (Grant No. 21QA1410400), the Science and Technology Commission of Shanghai Municipality (Grant No. 21S31901200), and the Youth Innovation Promotion Association of the Chinese Academy of Sciences (Grant No. 2020254).

Institutional Review Board Statement: Not applicable.

Informed Consent Statement: Not applicable.

Data Availability Statement: Not applicable.

Conflicts of Interest: The authors declare no conflict of interest.

References

1. Nel, A.E.; Madler, L.; Velegol, D.; Xia, T.; Hoek, E.M.V.; Somasundaran, P.; Klaessig, F.; Castranova, V.; Thompson, M. Understanding biophysicochemical interactions at the nano-bio interface. *Nat. Mater.* **2009**, *8*, 543–557. [CrossRef] [PubMed]
2. Chen, Y.P.; Wang, J.; Li, X.L.; Hu, N.; Voelcker, N.H.; Xie, X.; Elnathan, R. Emerging Roles of 1D Vertical Nanostructures in Orchestrating Immune Cell Functions. *Adv. Mater.* **2020**, *32*, 2001668. [CrossRef] [PubMed]
3. Biggs, M.J.P.; Richards, R.G.; Dalby, M.J. Nanotopographical modification: A regulator of cellular function through focal adhesions. *Nanomed.-Nanotechnol.* **2010**, *6*, 619–633. [CrossRef] [PubMed]
4. Padmanabhan, J.; Kinser, E.R.; Stalter, M.A.; Duncan-Lewis, C.; Balestrini, J.L.; Sawyer, A.J.; Schroers, J.; Kyriakides, T.R. Engineering Cellular Response Using Nanopatterned Bulk Metallic Glass. *ACS Nano* **2014**, *8*, 4366–4375. [CrossRef] [PubMed]
5. Vassey, M.J.; Figueredo, G.P.; Scurr, D.J.; Vasilevich, A.S.; Vermeulen, S.; Carlier, A.; Luckett, J.; Beijer, N.R.M.; Williams, P.; Winkler, D.A.; et al. Immune Modulation by Design: Using Topography to Control Human Monocyte Attachment and Macrophage Differentiation. *Adv. Sci.* **2020**, *7*, 1903392. [CrossRef]
6. Gui, N.; Xu, W.; Myers, D.E.; Shukla, R.; Tang, H.P.; Qian, M. The effect of ordered and partially ordered surface topography on bone cell responses: A review. *Biomater. Sci.* **2018**, *6*, 250–264. [CrossRef]
7. Yao, X.; Peng, R.; Ding, J.D. Cell-Material Interactions Revealed via Material Techniques of Surface Patterning. *Adv. Mater.* **2013**, *25*, 5257–5286. [CrossRef]
8. Yim, E.K.F.; Pang, S.W.; Leong, K.W. Synthetic nanostructures inducing differentiation of human mesenchymal stem cells into neuronal lineage. *Exp. Cell Res.* **2007**, *313*, 1820–1829. [CrossRef]
9. Dalby, M.J.; Gadegaard, N.; Tare, R.; Andar, A.; Riehle, M.O.; Herzyk, P.; Wilkinson, C.D.W.; Oreffo, R.O.C. The control of human mesenchymal cell differentiation using nanoscale symmetry and disorder. *Nat. Mater.* **2007**, *6*, 997–1003. [CrossRef]
10. Fu, J.P.; Wang, Y.K.; Yang, M.T.; Desai, R.A.; Yu, X.A.; Liu, Z.J.; Chen, C.S. Mechanical regulation of cell function with geometrically modulated elastomeric substrates. *Nat. Methods* **2010**, *7*, U733–U795. [CrossRef]
11. Claes, L.; Recknagel, S.; Ignatius, A. Fracture healing under healthy and inflammatory conditions. *Nat. Rev. Rheumatol.* **2012**, *8*, 133–143. [CrossRef] [PubMed]
12. Brown, B.N.; Ratner, B.D.; Goodman, S.B.; Amar, S.; Badylak, S.F. Macrophage polarization: An opportunity for improved outcomes in biomaterials and regenerative medicine. *Biomaterials* **2012**, *33*, 3792–3802. [CrossRef] [PubMed]
13. Mosser, D.M.; Edwards, J.P. Exploring the full spectrum of macrophage activation. *Nat. Rev. Immunol.* **2008**, *8*, 958–969. [CrossRef]

14. Li, K.; Liu, S.W.; Hu, T.; Razanau, I.; Wu, X.D.; Ao, H.Y.; Huang, L.P.; Xie, Y.T.; Zheng, X.B. Optimized Nanointerface Engineering of Micro/Nanostructured Titanium Implants to Enhance Cell-Nanotopography Interactions and Osseointegration. *ACS Biomater. Sci. Eng.* **2020**, *6*, 969–983. [CrossRef] [PubMed]
15. McWhorter, F.Y.; Wang, T.T.; Nguyen, P.; Chung, T.; Liu, W.F. Modulation of macrophage phenotype by cell shape. *Proc. Natl. Acad. Sci. USA* **2013**, *110*, 17253–17258. [CrossRef] [PubMed]
16. Zaveri, T.D.; Dolgova, N.V.; Chu, B.H.; Lee, J.Y.; Wong, J.E.; Lele, T.P.; Ren, F.; Keselowsky, B.G. Contributions of surface topography and cytotoxicity to the macrophage response to zinc oxide nanorods. *Biomaterials* **2010**, *31*, 2999–3007. [CrossRef] [PubMed]
17. Chen, S.L.; Jones, J.A.; Xu, Y.G.; Low, H.Y.; Anderson, J.M.; Leong, K.W. Characterization of topographical effects on macrophage behavior in a foreign body response model. *Biomaterials* **2010**, *31*, 3479–3491. [CrossRef]
18. Balaban, N.Q.; Schwarz, U.S.; Riveline, D.; Goichberg, P.; Tzur, G.; Sabanay, I.; Mahalu, D.; Safran, S.; Bershadsky, A.; Addadi, L.; et al. Force and focal adhesion assembly: A close relationship studied using elastic micropatterned substrates. *Nat. Cell Biol.* **2001**, *3*, 466–472. [CrossRef]
19. Sjostrom, T.; Dalby, M.J.; Hart, A.; Tare, R.; Oreffo, R.O.C.; Su, B. Fabrication of pillar-like titania nanostructures on titanium and their interactions with human skeletal stem cells. *Acta Biomater.* **2009**, *5*, 1433–1441. [CrossRef]
20. Lim, J.Y.; Hansen, J.C.; Siedlecki, C.A.; Hengstebeck, R.W.; Cheng, J.; Winograd, N.; Donahue, H.J. Osteoblast adhesion on poly(L-lactic acid)/polystyrene demixed thin film blends: Effect of nanotopography, surface chemistry, and wettability. *Biomacromolecules* **2005**, *6*, 3319–3327. [CrossRef]
21. Lim, J.Y.; Dreiss, A.D.; Zhou, Z.Y.; Hansen, J.C.; Siedlecki, C.A.; Hengstebeck, R.W.; Cheng, J.; Winograd, N.; Donahue, H.J. The regulation of integrin-mediated osteoblast focal adhesion and focal adhesion kinase expression by nanoscale topography. *Biomaterials* **2007**, *28*, 1787–1797. [CrossRef] [PubMed]
22. Krasteva, N.; Seifert, B.; Albrecht, W.; Weigel, T.; Schossig, M.; Altankov, G.; Groth, T. Influence of polymer membrane porosity on C3A hepatoblastoma cell adhesive interaction and function. *Biomaterials* **2004**, *25*, 2467–2476. [CrossRef] [PubMed]
23. Biggs, M.J.P.; Richards, R.G.; Gadegaard, N.; McMurray, R.J.; Affrossman, S.; Wilkinson, C.D.W.; Oreffo, R.O.C.; Dalby, M.J. Interactions with nanoscale topography: Adhesion quantification and signal transduction in cells of osteogenic and multipotent lineage. *J. Biomed. Mater. Res. A* **2009**, *91*, 195–208. [CrossRef] [PubMed]
24. Hart, A.; Gadegaard, N.; Wilkinson, C.D.W.; Oreffo, R.O.C.; Dalby, M.J. Osteoprogenitor response to low-adhesion nanotopographies originally fabricated by electron beam lithography. *J. Mater. Sci.-Mater. Med.* **2007**, *18*, 1211–1218. [CrossRef] [PubMed]
25. Makaremi, S.; Luu, H.; Boyle, J.P.; Zhu, Y.J.; Cerson, C.; Bowdish, D.M.E.; Moran-Mirabal, J.M. The Topography of Silica Films Modulates Primary Macrophage Morphology and Function. *Adv. Mater. Interfaces* **2019**, *6*, 1900677. [CrossRef]
26. Lee, H.S.; Stachelek, S.J.; Tomczyk, N.; Finley, M.J.; Composto, R.J.; Eckmann, D.M. Correlating macrophage morphology and cytokine production resulting from biomaterial contact. *J. Biomed. Mater. Res. A* **2013**, *101*, 203–212. [CrossRef]
27. Kianoush, F.; Nematollahi, M.; Waterfield, J.D.; Brunette, D.M. Regulation of RAW264.7 macrophage polarization on smooth and rough surface topographies by galectin-3. *J. Biomed. Mater. Res. A* **2017**, *105*, 2499–2509. [CrossRef]
28. McWhorter, F.Y.; Davis, C.T.; Liu, W.F. Physical and mechanical regulation of macrophage phenotype and function. *Cell Mol. Life Sci.* **2015**, *72*, 1303–1316. [CrossRef]
29. Singh, S.; Awuah, D.; Rostam, H.M.; Emes, R.D.; Kandola, N.K.; Onion, D.; Htwe, S.S.; Rajchagool, B.; Cha, B.H.; Kim, D.; et al. Unbiased Analysis of the Impact of Micropatterned Biomaterials on Macrophage Behavior Provides Insights beyond Predefined Polarization States. *ACS Biomater. Sci. Eng.* **2017**, *3*, 969–978. [CrossRef]
30. Malheiro, V.; Lehner, F.; Dinca, V.; Hoffmann, P.; Maniura-Weber, K. Convex and concave micro-structured silicone controls the shape, but not the polarization state of human macrophages. *Biomater. Sci.* **2016**, *4*, 1562–1573. [CrossRef]
31. Shao, D.D.; Li, K.; You, M.Y.; Liu, S.W.; Hu, T.; Huang, L.P.; Xie, Y.T.; Zheng, X.B. Macrophage polarization by plasma sprayed ceria coatings on titanium-based implants: Cerium valence state matters. *Appl. Surf. Sci.* **2020**, *504*, 144070. [CrossRef]
32. Lv, L.; Li, K.; Xie, Y.T.; Cao, Y.Z.; Zheng, X.B. Enhanced osteogenic activity of anatase TiO_2 film: Surface hydroxyl groups induce conformational changes in fibronectin. *Mater. Sci. Eng. C-Mater.* **2017**, *78*, 96–104. [CrossRef] [PubMed]
33. Vanterpool, F.A.; Cantini, M.; Seib, F.P.; Salmeron-Sanchez, M. A material-based platform to modulate fibronectin activity and focal adhesion assembly. *BioRes. Open Access* **2014**, *3*, 286–296. [CrossRef] [PubMed]
34. Li, K.; Hu, D.D.; Xie, Y.T.; Huang, L.P.; Zheng, X.B. Sr-doped nanowire modification of Ca-Si-based coatings for improved osteogenic activities and reduced inflammatory reactions. *Nanotechnology* **2018**, *29*, 084001. [CrossRef]
35. Abshire, M.Y.; Thomas, K.S.; Owen, K.A.; Bouton, A.H. Macrophage motility requires distinct alpha5beta1/FAK and alpha4beta1/paxillin signaling events. *J. Leukoc. Biol.* **2011**, *89*, 251–257. [CrossRef]
36. Massia, S.P.; Hubbell, J.A. An RGD spacing of 440 nm is sufficient for integrin alpha V beta 3-mediated fibroblast spreading and 140 nm for focal contact and stress fiber formation. *J. Cell Biol.* **1991**, *114*, 1089–1100. [CrossRef]
37. Selhuber-Unkel, C.; Lopez-Garcia, M.; Kessler, H.; Spatz, J.P. Cooperativity in Adhesion Cluster Formation during Initial Cell Adhesion. *Biophys. J.* **2008**, *95*, 5424–5431. [CrossRef]
38. Schvartzman, M.; Palma, M.; Sable, J.; Abramson, J.; Hu, X.A.; Sheetz, M.P.; Wind, S.J. Nanolithographic Control of the Spatial Organization of Cellular Adhesion Receptors at the Single-Molecule Level. *Nano Lett.* **2011**, *11*, 1306–1312. [CrossRef]

39. Gautrot, J.E.; Malmstrom, J.; Sundh, M.; Margadant, C.; Sonnenberg, A.; Sutherland, D.S. The Nanoscale Geometrical Maturatior of Focal Adhesions Controls Stem Cell Differentiation and Mechanotransduction. *Nano Lett.* **2014**, *14*, 3945–3952. [CrossRef]
40. Lv, L.; Xie, Y.T.; Li, K.; Hu, T.; Lu, X.; Cao, Y.Z.; Zheng, X.B. Unveiling the Mechanism of Surface Hydrophilicity-Modulatec Macrophage Polarization. *Adv. Healthc. Mater.* **2018**, *7*, 1800675. [CrossRef]
41. Hamidzadeh, K.; Christensen, S.M.; Dalby, E.; Chandrasekaran, P.; Mosser, D.M. Macrophages and the Recovery from Acute anc Chronic Inflammation. *Annu. Rev. Physiol.* **2017**, *79*, 567–592. [CrossRef] [PubMed]
42. Dan, H.C.; Cooper, M.J.; Cogswell, P.C.; Duncan, J.A.; Ting, J.P.Y.; Baldwin, A.S. Akt-dependent regulation of NF-κB is controllec by mTOR and Raptor in association with IKK. *Gene Dev.* **2008**, *22*, 1490–1500. [CrossRef] [PubMed]
43. Zhao, L.; Lee, J.Y.; Hwang, D.H. The phosphatidylinositol 3-kinase/Akt pathway negatively regulates Nod2-mediated NF-kappa B pathway. *Biochem. Pharmacol.* **2008**, *75*, 1515–1525. [CrossRef]
44. Park, Y.C.; Lee, C.H.; Kang, H.S.; Chung, H.T.; Kim, H.D. Wortmannin, a specific inhibitor of phosphatidylinositol-3-kinase enhances LPS-induced NO production from murine peritoneal macrophages. *Biochem. Biophys. Res. Commun.* **1997**, *240*, 692–696 [CrossRef] [PubMed]
45. Guha, M.; Mackman, N. The phosphatidylinositol 3-kinase-Akt pathway limits lipopolysaccharide activation of signaling pathways and expression of inflammatory mediators in human monocytic cells. *J. Biol. Chem.* **2002**, *277*, 32124–32132. [CrossRef] [PubMed]
46. Jimenez, C.; Portela, R.A.; Mellado, M.; Rodriguez-Frade, J.M.; Collard, J.; Serrano, A.; Martinez, A.C.; Avila, J.; Carrera, A.C. Role of the PI3K regulatory subunit in the control of actin organization and cell migration. *J. Cell Biol.* **2000**, *151*, 249–261. [CrossRef] [PubMed]
47. Mettouchi, A.; Klein, S.; Guo, W.; Lopez-Lago, M.; Lemichez, E.; Westwick, J.K.; Giancotti, F.G. Integrin-specific activation of Rac controls progression through the G(1) phase of the cell cycle. *Mol. Cell* **2001**, *8*, 115–127. [CrossRef]

Journal of
Functional Biomaterials

MDPI

Review

Calcium Phosphate-Based Biomaterials for Bone Repair

Xiaodong Hou [1,2,†], Lei Zhang [1,3,4,†], Zifei Zhou [1,3,4,*], Xiong Luo [1], Tianlong Wang [1], Xinyu Zhao [1], Bingqiang Lu [1], Feng Chen [1,*] and Longpo Zheng [1,3,4,*]

1 Center for Orthopaedic Science and Translational Medicine, Department of Orthopedics, Shanghai Tenth People's Hospital, Tongji University School of Medicine, Shanghai 200072, China
2 Department of Orthopedics, First Affiliated Hospital of Kunming Medical University, Kunming Medical University, Kunming 650032, China
3 Shanghai Trauma Emergency Center, Shanghai 200072, China
4 Orthopedic Intelligent Minimally Invasive Diagnosis & Treatment Center, Shanghai Tenth People's Hospital, Tongji University School of Medicine, Shanghai 200072, China
* Correspondence: zhouzf@tongji.edu.cn (Z.Z.); fchen@tongji.edu.cn (F.C.); dr.zheng@tongji.edu.cn (L.Z.)
† These authors contributed equally to this work.

Abstract: Traumatic, tumoral, and infectious bone defects are common in clinics, and create a big burden on patient's families and society. Calcium phosphate (CaP)-based biomaterials have superior properties and have been widely used for bone defect repair, due to their similarities to the inorganic components of human bones. The biological performance of CaPs, as a determining factor for their applications, are dependent on their physicochemical properties. Hydroxyapatite (HAP) as the most thermally stable crystalline phase of CaP is mostly used in the form of ceramics or composites scaffolds with polymers. Nanostructured CaPs with large surface areas are suitable for drug/gene delivery systems. Additionally, CaP scaffolds with hierarchical nano-/microstructures have demonstrated excellent ability in promoting bone regeneration. This review focuses on the relationships and interactions between the physicochemical/biological properties of CaP biomaterials and their species, sizes, and morphologies in bone regeneration, including synthesis strategies, structure control, biological behavior, and the mechanisms of CaP in promoting osteogenesis. This review will be helpful for scientists and engineers to further understand CaP-based biomaterials (CaPs), and be useful in developing new high-performance biomaterials for bone repair.

Keywords: calcium phosphate; bone regeneration; hydroxyapatite; biomineralization; osteogenesis

1. Introduction

Bone as a mineralized tissue has an irreplaceable role in supporting and protecting the body of human beings. Defects of bone caused by trauma, aging, inflammation, infection, and tumors seriously affect people's health and normal life [1]. A critical bone defect, which refers to a defect greater than 2 cm in length or greater than 50 percent of the circumference of the defect, cannot completely regenerate by self-growth and requires the use of biomaterials to guide its repair [2]. Millions of bone grafting operations are performed every year in the world for the treatment of critical bone defects, resulting in a huge economic burden to the families of patients and the whole of society [3]. Therefore, the development of high-performance biomaterials for bone repair is of great scientific significance and clinical application value.

At present, the most commonly used methods for treating bone defects include bone transplantation, membrane-guided regeneration, Ilizarov technology, and bone tissue engineering [4–8]. However, these methods are insufficient in meeting the requirements for clinical treatment of bone defects. Autologous bone grafting, the clinical gold standard for treating bone defects, is usually limited by the quantity of available tissue and the risk of secondary surgery and infection, blood loss and operation time [3]. Allografts and xenografts are important alternative options of autografts for orthopaedic applications in terms of osteogenic, osteoinductive, and osteoconductive properties, and there are two main categories, including cellular bone matrices (CBM) and peptide enhanced xeno-hybrid bone grafts developed as commercial products for clinical use. Cellular bone matrices have four necessary and beneficial elements for bone growth and healing: osteoinduction, osteoconduction, osteogenic activity and angiogenic activity [9]. However, CBM have several challenges with respect to intrinsic biological characteristics, such as viable cell sources, donor age at the time of graft harvest, and cell survival after transplantation, which may cause differences in expected outcomes due to different batches of the same product. Xeno-hybrid bone matrices are appealing, innovative, osteoconductive and osteoinductive bone substitutes to autografts. The compatibility of xeno-hybrid bone matrices in favoring cellular attachment, osseointegration, bone remodeling and satisfactory mechanical performance has been attested by numerous clinical studies [9]. However, further independent clinical studies are required to confirm these promising results and to promote their application. It is worth mentioning that although there are potential risks of infection when using allografts, the allografts are procured, processed, and distributed only by Tissue Banks, which operate under strict guidelines and sterile conditions in Class A environments, which helps to minimize the abovementioned issues [10]. The technology of tissue engineering represents an emerging strategy for repair of bone defect [11]. However, it is still a big challenge to construct functional bone tissue in vitro, due to the proliferation and differentiation of seeding cells, bioactivity of growth factors and physicochemical and biological properties of scaffolds [12,13]. Recently, in-situ tissue engineering has been proposed for autologous tissue regeneration, which is based on tissue-specific scaffolds, by regulating the microenvironment and in vivo recruiting stem and progenitor cells [14,15]. Therefore, the preparation of functional scaffolds with ideal biocompatibility, bioactivity and biodegradability is the critical factor that limits the rapid development of in-situ tissue engineering for bone defect repair.

Synthetic bone scaffolds have been increasingly applied in the field of bone repair. Compared with the autologous bone grafts, although there are some poor properties of osteinductive and osteogenic activities, synthetic bone scaffolds with abundant sources provide a wide variety of choices in structure, chemical/mechanical properties and biological functions to meet specific requirements in bone regeneration [15]. Considering the limitations above-mentioned, artificial bone substitutes have attracted tremendous attention and have been rapidly developed. Among the varied biomaterials used in bone repair, calcium phosphate (CaP)-based biomaterials occupy a particular position as a result of their resemblance to the chemical components and structures of natural bone tissue. CaP is not a specific material but represents a big family of materials that are compounds formed by the reaction of calcium ions and phosphate ions. The apatite reported by Werner in 1788 was the earliest discovered member of CaP [16]. By 1926, Jong revealed the relationship between apatite and the inorganic minerals of bone [17]. Therefore, CaP-based biomaterials (CaPs) were proposed for use as therapeutic agents for bone regeneration [18]. In 1971, Monroe was the first to report the use of CaP ceramics, which are white translucent polycrystalline ceramics that contain hydroxyapatite (HAP) [19]. Since that time, CaP ceramics have been developed greatly for the application of bone repair [20,21]. A CaP bone cement (CPC) was created by the hydrolysis of TCP and was used for the first time in the early 1920s as a treatment for bone repair [18]. Since then, CPC has been prepared with many different chemical formulas and application as described by Cama [22]. Mineralized collagen with orderly, organized collagen and HAP is the basic unit of natural bone tissues and is involved in building complex biomineralized systems with hierarchical structures [23]. Hence, researchers in the field of biomaterials are interested in biomimicking mineralized collagen for developing bone substitute materials by utilizing the biomimetics strategy [24,25]. In 2003, biomimetic mineralized collagen nanofibrils were designed and prepared by Cui et al., which are similar in both composition and structure to natural bone [26]. As of now, various methods have been developed for preparing mineralized collagen, and these products have excellent bioabsorbability and osteoconductive properties, leading to their potential in promoting bone regeneration [27].

With the advancement of materials science and technology, synthetic CaP materials, as promising biomaterials that provide similar bone environments for cell attachment, proliferation, and differentiation for bone regeneration, have attracted more and more interest and have shown excellent biocompatibility, osteoconductivity and osteoinductivity. However, synthesis methods, structural regulation and functionalization of CaPs are still long-term processes that need to be investigated to meet the requirements of different applications [28]. This review presents an up-to-date overview of advances in CaP biomaterials with different crystal phases and structures, strategies for fabricating biomimetic hierarchical nano-/microstructures and highlights their applications in bone regeneration (Figure 1 and Table 1). The review will help to further understand the relationships among the physical, chemical and biological properties of CaP biomaterials, and thus guide the preparation of the next generation of CaP biomaterials for bone repair.

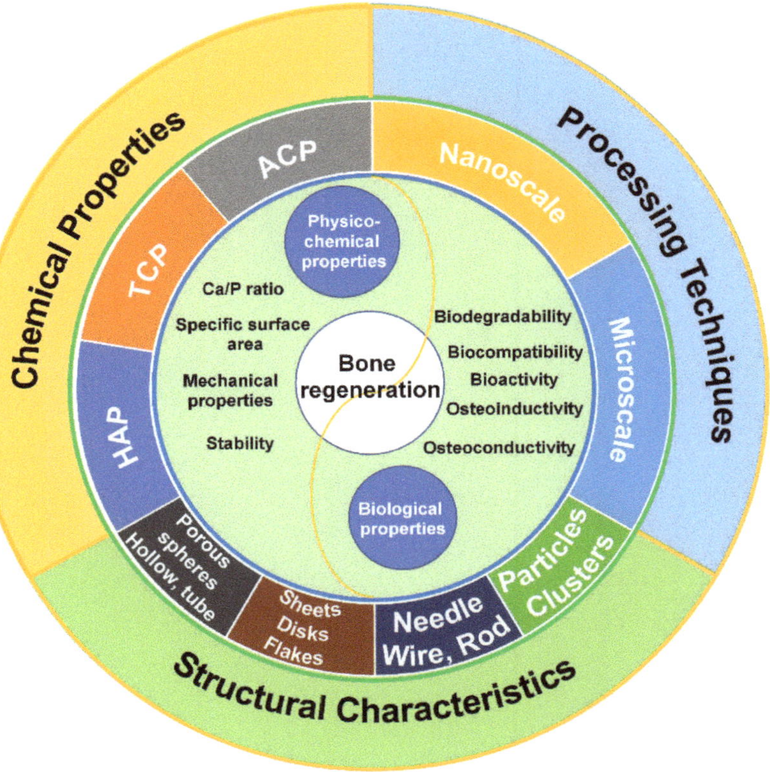

Figure 1. Factors affecting calcium phosphate-based biomaterials for bone regeneration, including chemical properties, processing techniques and structural characteristics.

Table 1. Summary of advantages and disadvantages of CaPs.

	Advantages	Disadvantages	Reference
Species			
HAP	Advanced osteoconductivity and osteoinductivity	Poor mechanical properties, and biodegradability	[29–31]
α-TCP	Advanced osteoconductivity and osteoinductivity, easy be resorpoted	Poor stability	[32,33]
β-TCP	Advanced osteoconductivity and osteoinductivity, more stable than α-TCP	Lower biodegradability than α-TCP	[31,34]
ACP	Excellent biodegradability, large specific surface, pH-responsive degradation	Lower surface energy than OCP and HAP, poor stability	[35,36]
OCP	Acts as the initial deposition site for bone, beenter osteoconductivity and osteoinductivity than HAP and ACP	Unstable, poor mechanical properties	[37]
DCPA/DCPD	Good biocompatibility, biodegradability and osteoconductivity	Poor stability, easy to casue inflammatory response by the degradation products	[38,39]
TTCP	Advanced biodegradability, biocompatibility and stability.	Cannot be synthesized in aqueous environment, easy to hydrolysis to HA	[40]
Size			
Microscales	Higher surface charge and excellent molecular adsorption properties	Lower biodegradability	[41]
Nanoscales	Improving the sintering ability of ceramics, and mechanical properties of implants, higher absorbability, easer to penetrate cell membrane.	Difficulty in synthesizing nanomaterials of specific sizes	[41]
Hierarchical nano/micro structures	Similarity of nature bone structure, better cell adhesion ability and bioactivity	Difficulty in controlling	[42]

2. Methods

The reports and information were gathered using the bibliographic databases PubMed and Web of Science. In addition, a search on Bing was performed to supplement the information. The following term combinations were searched: "calcium phosphate-based materials name" and ("bone regeneration" or "bone repair" or "bone healing"), "calcium phosphate-based materials name" including "calcium phosphate", "CaP", "hydroxyapatite", "HA", HAP", "tricalcium phosphate", "TCP", "amorphous calcium phosphate", "ACP", "octacalcium phosphate", "OCP", "dicalcium phosphate anhydrous", "DCPA", "dibasic calcium phosphate dihydrate", "DCPD", "tetracalcium phosphate", "TTCP", "dicalcium phosphate monohydrate", "DCPM", "cement" and "ceramic". Considering only innovative reports, the range of the publishing data was set as "2010–2022". However, some classical and revolutionary works outside this window were also included. This study reviews and discusses representative literature.

3. Chemical Properties of Calcium Phosphate

3.1. Species of Calcium Phosphate

CaP is not only a family of natural minerals, but also includes biominerals in humans, which are the main inorganic component of hard tissue (bone and teeth) [31]. In the past decades, a wide variety of CaP-based biomaterials have been used in bone regeneration studies and clinical applications. As is well known, CaP biomaterials promote cell adhesion and growth, which then induces the formation of new bone minerals via their interaction with extracellular matrix proteins [43]. In the application of bone regeneration, the bioactivity of CaPs is critical and usually varied depending on their species [44]. The bioactive features of CaPs is related with to the degradation properties of CaP [45]. Due to different Ca/P ratios, the different species of CaP biomaterials result in variations in in vitro and in vivo calcium and phosphate ion release. Consequently, the pH of the local microenvironment of bone is affected by the released calcium and phosphate ions, which then influence the viability of osteoblasts and osteoclasts [46,47]. Moreover, the increased concentration of calcium and phosphate ions can promote the formation of bone minerals, as well as affect the expression of osteogenic differentiation-related genes (e.g., Col-I, ALP, OPN, OCN, RunX2 and BMPs) of bone cells [31,48].

Calcium exists wildly in natural bone minerals and is a key ion in forming the bone matrix [49]. Ca^{2+} is also capable of forming and maturing bone tissue by calcification. Ca^{2+} influence the bone cell maturation and bone tissue regeneration by regulating related cellular signaling pathways [50,51]. For instance, Ca^{2+} activating ERK1/2 causes pathway activation of osteoblastic-related bone formation [52]. In addition, an increased life span of osteoblasts has been observed with the activation of the PI3K/Akt signal axis by Ca^{2+} [53]. Meanwhile, phosphate ions degraded from CaP are present in large quantities in the human body, and can be utilized in various physiological systems, including construction of proteins, nucleic acids, and adenosine triphosphate [54]. Approximately 80% of phosphate ions in the body occur with calcium ions in the form of CaP in bone, which affects the development of bone tissue [55]. It is well known that the differentiation and growth of osteoblasts are regulated by phosphate ions by IGF-1, ERK1/2, BMP, and other pathways [56,57].

The osteoconductivity and osteoinductivity of CaP materials are closely related to their physical and chemical characteristics, such as, solubility, stability, and mechanical strength [31], are determined by the species of the CaP materials, Therefore, the selection of one kind of CaP biomaterial from their family according its characteristics is important in preparing biomaterials for the use of bone regeneration. A large number of CaP biomaterials have been used in bone regeneration and other biological research, including tricalcium phosphate (TCP), hydroxyapatite (HAP), amorphous calcium phosphate (ACP), octacalcium phosphate (OCP), dicalcium phosphate anhydrous (DCPA), dibasic calcium phosphate dihydrate (DCPD), and tetracalcium phosphate (TTCP) [42]. Basic information concerning these CaP biomaterials are displayed in the Table 2 [33]. The crystal phases of calcium phosphate were discovered before the 20th century (amorphous phases were discovered in the 1950s) and accurately characterized during the 20th century (OCP was defined in 1957 [58]). After that, no new crystal phases (non-doped, non-substituted, only Ca, P, O, H) have been reported. Dicalcium phosphate monohydrate ($CaHPO_4 \cdot H_2O$, DCPM) was obtained by controlling the transformation of a special amorphous calcium phosphate ($CaHPO_4 \cdot xH_2O$, ACHP) in a water-deficient environment (water/methanol mixed solvent, or in humid air) by Lu et al. in 2020 [59]. The discovery of DCPM brought a new member to the calcium phosphate family. However, applications of DCPM in bone repair, and even in biomedicine, has not yet been carried out.

Table 2. The basic information of usual calcium phosphate materials.

Name	Formula	Ca/P	Solubility at 25 °C (g/L)
HAP	$Ca_{10}(PO_4)_6(OH)_2$	1.67	~0.0003
α-TCP	α-$Ca_3(PO_4)_2$	1.5	~0.0025
β-TCP	β-$Ca_3(PO_4)_2$	1.5	~0.0005
ACP	$Ca_xH_y(PO_4)_z \cdot nH_2O$, n = 3–4.5, 15–20% H_2O	1.2–2.0	/
OCP	$Ca_8(HPO_4)_2(PO_4)_4 \cdot 5H_2O$	1.33	~0.0081
DCPA	$CaHPO_4$	1.0	~0.048
DCPD	$CaHPO_4 \cdot 2H_2O$	1.0	~0.088
TTCP	$Ca_4(PO_4)_2O$	2.0	~0.0007

HAP: hydroxyapatite, α-TCP: α-tricalcium phosphate, β-TCP: β-tricalcium phosphate, ACP: amorphous phosphate calcium. The solubility date at 25 °C of ACP cannot be measured precisely. However, the comparative solubility in acidic buffer is ACP >> α-TCP >> β-TCP >> HAP.

3.2. Hydroxyapatite

HAP is the most abundant crystal phase of biominerals in human bones, and accounts for ~70% of the dry weight of bone tissue [60]. Among all CaP materials, HAP is only inferior to fluorapatite (FAP) in terms of stability and insolubility. The chemical formula of HAP is $Ca_5(PO_4)_3(OH)$. However, HAP is usually referred to as $Ca_{10}(PO_4)_6(OH)_2$ to indicate the hexagonal unit cell of HAP [61]. There are two approaches for the formation of HAP, including the natural formation process and artificial synthesis. The hexagonal crystal structure of naturally formed HAP usually has defects, which can be filled by vacancies or other ions [33]. However, structural defects in synthesized HAP may depend on synthesis procedures or conditions. Monoclinic and hexagonal crystals are the two crystal phases of synthesized HAP; the monoclinic crystal phase can change to the hexagonal crystal phase when the temperature is higher than 250 °C. The hexagonal crystal structure of HAP is the predominant phase found in the biological environment as a result of its high stability [62]. HAP is considered the most stable phase of CaP and the final mineral phase in bone, whereas the other CaP phases (e.g., ACP and OCP) in bone are precursors, or sub-precursors, that transform into HAP under in vivo or aqueous environments with high pH [31,33]. The phase transformation of several CaPs usually occurs under different conditions [63]. The equilibrium of phase transformation between various CaP phases is related to temperature and the ratio of CaO and P_2O_5. The physicochemical and biological properties of HAP significantly change with the Ca/P ratios and the replacement of ions or vacancies in HAP crystal structure. For example, the mechanical properties of HAP are enhanced with increasing Ca/P ratio, and reach a maximum when the stoichiometric ratio is 1.67. Interestingly, once the Ca/P ratio exceeds 1.67, the strength of HAP decrease [64]. The defects of HAP crystal structure can be replaced with F^-, Cl^-, CO_3^{2-}, Mg^{2+}, Sr^{2+}, and other ions. As a result of Mg^{2+} replacement, the size and density of HAP nanostructure particles, which contribute to the specific mechanical properties of bones, may be altered [65]. Furthermore, the crystallization of HAP is inhibited by Mg^{2+}, and results in the formation of fewer large crystals and a greater number of apatite nuclei. The significance of this is that nanocrystalline bone apatites are necessary for the proper bone formation–resorption turnover that occurs via bone cells [66]. The replacement of F^- ions can increase stability, while Mg^{2+} enhances biological activity, compared with pure HAP [66]. Several studies have shown that Mg^{2+} has the ability to influence bone metabolism, regulate the activity of osteoblasts and osteoclasts, and to stimulate new bone growth [67,68]. Therefore, artificial Mg substituted HAP in different forms has been carried out, and has displayed advanced bioactivity [69,70]. Furthermore, Mg-based CaP materials can result in neuralization and the synthesis and release of CGRP to promote osteogenesis [71–73].

HAP has been used clinically in bone regeneration since the 1980s, as implants and coatings of other implants [61,74]. HAP has good biocompatibility, bioactivity, and osteoconductive properties. In the presence of Ca^{2+} and $PO_4{}^{3-}$ ions, the surface of HAP can act as a nucleation site for the initiation of biomineralization [75]. Therefore, HAP is used widely for dental surgery, long bone defects, bone nonunion, vertebral fusion operation and maxillofacial repair [76]. The biocompatibility, osseointegration, and bioactivity of metal implants are improved by coating their surfaces with HAP, which enhances the bone contact area and cell adhesion properties of the implants [30]. Moreover, HAP can promote the biomineralization of macromolecule-based scaffolds. HAP nanoparticles penetrate into the demineralized collagen scaffold and serve as mineralization seeds that promote the occurrence of remineralization of the collagen matrix [77].

HAP has high chemical stability, but has weakness in mechanical properties. Stress along the Z-axis direction of HAP crystals creates brittleness [78]. It is worth mentioning that wear resistance, the friction coefficient and hardness of dense HAP are similar to natural those of mineralized tissues [32]. The fatigue resistance of dense HAP is superior to porous HAP [64]. As a result, HAP is not used as a load-bearing implant due to its poor mechanical properties, but is usually implanted with granules and porous scaffolding [29]. It is still a huge challenge to improve the mechanical properties of HAP. Metal oxides including zirconia, alumina and titania are common reinforcing agents [79]. However, the biocompatibility and biodegradation properties of HAP-based biomaterials are compromised by the addition of these reinforcing agents, which are bioinert or none-biodegradable [80]. Constructing composites with a polymer is an effective way to improve the mechanical properties of HAP. Natural polymers such as chitosan, hyaluronic acid, silk fibroin and gelatin are common components for fabricating hybrid scaffolds [81]. For instance, the hydroxy propyl methyl cellulose of chitosan has been crosslinked to fabricate chitosan/HAP sponge-like scaffolds which have excellent compressive strength, elasticity and degradability [82]. Considering the importance of mechanical properties for bone repair, particularly in load-bearing bones, further research is necessary to improve the mechanical properties of HAP [83,84].

The biological performance of artificial bone implants is extremely important. HAP is considered to have good biocompatibility and bioactivity in osteoconductivity, but has poor osteoinductivity [75]. Therefore, it is usual to combine HAP with other materials to improve its osteoinductivity. Beta-tricalcium phosphate (β-TCP), another common kind of calcium phosphate used in bone regeneration, has better osteoinductivity than HAP. This biphasic calcium phosphate (BCP) material has been synthesized by combining HAP and β-TCP to take advantage of the properties of both and obtain better bioactivity for bone regeneration [85]. The BCP material possesses superior bioactivity, biodegradability, osteoinductivity, and mechanical properties than HAP or β-TCP alone, and has greater ability to stimulate osteogenic differentiation of BMSCs [86]. Hence, bone grafts and dental materials are commonly prepared with BCP material [87]. Zhu et al. constructed BCP bioceramics with micro-whiskers and a nanoparticle hybrid structure which may be applied in research on load-bearing bone tissue regeneration to provide mechanical support [88].

HAP is an advanced material for preparing bone grafts owing to its similarity to natural minerals and excellent biocompatibility and osteoconductivity. However, the preparation performance regulation of hydroxyapatite materials for bone regeneration remains a long-term and challenging endeavor. First of all, basic research on hydroxyapatite in bone tissue remains largely unexplored. For example, there is a lack of understanding regarding the factors involved in the formation of hydroxyapatite, including precursors and crystal growth regulatory factors, in the process of bone tissue biomineralization. In addition, the interaction between hydroxyapatite and collagen molecules, particularly how the regular hydroxyapatite-collagen complex is formed, needs to be further studied. On the other hand, there are still many technical and scientific problems associated with the synthesis of hydroxyapatite and the preparation of scaffolds. For instance, the mechanism of hydroxyapatite crystal growth needs to be further explored to control the scale, and a controllable fabrication strategy for ordered biomimetic structures must be developed to fabricate scaffolds with good mechanical properties and controllable porous structures.

3.3. Tricalcium Phosphate

TCP, as one of the most studied calcium phosphate materials, contains two crystalline phases (α-TCP and β-TCP). There are several phases of CaP materials that have similar compositions to TCP, and the term TCP here is used for the phase with a chemical composition of $Ca_3(PO_4)_2$ and a Ca/P ratio of 1.5. Pure crystalline α-TCP cannot be precipitated in aqueous solutions since it is very poorly soluble, unlike β-TCP [89,90]. There are three approaches for synthesizing β-TCP, including solid-state reaction, thermal conversion, and precipitation. Usually, crystalline β-TCP is prepared at a high temperature of ~800 °C such as by thermal decomposition of calcium deficient hydroxyapatite (CDHA), and the other is the solid-state interaction between acidic CaP (i.e., DCPA) and alkaline (i.e., CaO) [33]. As well, it has been shown that β-TCP precipitates in organic solutions, such as ethylene glycol, methanol, tetrahydrofurane, and ethyl propionate [91–93]. Tang et al. synthesized β-TCP at a relatively low temperature at about 150 °C in organic solvents (e.g., ethylene glycol) [94]. Moreover, β-TCP transforms into the α-TCP at higher temperatures (above 1125 °C), so α-TCP may be considered as the high-temperature phase of β-TCP [38].

TCP has excellent stability and can be stored in a dry environment at room temperature for a long period of time. β-TCP is more stable than α-TCP according to a density functional study [95]. α-TCP has superior reactivity and specific energy in an aqueous solution than β-TCP, and is capable of being hydrolyzed to CDHA [33]. In clinical applications, β-TCP has higher osteoconductivity and osteoinductivity than HAP and is primarily used in bone cements and bioceramics [34,96], while α-TCP is normally used in cements, since it is subject to a phase conversion to HAP upon water contact [97,98]. It should be noted that the rate of resorption of pure α-TCP is higher than new bone formation, which leads an imbalance between the process of bone formation and implant degradation [38]. Therefore, α-TCP is usually used as a component in CaP cements with other CaP materials [33]. In contrast, β-TCP has a relatively lower resorption rate than α-TCP, and has good prospects for application in bone regeneration [32]. The nano-porous structure of β-TCP allows for excellent biomineralization and cell adhesion; these properties can stimulate osteoblast and BMSCs proliferation [99]. In addition, compared to HAP, β-TCP has better biodegradability and resorption rate, which can increase the biocompatibility of the implants for bone regeneration [31].

3.4. Amorphous Calcium Phosphates

Amorphous calcium phosphates (ACPs) are a special phase of CaP with various chemical compositions. ACPs have long-range order but short-range disorder regarding their crystal properties [33]. Initially, ACPs were discovered during the preparation of HAP in vitro; therefore, ACPs were considered as precursors of HAP [100]. A study in 1972 found that ACPs were the first phase to form and were transformed into octacalcium phosphate (OCP) during the synthesis of HAP in vitro; the final phase conversion occurred from OCP to HAP [101]. Glimcher et al. believed that ACPs may be the precursor stage of bone formation due to the presence of uniform intra-collagen mineralized particles found in collagen mineralization in vitro through ACPs [102]. ACPs are classified into two groups based on their preparation temperature, namely low-temperature ACPs and high-temperature ACPs [36]. Low-temperature ACPs usually occur as precursors during the precipitation process of other CaP compounds. Since the surface energy of ACPs is lower than that of OCPs and HAPs, ACPs are thought to form at the onset of precipitation [36].

The chemical composition of ACPs depends on pH value and the concentration of calcium and phosphate ions in aqueous solution. The recrystallization of ACPs occurs with increased concentration of Ca^{2+} and PO_4^{3-}. In addition, ACPs may recrystallize slowly or transform into CaP materials with a higher crystalline degree, such as CDHA, in a reaction system with a continuous and mild stirring rate, especially when at higher reaction temperatures [33]. In studies concerning the influence of pH, researchers have discovered that the Ca/P ratio of ACPs increases from 1.18 to 1.53 as the pH value of the system changes from 6.6 to 11.7 [103,104]. Up to now, the atomic distribution in ACPs is still not entirely clear, which is an important research topic in studies of biominerals [36]. Freshly precipitated ACPs usually display spherical-like structures with diameters between 20 and 200 nm as seen by electron microscopy [33]. Some researchers believe that the basic structural unit of ACPs is thought to be a spherical cluster structure with a diameter of 0.95 nm. The ACP chemical formula is $Ca_9(PO_4)_6$ [36,105].

ACPs constitute the initial phase of HAP and are essential components in the process of bone regeneration and bone mineralization due to their particular physical properties and structure [106]. ACPs possess superior biological properties, such as osteoconductivity and biodegradability, leading to a variety of applications including CaP bone cements, biological tissue engineering scaffolds, bone repair biomaterials, and dental implants [107,108]. In addition, the nano-sized clusters in the ACPs have characteristics of large specific surface areas and pH-responsive degradation, which makes them ideal drug delivery carriers for studies relating to drug loading and controlled release [35].

The preparation of ACPs is regulated by small molecules such as ATP, which can effectively inhibit the phase transformation of ACP [109]. As a result of an ATP-assisted preparation strategy, the product is an ACP composite nanoparticle containing ATP and ADP molecules. Furthermore, the compound has good biocompatibility and osteogenic activity, and can up-regulate the expression of osteogenic genes in BMSCs. An injectable hydrogel prepared by combining ACP compounds with fibrinogen displayed excellent promoting effects in in vivo bone regeneration [110].

An ALP-catalyzed hydrolysis reaction was used to generate EACP nanominerals in an alkalescent aqueous solution similar to mitochondrial surroundings [109]. The mechanism of EACP promoting bone healing was demonstrated in that the ADP/AMP biomolecules and Ca^{2+} ions released from EACP can increase the activation level of AMPK and promote autophagy and osteogenic differentiation in hBMSCs. Additionally, a number of theories suggest that ACP plays an important role in the biomineralization process as a precursor of apatite formation [111–113]. There is also evidence that the biomineralization route involves the formation of the mineral phase within matrix vesicles that are associated with small crystals of calcium phosphate mineral [114], which are usually an amorphous phase involved in the formation of these vesicles [115]. Meanwhile, disordered collagen fibrils may contribute to the stabilization of ACP, resulting in both amorphous and crystalline bone mineral [116]. At present, the collagen fibrils as the temptation for bone mineral growth are attracting a great deal of interest in biomineralization research [117]. However, the specific mechanisms by which ACP promotes bone regeneration are highly controversial and require further investigation.

3.5. Application of Other CaP Phases

Tetracalcium phosphate (TTCP) is the most basic CaP phase, which is in a metastable state and gradually hydrolyzed to HAP and $Ca(OH)_2$ in a humid environment or aqueous solution [38]. TTCP often occurs as a by-product of HAP plasma coating, which is a mixture of α-TCP, TTCP and CaO from the high temperature phase [40]. The chemical synthesis process of TTCP can only be carried out in dry air or vacuum environment. It is synthesized by a solid-phase reaction at over 1300 °C. In the presence of water vapor, TTCP is decomposed into HAP [38]. There are three types of TTCP bone cement: single component, multi component, and polymer. In biological applications, TTCP is usually used as a component for preparing self-curing bone cements, biological composites or root canal sealants [118–120]. By combining TTCP with DCPA or DCPD (Ca/P = 2.0), bone cements with the stoichiometry composition of HAP can be produced [40]. A set cement with the best mechanical properties was obtained using an equimolar mixture TTCP and DCPA with a particle size (diameter) ratio of approximately 10: 1 [40]. However, another study found that cement with a diameter ratio (TTCP:DCPA) of 20:1 had the highest compressive strength [121].

TTCP bone cements show advanced biological performance. Qin et al. fabricated three-dimensional porous TTCP scaffolds via selective laser sintering technology (SLS) [122]. After immersion in SBF for one day, nanoapatite was produced on the surface of TTCP scaffolds. The scaffold surface was completely covered with apatite after three days, indicating good biological activity. Furthermore, the biocompatibility of TTCP scaffolds was evaluated by cell culture, which confirmed their high biocompatibility. An evaluation of the histological effects of the TTCP cement applied to the pulp of rat upper incisors demonstrated great advantages over calcium hydroxide ($Ca(OH)_2$) cement [123]. Tsai et al. investigated a single component TTCP cement (containing $(NH_4)_2HPO_4$ as the liquid) in rabbit femurs for 24 weeks in vivo [124]. Following implantation, histological examination of the recovered implants demonstrated good cement-bone host bonding, with cement resorption, new blood vessels, osteoocytes, and osteoblast-like cells identified. At the end of 24 weeks, only a small amount of residual bone cement was found, and the cortical bone was almost completely remodeled.

Octacalcium phosphate (OCP), as a precursor to HAP crystal formation, along with ACP and DCPD, play an important role in bone formation and biomineralization [37,125]. A very similar structure exists between OCP and HAP but OCP is more unstable than HAP and is hydrolyzed to HAP [37]. The mechanism of hydrolysis of OCP is still not completely clear. Two hypotheses, dissolution-reprecipitation mechanism [126] and ion diffusion-crystallization conversion [127], are proposed to explain the hydrolysis of OCP. Eliminating the HPO_4^{2-} from the OCP water layer has been confirmed as a necessary step for phase transformation, and is believed to be the rate-determining factor of the conversion [128]. The transformation of OCP was observed under in vitro and in vivo conditions. Upon being placed in water with a starting pH of 7.2, the mixture of OCP and HAP was examined after 1 h, and at 12 h the structural transition was completed [129]. The OCP was completely hydrolyzed to CDHA within 6 h in deionized water [130]. pH also affects the transformation rate. For example, Suzuki et al. found that OCP hydrolysis was postponed at pH 11 compared with pH 7.4 [131]. Interestingly, the hydrolysis of OCP into HAP is very slow in in vivo conditions. Implanted OCP in a rat calvarial defect hydrolyzed very slowly after 21 days [132]. In SBF at a temperature of 36.5 °C and a pH of 7.25, the hydrolysis of OCP to HAP took place to a small extent over the course of 28 days [133]. In addition, OCP may be non-stoichiometric, and its structure may be calcium-deficient (Ca/P = 1.26) or calcium-excessive (Ca/P = 1.48) [134].

OCP has good osteoinductivity and is widely used in bone repair research, including the coating of metal grafts, the use of CaP bone cement, and the construction of composite bone repair scaffolds [135–137]. OCP/Col composite scaffolds constructed from OCP particles and collagen have a higher osteoconductivity than OCP alone, and the osteoconductivity is positively correlated with the dose of OCP [138,139]. By providing a nuclear structure, OCP acts as an initial deposition site for bone, and its conversion to HAP plays a significant role in bone formation, which may explain its beneficial effects on bone growth [132,140,141]. By implanting the precursors of HAP, such as OCP, ACP, and DCPA, along with HAP particles in the subperiosteal region of the mouse calvaria, bone tissue appeared with OCP in approximately one week. At about 3 weeks, bone tissue appeared in ACP and DCPA, and was later found in HAP particles (5 weeks), which further indicates that OCP has good ability to promote bone formation [140]. Moreover, osteoblasts that can initiate bone formation were found on the surface of OCP particles in the OCP group. Ultrastructural SEM examination confirmed that osteoblasts were directly attached to OCP to form bone matrix, and that filaments were formed around OCP. There were many similarities in the composition of the granulated and granular complexes around with OCP to the bone nodules formed during intramembranous osteogenesis [142].

The application of OCP in bone graft biomaterials and bone regeneration has a promising future due to its good osteoconductivity and osteoinductivity. It is of paramount importance to explore and understand the biological mechanism of good osteoinductivity of OCP, as well as the influence of Ca/P stoichiometry and microstructure on its intrinsic biological activity [143,144].

Dicalcium phosphate anhydrous (DCPA) and dibasic calcium phosphate dihydrate (DCPD) are acidic CaP materials. DCPA is an anhydrous crystalline form of DCPD. Since there are no hydrated molecules, the solubility of DCPA is lower than DCPD. Both can be precipitated from an aqueous solution at 100 °C. The difference between DCPA and DCPD is that DCPA does not form in vivo, as confirmed by no DCPA being formed in normal or pathological calcification nodus [33]. DCPA is often mixed with other calcium phosphate materials to prepare bone cement, and it is also used to provide calcium and phosphorus in foods and toothpastes [38,145,146].

DCPD is the dihydrate crystalline state of DCPA [38]. By adjusting pH in the range of 3–4 at room temperature, DCPD can be produced by the neutralizing reaction of Ca $(OH)_2$ and H_3PO_4. Metathesis reactions using calcium-containing phosphates in a slightly acidic environment can also synthesize DCPD [145]. Due to its biocompatibility, biodegradability, and osteoconductivity, DCPD is often used as a component of bone cements and toothpaste to promote bone and tooth mineralization [38,147]. It is worth noting that DCPD was found to be converted to calcium deficient hydroxyapatite (CDHA) in vivo [148]. This conversion process released many acidic substances when excessive DCPD was implanted in vivo, causing a severe inflammatory response [39].

Dicalcium phosphate monohydrate (DCPM) as a crystal phase has a Ca/P ratio of 1:1, and is a new metastable CaP with structural water without DCPD and DCPA [59]. DCPM is formed using ethanol and water mixtures that maintain a low level of hydration and inhibit the formation of DCPD. X-ray powder diffraction (XRPD), was used to determine the crystal structure of DCPM, conformed with a = 8.0063(4) Å, b = 6.7954(5) Å, c = 7.7904(5) Å, $\alpha = \gamma = 90°$, β = 91.548(4) Å. In addition, after immersion in water for only one hour, this new crystalline form of calcium phosphate monohydrate transforms into hydroxyapatite, which is the stable form of calcium phosphate found in human bones. This represents a two-fold increase in speed as compared to the dicalcium phosphate dihydrate (DCPD) phase, which is usually used in bone cements today. Furthermore, DCPM can be stabilized by organic molecules such as citrate salts, which are abundant in the human body, and can adsorb a large quantity of small molecules. Consequently, DCPM is an interesting option to encapsulate and release drugs to enhance bone healing and remineralization. However, there is still much work to be done on the features and applications of DCPM in bone repair and biomedicine.

And some commercial products of CaPs were displayed in the Table 3.

Table 3. The commercial products of CaPs.

CaP Materials	Product Name	Producer
HAP	Actifuse ApaPore Bonetite Bone Source Bioroc Cerapatite Ostim Synatite	ApaTech, UK ApaTech, UK Pentax, Japan Stryker orthopaedics, NJ, USA Depuy-Bioland, France Ceraver, France Heraeus Kulzer, Germany SBM, France
β-TCP	adbone® TCP Biosorb Cerasorb Conduit Osferion SynthoGraft Vitoss	Medbone, Portugal SBM S.A., France Curasan, Germany DePuy Spine, USA Olympus Terumo Biomaterials, Japan Synthograft, MA, USA Orthovita, PA, USA
HAP + β-TCP	BCP Graftys BCP MBCP OsSatura BCP	Medtronic, MN, USA Graftys, France Biomatlante, France Integra Orthobiologics, CA, USA
HAP + α-TCP	Skelite	Millennium Biologix, ON, Canada
CDHA	Osteogen	Impladent, NY, USA
ACP + DCPD	Biobon (α-BSM)	Etex, MA, USA
DCPD + β-TCP	ChronOS	DePuy Synthes, PA, USA
TTCP + DCPA + saline	BoneSource HAC	Stryker Instruments, MI, USA
α-TCP + TTCP + $CaHPO_4$ + HAP	BIOPEX	Taisho Pharmaceutical, Japan
HAP + collagen	Healos Fx	DePuy Spine, USA
HAP + PLLA	SuperFIXSORB30	Takiron, Japan
HAP + Polyethylene	HAPEX	Gyrus, TN, USA
β-TCP + PMMA	Cal-CEMEX	Tecres Spa, Italy

4. Effects of Sizes and Structural Characteristics of CaP Materials

4.1. Sizes of CaP Materials

Vertebrates' bones are composed of multiple levels and sizes of units, ranging from nanometers to micrometers, with precise yet complex arrangements (Figure 2 [149]). In terms of microstructure, the trabeculae thickness ranges between 50 and 300 μm, with their orientation depending on the distribution of load in the bone [150]. Mineralized collagen fibers form lamellae whose width is approximately 3–7 μm [150]. Mineralized collagen fibrils with a diameter about 100 nm are formed on the lamellae at the nanoscale level. These fibrils, 300 nm long and 1.5 nm thick, are the basic building blocks of the bone material and consist of collagen molecules [33]. Crystals grow with an approximate repeat distance of 67 nm on the fibrils [151]. This hierarchal structure of biomineralized fibrils and trabeculae from the nanoscale to the microscale are critical to the isotropic properties in the bone, and enhance load-bearing capability. Therefore, synthesizing different sizes of CaP materials, constructing the biomimetic hierarchal grafts, and understanding the biological performance of various sizes of CaP crystals should improve bone defect treatment.

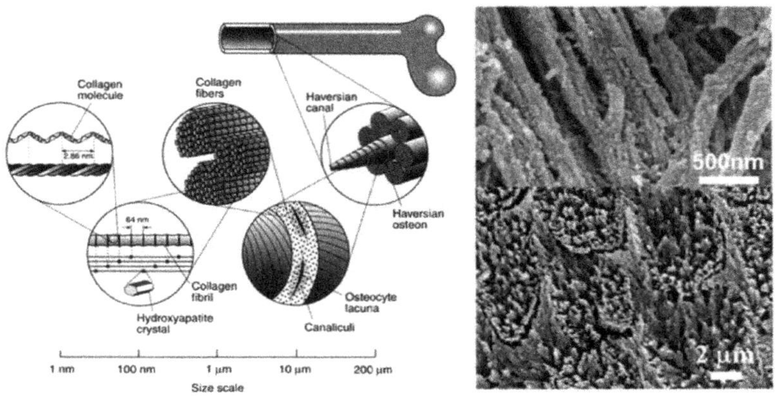

Figure 2. Bone (left) is a complex, hierarchically structured biological material in which the building components are precisely arranged at scales spanning six of orders of magnitude. The image on the left shows sketches of the structural elements of cortical/compact bone (which comprises the harder, outer layer of the cross-section of bone, surrounding the softer trabecular/spongy/cancellous bone) at different scales. The image on the upper right side shows the nanostructure of mineralized collagen fibers in bone. HAP particles are incorporated within the organic matrix. The image on the below right displays the fine structure of dental enamel, the hardest substance in the body, which is composed of an almost pure mineral with elongated HAP nanofibers connected into bundles and forming equally uniaxially directed enamel rods. Reprinted with permission from Ref. [149]. 2011, Uskokovic, V.

Over the past decades, many strategies have been developed to control the size of CaP crystals, including chemical precipitation, sol-gel process, microemulsion, a hydrothermal method, solution combustion synthesis and an electrospinning method. In addition, microwave, ultrasonic and gravity precipitation methods, as well as the use of reagents such as proteins, polymers and chelated reagents, regulates the size of CaP crystals. Synthesis methods for controlling the size of CaPs are the initial and key steps to understanding the physiochemical properties and biological performance of different sizes of CaPs. The specific methods and details of these strategies are described in [152]. Several classical and effective strategies are introduced below.

The chemical precipitation approach is the simplest method for preparing CaP crystals. The size of CaP crystals is usually related to the reaction time and temperature [153]. However, the difficulty of precisely obtaining the size of CaP crystals is a big problem using this method. The microemulsion method is highly effective for regulating the size of CaP crystals due to the uniform and narrow channels of the microreactors strictly restricting the nucleation and growth of crystals [154,155]. The surfactants in the microemulsion play a critical role in modulating the CaP crystal size. When the molar ratio of organic solvents to surfactants increases, the length of HAP nanorods increases also [156]. In addition, changing the molar ratio between water and surfactants ($W_0 = H_2O$/surfactant) can also regulate CaP crystal size. HAP products are changed from nanospheres with a diameter about 25–40 nm to the needle-like crystals with 4–8 nm in diameter and 80–100 nm in length when the W_0 is increased from 5 to 8, and when the increase of W_0 was from 5 to 15, the HAP products were changed to the rod-like crystals (10–17 nm in diameter and 24–50 nm in length) [157].

The use of chelators is another method for controlling the size of CaPs [152,158]. For instance, HAP nanorods (120 nm in average length and 25 nm in diameter) were obtained in the presence of citric acid, while nanowires (0.7–1 μm in length and 30–40 nm in diameter) were produced in the absence of citric acid (Figure 3) [158]. Regulating the concentrations of precursors can, to a certain degree, tailor the size of CaP crystals via precursor transformation routes. Figure 4 shows that the size and aspect ratio of the β-TCP platelets increases as the concentration of the precursor increases [94,159].

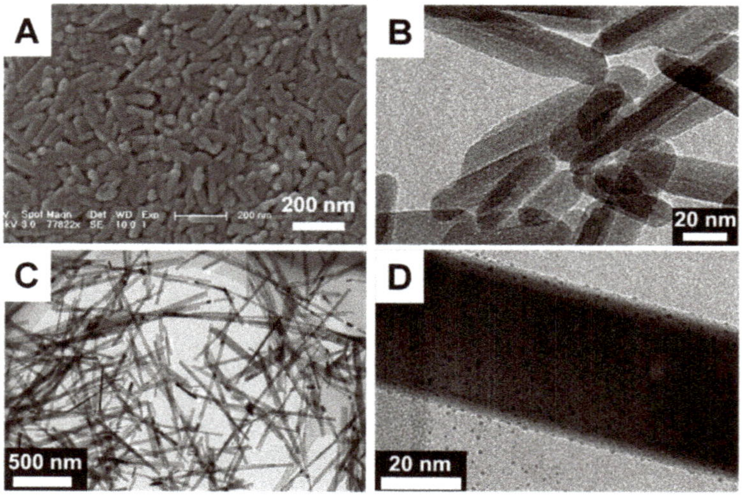

Figure 3. SEM and TEM images of HAP crystals with different sizes after in the addition (**A**,**B**) and in the absence of citric acid (**C**,**D**). Reprinted with permission from Ref. [158]. 2019, Zhang, C. M.

It is known that the properties of CaPs are closely related to their size, which may determine their application potential [42,160,161]. The nanoscale CaP particles used for bioceramics display improved sintering ability in ceramics preparation and prevent microcracks caused by extreme sintering temperature, owing to high surface energy [29]. Implants made from HAP nanoparticles have superior mechanical properties compared to micro-sized HAP crystals [156], and offer an effective way to enhance the mechanical properties of CaP-based grafts. Furthermore, CaP materials with nanoscale crystals have greater absorbability compared with microscale crystals, and are therefore more suitable for use in bone tissue regeneration [156]. In addition, nanoscale CaP crystals are usually used in cell targeting [162], drug/protein/gene delivery systems [163,164] and gene transfection/silencing [165,166].

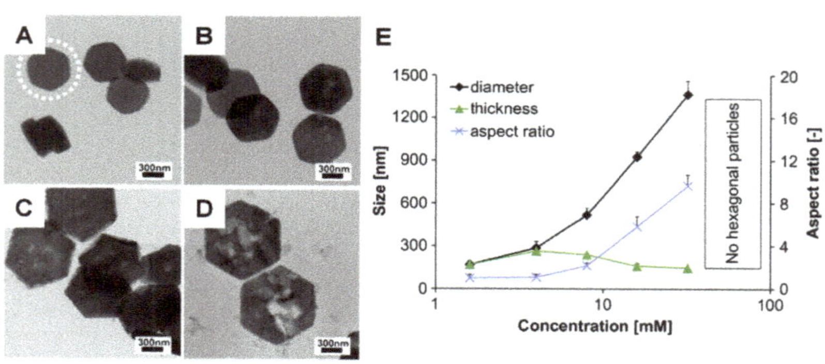

Figure 4. TEM images of β-TCP platelets with different precursor amounts of (**A–D**) 10, 15, 26 and 40 mg, respectively. (**E**) Effect of precursor levels on the dimensions of hexagonal β-TCP platelets. Reprinted with permission from Ref. [94]. 2009, Tao, J. H. and Ref. [159]. 2013, Galea, L.

For biological effects, the size of CaP materials plays a critical role in regulating osteoblast proliferation, cellular activity, apoptosis, and osteogenic differentiation [31,42,161]. And some in vitro/ vivo experiments of CaPs and their markable outcomes were displayed in Table 4. The HAP particles at the nanoscale, particularly those with a diameter of 20 nm, have been shown to promote cell proliferation and bioactivity, and inhibit cell apoptosis compared with those at the microscale [41,167]. These positive effects are probably attributable to the better penetration abilities of smaller HAP nanoparticles (Figure 5) [41]. The size of CaP materials affects bone regeneration by inducing macrophage polarization [168]. HAP crystals at the nanoscale induce M2 macrophage polarization [169], while at the microscale polarize into M1 macrophages [170].

Table 4. In vitro and in vivo experiments of CaPs.

First Author	CaPs	In Vitro	In Vivo	Outcomes	Reference
Mahon OR	HAP nanoparticles	hBMSCs, HUVECs	Rat	Promoting M2 macrophages polarization and angiogenesis; Specifically enhancing IL-10 production	[171]
Ji C	β-TCP scaffold	rBMSCs, HUVECs, RAW264.7	Rat	Specifically enhancing the expression of osteoclast differentiation and extracellular space pathway genes to promote the process of bone remodeling	[172]
Raymond Y	α-TCP scaffold	MG-63	Rabbit	Hydrothermal process promising a more favorable microstructure, nanoporosity, and nanopore size; significantly enhancing bone formation	[173]
Zhou Z	ACP/GelMA scaffold	rBMSCs	Rat	Inducting the ALP into the biomimetic strategy to produce mineralized ACP nanoparticles; enhancing the proliferation of BMSCs and upregulating the osteogenic differentiation owing to bioactivity of ALP	[174]
Kurobane T	OCP/gelatin scaffold	HUVECs	Rat	Stimulating the angiogenesis then enhancing the bone regeneration. Exploring the relationship between OCP dose and angiogenesis	[175]
Sheikh Z	DCPA cement	-	Rabbit	Complete resorption and more bone formation than DCPD cement. Bone formation and resorption in DCPA cement are site specific	[176]
Ko CL	DCPA/DCPD -rich cement	mBMSCs	Rabbit	Having higher cell viability, ALP activity, and ALP quantity. Showing lesser residual implant and higher new bone formation	[177]
Tsai CH	TTCP cement	-	Rabbit	The new bone formed started at the center of TTCP cement at 12 weeks. The resorption of grafts, bone ingrowth and remodeling activities were completed at 24 weeks.	[124]

hBMSCs: human bone mesenchymal stem cells; rBMSCs: rat bone mesenchymal stem cells; mBMSCs: mice bone mesenchymal stem cells; HUVECs: Human Umbilical Vein Endothelial Cells.

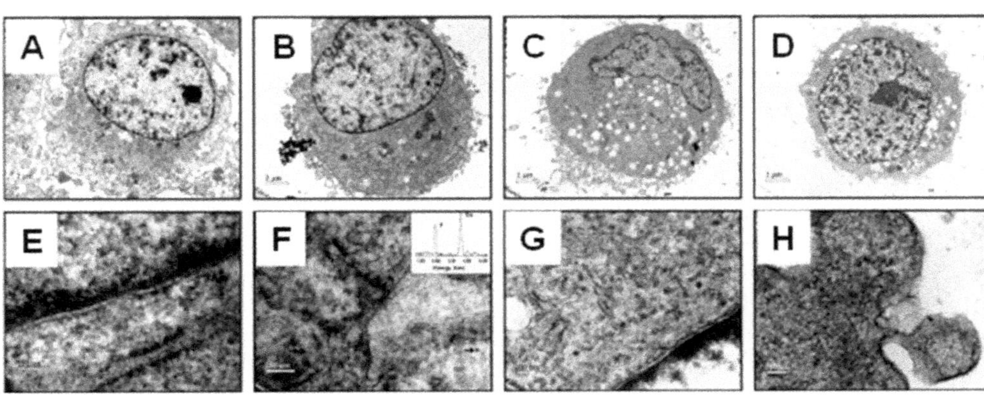

Figure 5. TEM images of MG-63 cells incubated with glass: (**A**,**E**) HAP particles at 20 nm (**B**,**F**), 80 nm (**C**,**G**) and the microscale (**D**,**H**). The black arrows in F indicate 20 nm HAP particles endocytosed by the cell. Reprinted with permission from Ref. [41]. 2009, Shi, Z. L.

It is noteworthy that the component sizes, the surface and pore scale levels of the CaP grafts and hybrid scaffolds are closely related and are important to bone regeneration, considering that the grafts and scaffolds are the main forms of bone implants. Li et al. demonstrated that compared to CaP bioceramics fabricated by microcrystalline structures, bioceramics using nanocrystalline structures have many advantages in the form of unique surface topography, good bioactivity, excellent osteoinductivity and proper biodegradability. Accordingly, the CaP nanoceramics have significantly superior biological performance in promoting osteogenic differentiation and bone formation (Figure 6) [161,178].

4.2. Pore and Surface Characteristics of CaP

Another key factor influencing the biological response of cells and the effects of bone repair is the pore size of scaffolds. Macropore and micropore structures exist in CaP ceramics or scaffolds. The macropore structure is designed to promote osteoinduction [42]. Numerous studies have confirmed the positive effects of macropores on cellular growth and tissue formation [179–181]. Macropore sizes in range of 300–500 μm are recommended because this scale ensures the transportation of nutrients and metabolites, osteogenesis and vascular ingrowth [42,182]. Macropores ranging from 300 to 500 μm may possess optimal surface tension, which has been demonstrated as an important determinant of cell adhesion and tissue growth, to the mechanoreceptors of cells [183].

Additionally, micropore structure, generally characterized by pores smaller than 50 μm, is critical for the CaP scaffolds. The micropore structure is thought not only to improve bone ingrowth, but also to create additional space for bone formation [184, 185]. Furthermore, a larger surface area, attributed to the presence of micropores, can enhance protein adsorption, ion exchange and mineralization [186]. Moreover, bone tissue regeneration is accelerated due to the capillary effect induced by micropores, which further improves the homogeneity of bone distribution in the scaffolds [187]. Interestingly, some studies have found that new bone tissue formation was observed in HAP ceramics with micropore structures that were subcutaneously and intramuscularly implanted in dogs, and did not occur in ceramics without micropores [184,188]. Micropore structure is becoming increasingly important as research advances. Bohner et al. found mineralized cell/collage-rich tissues only in scaffolds with micropores having threshold sizes greater than 1–10 μm, and were not related to macropore size [185]. Regulating micropore size from 1.58 ± 0.65 μm to 0.65 ± 0.25 μm in TCP ceramics resulted in more abundant bone tissue, which further confirmed the importance of micropore structure [189].

The surface structures of scaffolds or ceramics at the microscale to nanoscale can influence cell adhesion, spreading, cytoskeletal distribution, and gene expression (Figure 7) [42,190]. Holthaus et al. [191] found that microgrooves with 60–100 μm widths had more cells (35–45%) growing than those of with widths of 20–40 μm (16–25%). In contrast, the opposite phenomenon was observed with respect to cell orientation. The number of aligned cells along the microgrooves was higher at 20–40 μm (64–79%) widths than at widths of 60–100 μm (29–47%). According to the study, the size of microgrooves similar to cell dimensions may be beneficial in guiding cell alignment and adhesion. Notably, the sensitivity of different cells to microstructure sizes varies. For example, microgrooves with 8 μm widths responded strongly for cell alignment of both myoblasts and osteoblasts, while those with the widths of 24 μm only affected myoblasts [192]. The depth of the microgrooves influences cell adhesion by single cell guidance. The adhesion force of an individual osteoblast cell to the matrix decreases with increasing depth of the microgroove from 3 μm to 5.5 μm [193].

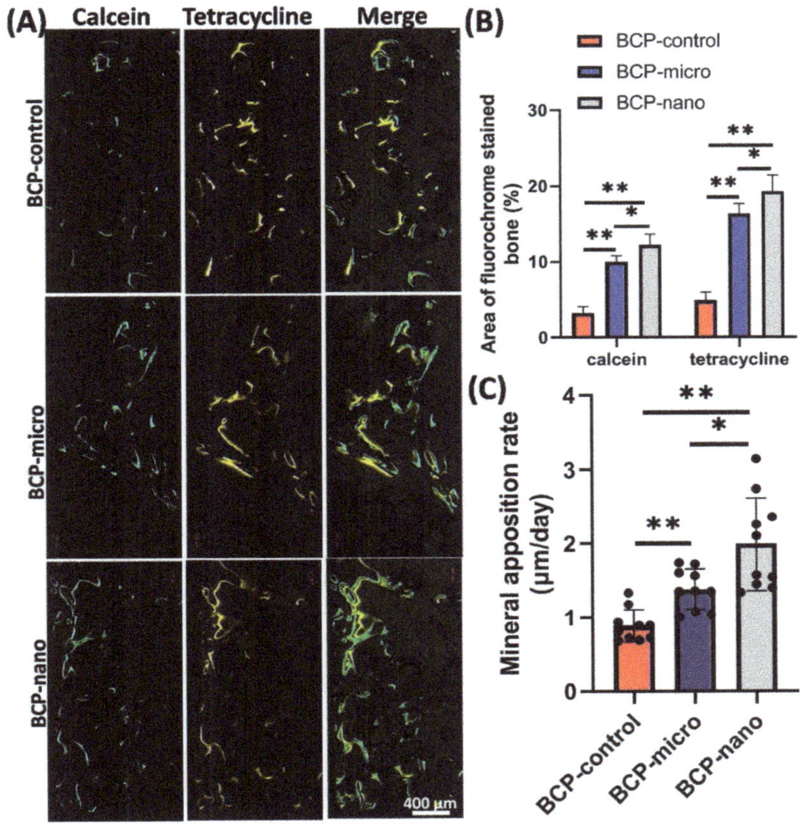

Figure 6. Fluorochrome-labeling analysis of new bone formation and mineralization with different BCP bioceramics at the microscale (BCP-control: irregular shape; BCP-micro: spherical shape) to nanoscale (BCP-nano: spherical shape) grain size. And (**A**) displayed the calcein-labeled newly formed bone at week 8 (column 1, green), tetracycline at week 10 (column 2, yellow), and merged images of the two fluorochromes (row 3). (**B**,**C**) analysis of the fluorochrome-labeled new bone area and mineral apposition rate. Reprinted with permission from Ref. [161]. 2022, Li, X.

Nanoscale surface structures of scaffolds and ceramics have been widely applied and have received much attention since the development of nanofabrication techniques [194,195]. Surface structures of scaffolds/ceramics on the nanoscale may affect the cells' response due to the ECM being composed of nanosized collagen fibrils and cellular receptors and filopodia which are also at the nanoscale [42]. In addition, many studies have been conducted on mechanisms responsible for osteogenic differentiation of BMSCs in response to surface structures and sizes [196–198]. However, more studies are needed to understand how nanoscale surface structures at different levels influence cell behavior and responses. The specific relationships between the mechanisms and the surface scale are not clear but have been discussed extensively in [42].

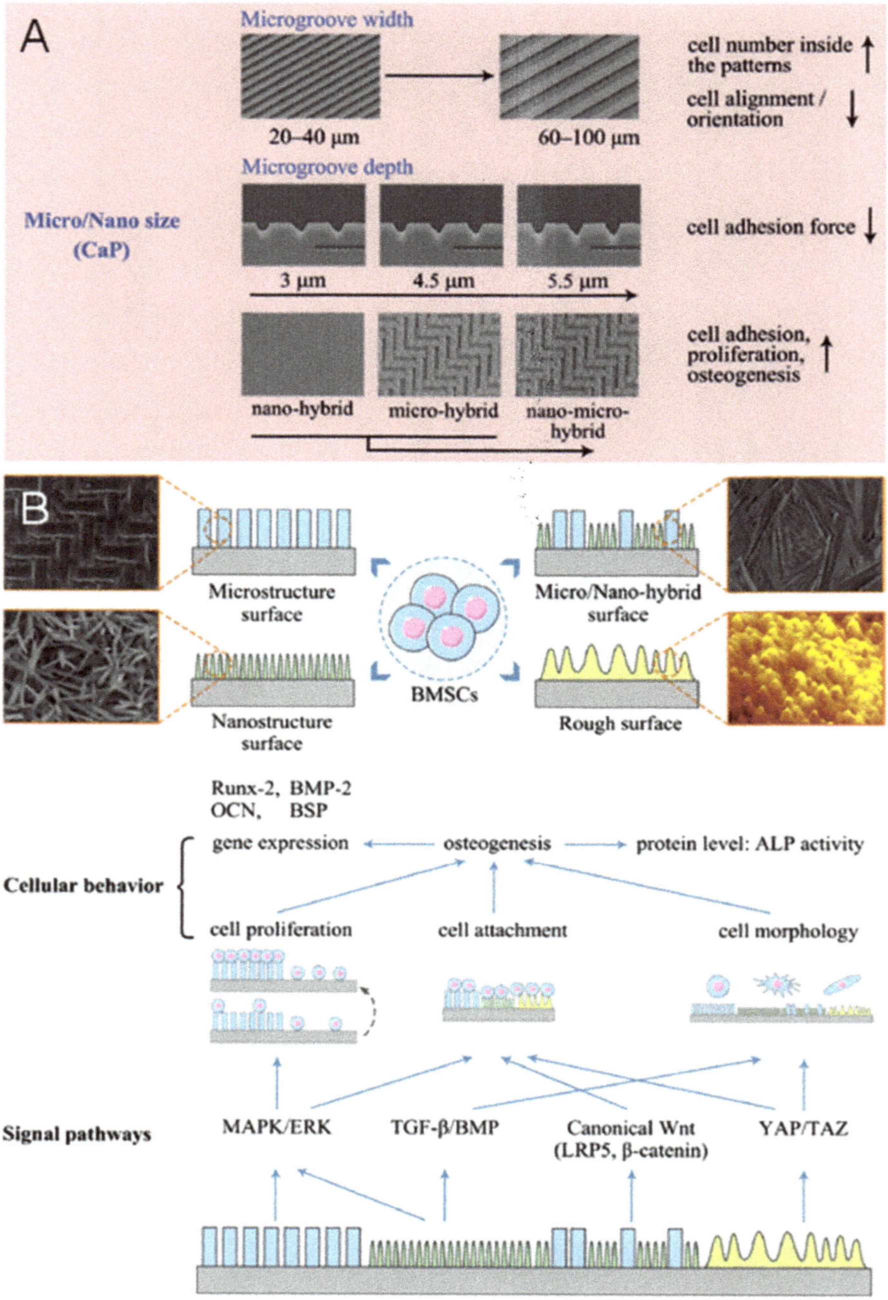

Figure 7. (**A**) Biological effects of surface structure sizes of grafts and scaffolds on cell behavior. (**B**) Different signaling pathways participated in osteogenic differentiation of BMSCs responding to various surface structures. Reprinted with permission from Ref. [42]. 2020, Xiao, D. Q.

5. Regulating Morphologies of CaP Materials

There is a wide variety of shapes and sizes among the natural CaPs. During bone mineralization in vivo, the average crystal size of minerals is smallest at the beginning of formation, and then gradually grows with maturity, leading to various sizes and shapes of crystals [199]. In the past, CaPs with nanoparticle shapes were believed to have better properties. However, minerals with needle-shaped and rod-shaped crystals were found in bone mineralized collagen [200], leading to a controversy regarding the effect of mineral crystal morphology on its properties. Sphere-shaped HAP nanoparticles are beneficial for the proliferation and migration of osteoblast compared to the rod-shaped nanoparticles [201]. Therefore, the morphologies of CaPs play an important role in biological responses and bone tissue regeneration.

With the development of preparation techniques and improved understanding of the mechanism of CaP formation, various CaPs with different morphologies can be produced in vitro by controlling the conditions. The commonly used preparation methods of CaP materials include co-precipitation, emulsion, hydrothermal, microwave-assisted, hydrolysis, solution combustion synthesis, and the sol-gel method [202–207]. CaPs with various morphologies can be prepared by adjusting pH, temperature, organic additives, the Ca/P ratio, and ion saturation of the reaction system [149,208].

pH affects precipitation and crystallization of HAPs by influencing the balance between the hydrogen-containing anions and orthophosphates [152]. Generally, based on temperature, they can be divided into two categories: high temperatures (~800 °C) and low temperatures. For example, TCP is required to be synthesized in a high temperature environment, but this high temperature solid-phase reaction method cannot produce uniform CaP nanoparticles [149]. Figure 8 displays the various shapes of β-TCP particles which have been prepared at different pH values and temperatures [159].

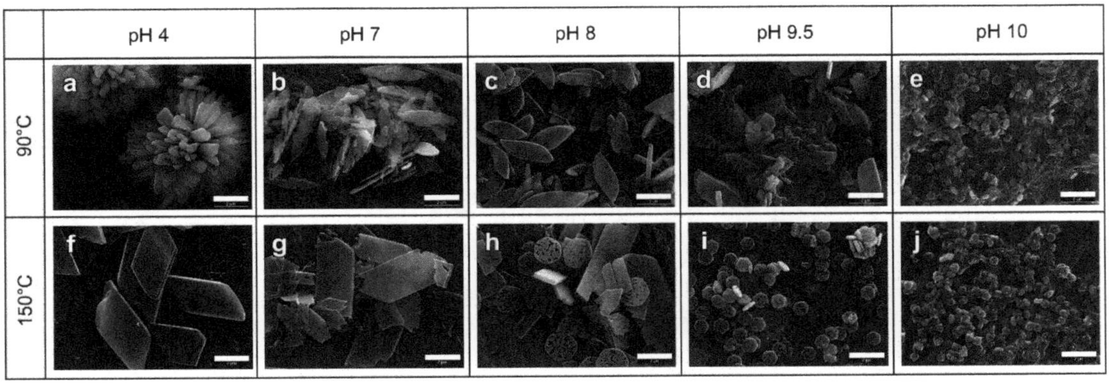

Figure 8. SEM images of β-TCP particles obtained at 90 °C (**a–e**) and 150 °C (**f–j**), and pH 4 (**a**,**f**), pH 7 (**b**,**g**), pH 8 (**c**,**h**), pH 9.5 (**d**,**i**) and pH 10 (**e**,**j**). Scale bar is 2 μm on all images. Reprinted with permission from Ref. [159]. 2013, Galea, L.

Organic additives, such as, hexadecyl trimethyl ammonium bromide [209], poly (acrylic acid) [210] and allylamine hydrochloride [211], affect the morphology of CaP materials. Organic additives regulate the morphology of CaP by the electrostatic interaction between the surface of the crystals and additives, and by regulating the zeta potential of crystals [212,213]. For instance, large HAP nanoparticles with high length-diameter ratios are obtained when adding poly(L-lysine), while using poly (L-glutamic acid) with higher charges results in the HAP nanoparticles with a smaller size [214]. On the other hand, the Ca/P ratio and reaction time are crucial factors in determining the morphology of CaP materials [208,215]. By varying the Ca/P ratio and reaction time, Zhang et al. prepared HAP microtubes and HAP nanowires [216].

Throughout the past decade, CaPs have been reported with a range of morphologies, including particles, spheres, rods, needles, wires, sheets, flakes, strips, porous structures, and hollow structures, all with different sizes ranging from the nanoscale to macroscopical [152,217]. These shapes can be classified into four groups according to dimensions [152]; Figure 9 shows different morphologies of CaPs.

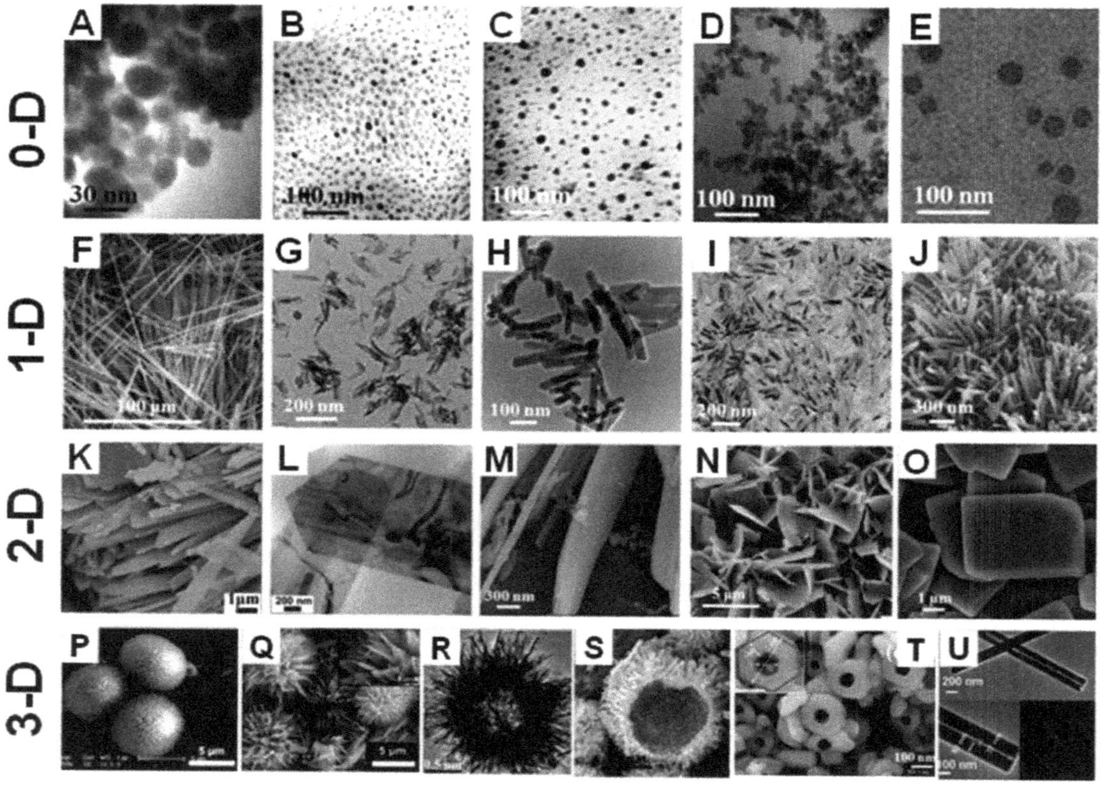

Figure 9. Different morphologies of CaP materials from 0-D to 3-D. (**A–E**), 0-D CaP crystals prepared by a milling method (**A**) [218], sol-gel method (**B**,**C**) [219], and microemulsion method (**D**,**E**) [220,221]. (**F–J**) 1-D shaped CaP crystals show HAP whiskers (**F**) [222], dicalcium phosphate dihydrate (DCPD) whiskers (**G**) [223], HAP nanorods (**H**,**I**) [224,225] and HAP nanowires (**J**) [226]. (**K–O**) displays 2-D shaped CaP crystals including HAP (**K–N**) [226–229] and DCPA (**O**) [230] nanosheets. HAP microspheres (**P**) [158] and flowers (**Q**) [158] obtained by self-assembly with nanorods and microsheets, respectively. (**R**,**S**) [231,232] Typical images of hollow HAP microspheres. (**T**,**U**) [233,234] HAP nanotubes.

Clusters, particles, and quantum dots are typically zero-dimensional (0-D) structured materials. Zero-dimensional-shaped CaPs have wide applications in delivery systems and the fabrication of CaP bioceramics and composites [235–238]. One-dimensional (1-D) CaP crystals represent structures with a length that is significantly greater than the cross-sectional dimension, e.g., needles, rods, wires and fiber-like structures. Two-dimensional (2-D) shaped CaP crystals contain sheets, disks, flakes and platelets, which have excellent molecular adsorption abilities and mechanical properties. It is well known that the physical properties of inorganic components, including size and morphology, affect the mechanical properties of the organic-inorganic composites [239,240]. As a result, 1-D and 2-D-shaped

CaP powders are commonly used as raw components for improving the mechanical properties of bio-composites, where the latter are usually considered to be the most effective stiffeners in isotopic composites [241].

In recent years, three-dimensional (3-D) CaP materials have attracted a great deal of attention owing to their superior biological performance and spectrum of biomedical applications [152]. Novel 3-D CaP architectures reported in the literature contain porous and mesoporous spheres, hollow and tube structures, ordered and patterned arrays, and porous scaffolds [242–245]. To fabricate 3-D CaPs, self-assembly and biomineralization methods are employed. These strategies use nanoparticles, nanorods, nanobelts, and nanosheets as building blocks to construct three-dimensional materials with various morphologies [246–248]. Furthermore, amino acids, proteins, and surfactants are commonly utilized and are most effective assistance in the technology of self-assembly and biomineralization for controlling the morphology of 3-D CaP functional materials [152].

6. Biomimetic Calcium Phosphate Scaffolds for Bone Regeneration

Highly organized arrays of HAP crystallites on the nanoscale level, and intricate bundles of aligned crystallites on the microscale level, are found in human bones and teeth, and are critical in determining many of the advanced biological and mechanical properties of bones and teeth [249,250]. Therefore, fabricating CaP biomaterials that mimic the structures of bones and teeth is a new strategy to improve the performance of biomaterials. Three-dimensional grafts and scaffolds with nano-/microstructured surfaces exhibited better biological properties owing to their similarity to human bones and teeth, and provided promotion of osteointegration and subsequent bone tissue regeneration [251–253].

Traditional methods to fabricate 3-D grafts with nano-/microstructured surfaces are assisted by organic solvents and reagents for directing structure. HAP columnar structures were formed elongated in the *c*-axis on the surfaces of HAP bioceramics and on the substrates of metals by a molecular template directing biomineralization method in a SBF system (Figure 10A–C) [254,255]. Active groups and components such as PO_4^{3-}, -NH_2, -COOH and polydopamine have been developed to promote the biomineralization of CaP crystals because of their excellent ability of capturingCa^{2+} ions, leading to the formation of nuclei that can induce CaP crystal growth [244,256].

Mineralization via SBF soaking usually results in low crystallinity [257]. Large amounts of surfactant and additives, some of which are hazardous to health and the environment, are need to be added to SBF to assist mineralization [258]. In recent years, 3-D printing technology has allowed fabrication of 3-D grafts with nano-/microstructured surfaces [253,259]. Wu et al. [260] mimicked the nano-/microstructural hierarchy of natural wood to fabricate biomimetic hierarchical porous scaffolds by 3D printing technology (Figure 11). It was found that the first-level macropores of the biomimetic, natural, wood-like, hierarchical structure scaffolds had good performance in promoting bone tissue ingrowth, whereas the second-level micro/nanoscale pores performed well in transporting nutrients and metabolites. These scaffolds with nano-/microstructured surfaces exhibited excellent features in osteoinductivity and bone tissue regeneration.

Microporous CaPs scaffolds with nano-/microstructured surfaces have an excellent capacity to transport cells, resulting in a high interest in the field of bone tissue engineering [261]. Three-dimensional microporous HAP bioceramic scaffolds have been constructed via hydrothermal method by using α-TCP bioceramic scaffolds as precursors without any assistance from structure-directing reagents and organic solvents [261]. The 3-D ordered scaffolds have highly interconnective macropores and various surface topographies such as nanosheets, nanorods and microrods that can be tailored by regulating the reaction medium of the NaH_2PO_4, Na_3PO_4 and $CaCl_2$ aqueous solution (Figure 12) [251].

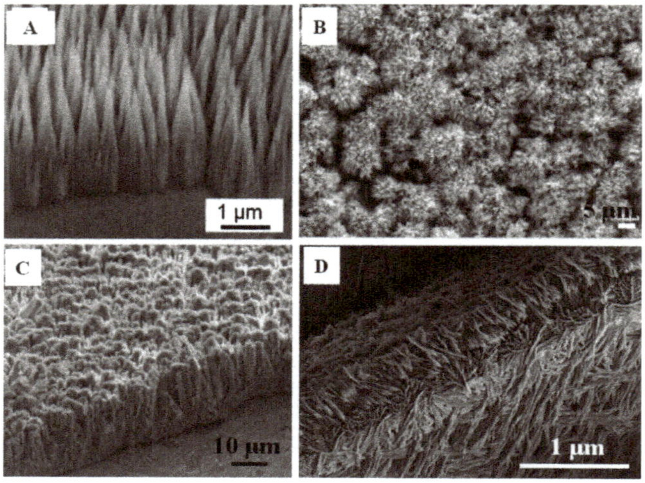

Figure 10. SEM images of bone grafts with 3-D architecture nano-/microstructured surfaces. (**A**) HAP columnar elongated along the c-axis on the HAP disks induced by aspartic acid in SBF; (**B**) highly packed and aligned FHAp coating with enamel-like structure on Ti plates via biomimetic growth in SBF; (**C**) bioinspired crystallization of continuous HAP films on titanium surface induced by Langmuir monolayer of zein protein; (**D**) multilevel hierarchically ordered artificial biomineral HAP ceramic with macroscopical size more than 1 cm using DCPD precursor transformation method. Reprinted with permission from Ref. [152]. 2014, Lin, K.

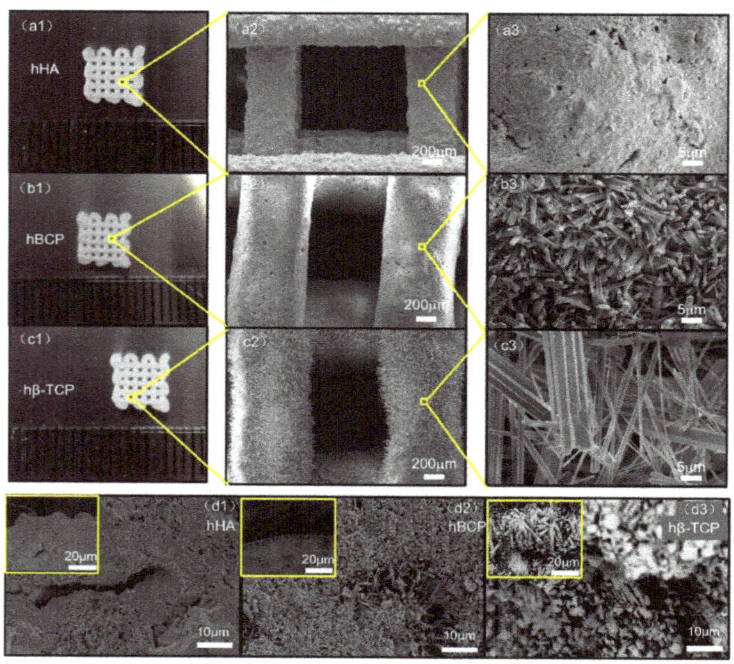

Figure 11. SEM images of 3D-printed biomimetic hierarchical porous calcium phosphate bioceramic scaffolds (hHA, hBCP and hβ-TCP). External macroscopic and microscopic morphologies of hHA (**a1–a3**),

hBCP (**b1–b3**), and hβ-TCP (**c1–c3**). Internal microscopic morphologies of hHA (**d1**), hBCP (**d2**), and hβ-TCP (**d3**). Smaller images in the upper left of (**d1–d3**) showing scaffold cross-sections of hHA, hBCP, and hβ-TCP, respectively. Reprinted with permission from Ref. [260]. 2020, Wu, L.

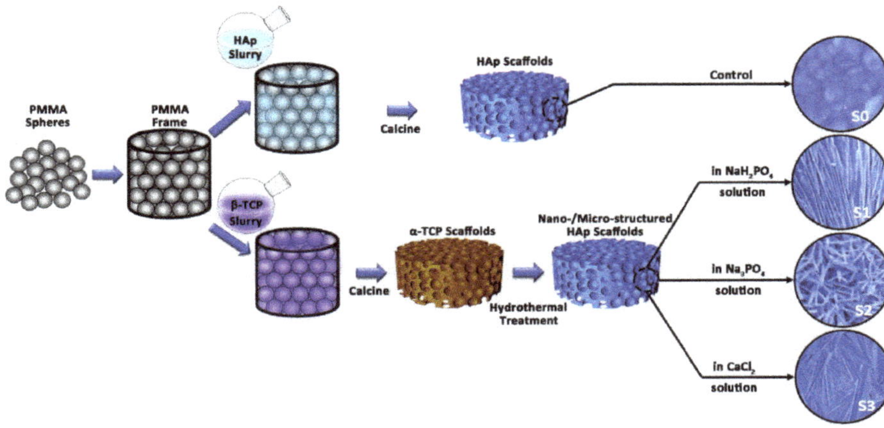

Figure 12. The schematic of the fabrication process for HAp bioceramic scaffolds with different surface topographies: (S0) smooth and flat surface, (S1) nanosheet surface, (S2) nanorod surface, (S3) micro-nano-hybrid surface. Reprinted with permission from Ref. [251]. 2013, Xia, L. G. and Ref. [152]. 2014, Lin, K.

And the HAP bioceramics with biomimetic nano-/microstructured hybrid scaffolds demonstrated excellent adhesion, proliferation, and osteogenic differentiation capabilities of BMSCs. In addition, excellent protein adsorption due to their surface characteristics contributed to the bioactive performance of the composite scaffolds, which resulted in enhanced bone regeneration. Furthermore, the mechanism of the biomimetic hybrid scaffolds with nano-/microstructured surfaces was investigated. It was found that integrins were activated initially by the attachment of cells to scaffolds, and that the BMP2 signaling pathway and related Cx43-based cell-cell interactions were activated subsequently, in addition to an interaction between BMP2 and Cx43 that facilitated osteogenic differentiation (Figure 13) [198].

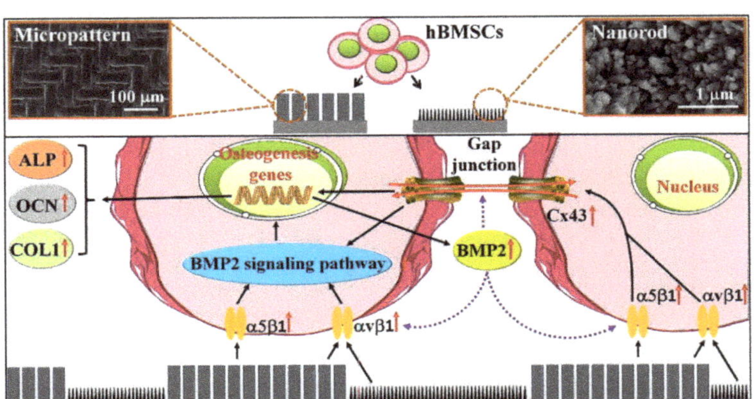

Figure 13. Mechanism of HAP hybrid scaffolds with nano/microstructured surface on the osteogenic differentiation. Reprinted with permission from Ref. [198]. 2018, Zhao, C. C.

7. Drug Delivery Applications and the Potential of CaPs

CaPs, especially porous nanostructured CaP materials, are widely used as drug delivery carriers due to their suitable architecture, large surface area and stability in biological fluids. As well-known, CaPs dissolve at a slightly acidic pH, which makes controlled delivery of drugs into cells possible [262]. Furthermore, the production of the ionic, non-toxic constituents Ca^{2+} and PO_4^{3-} after dissolution of CaPs can prevent particle accumulation and induce the release of drugs into the cells [263]. As a result, CaPs are commonly used as carriers for the delivery of commercial drugs, bioactive molecules and genetic materials [264–266].

The drug delivery behavior of CaPs is influenced by characteristics including crystallinity, microstructural properties, surface area and charge, particle size, and morphology [267]. Higher levels of drug complex loading were observed for HAP nanoparticles with lower crystallinity and higher surface area when compared to similar nanoparticles with higher crystallinity [268]. Lower release rates were observed with CaP particles of higher crystallinity, whose solubility is affected by their crystallinity, resulting in a decrease in drug release rate [269]. Additionally, the morphology of the CaPs results in different surface areas, resulting in different drug loading efficiency. In a study by Palazzo et al., therapeutic drugs were absorbed onto plate-like HAP particles approximately 1.3 times more than onto needle-like particles [270]. In addition, spherical CaP nanoparticles provide more effective drug loading and release properties than particles with flaky, brick-like, or elongated orthogonal morphologies [271]. In terms of particle size, particles with a size of 20 ± 5 nm are best accepted by osteoblast cell lines [272]. Hence, the CaP nanoparticles are favored in the drug delivery system.

A number of therapeutic factors have been delivered by CaP nanoparticles, including antibiotics, anti-inflammatory drugs, and growth factors for bone healing [267]. A CaPs-based antibiotics delivery system is mainly used for the treatment of bone defects that are infected or caused by infection. Moreover, by following this approach, high concentrations of antibiotics are only found at the anatomical sites of interest, thereby minimizing the toxic effects of antibiotics [273]. In addition, growth factors, including the family of BMPs, transforming growth factor-beta (TGF-β), platelet-derived growth factor (PDGF) and vascular endothelial growth factor (VEGF), are critical in the processes of bone repair [274]. Hence, the addition of growth factors can further enhance the osteogenic ability of CaPs. BMP-2 interacts with HAP via the functional groups -OH, -NH2, and -COO [275]. The continuous releasing period can be delayed to 15 days in vitro when delivered by HAP nanoparticles [276]. It is important to note that when these growth factors are loaded onto CaPs, attention must be paid to preventing denaturation of the protein, reducing its functionality.

Gene delivery participates in the promotion and facilitation of bone regeneration [277]. Furthermore, a new approach to tissue regeneration is being brought about by the most recent discoveries and advancements in gene delivery [267]. This makes calcium phosphate nanoparticles an attractive option for bone regeneration.

8. The Mechanism of Calcium Phosphate Promoting Osteogenesis

CaPs have applied to bone tissue regeneration engineering for decades, and show advanced osteoconductivity, osteoinductivity and bone healing effects. However, the mechanisms of CaPs in promoting bone regeneration is still a mystery. Many mechanisms have been proposed and confirmed such as osteogenesis, vascularization, neuralization, inflammatory and immunology; some classical mechanisms are reviewed in this work.

8.1. Osteogenic Differentiation

The mobilization and recruitment of BMSCs from the bone marrow through the peripheral circulation play an important role in the repair of bone defects [278]. At the cellular level, bone tissue regeneration is determined by the induction and promotion of directed differentiation of BMSCs into osteoblasts following recruitment to the site of the bone defect. With the degradation of the CaP material, Ca^{2+} and PO_4^{3-} are gradually released, leading to a local ion concentration above the physiological level. As a result of the increased ion concentration, osteoblast proliferation and differentiation and the subsequent process of bone formation are significantly affected [279]. As well, Ca^{2+} is an important homing signal that facilitates multicellular processes such as bone remodeling and wound healing by bringing together different types of cells [280]. A high calcium concentration, for example, has been shown to stimulate the migration of BMSCs [281], pre-osteoblasts [282], and osteoblasts [283] to the site of bone resorption and their maturation into new bone-producing cells. Moreover, the influx of Ca^{2+} increases intracellular Ca^{2+} levels, thereby increasing the polarization of the membrane front, which is crucial for determining the direction of sustained cell migration [284,285].

Apart from this, extracellular Ca^{2+} plays an important role in maintaining osteoblast proliferation and differentiation near the site of bone resorption via calcium/calmodulin signaling [286]. Addition of elevated Ca^{2+} levels to osteoblastic cell cultures has been shown to have an effect on bone cell fate independently of systemic calciotropic factors in a concentration-dependent manner [287,288]. The presence of Ca^{2+} also stimulates the expression of osteogenic growth factors, including parathyroid hormone-related peptide [289], BMP-2, and BMP-4 [290]. In addition, implants enriched with Ca^{2+} significantly improve in vivo osseointegration and bone formation; for example, in a CaP glass, titanium substrate implanted with Ca^{2+}, a collagen gel exposed to Ca^{2+}, and CaP-coated implants [291–293]. In an encouraging development, Ca^{2+} has recently been implicated as an important messenger involved in the non-canonical Wnt/calcium signaling pathway for bone formation independent of the β-catenin pathway [294]. In this pathway, calcium-sensitive enzymes such as Ca^{2+}-CaMKII, protein kinase C, and calcineurin are activated by an intracellular release of calcium.

In addition, PO_4^{3-} also contributes to the proliferation and differentiation of osteoblasts. Several osteogenesis-related biological processes are affected by PO_4^{3-} as an important signaling molecule, such as osteogenic associated gene expression (e.g., osteopontin) [295] and bone-related protein secretion (e.g., matrix Gla protein (MGP)) [56]. Furthermore, inorganic phosphate (Pi) has been shown to play a crucial role in the physiological mineralization of bone matrix, which is mediated by the enzyme ALP [296]. As well, BMP-2 stimulates Pi transport by osteogenic cells primarily through sodium-dependent phosphate transporters to induce bone matrix calcification [297].

It should be noted that the structures of CaP biomaterials also activate biological signals associated with adhesion, proliferation, and differentiation of cells, such as dimensions, geometries, porosity/microporosity, grain size, and surface topography. Cell sensor and adhesion-related signal pathways play a primary and key role in this biological process. The ERK and p38 MAPK signaling pathways became phosphorylated when BMSCs are cultured on micro/nano structured HAP scaffolds, and when the signaling pathways are blocked by inhibitors, osteogenic differentiation is attenuated [251]. In addition, the MAPK/ERK signaling pathway has been shown to play a significant role in the regulation of cell functions such as proliferation and differentiation [197]. Consequently, BMSCs may have sensed micro/nanostructured surfaces through focal adhesion formation and subsequently activated ERK and p38 signaling pathways, resulting in the upregulation of relevant genes and the formation of osteoblasts. In addition, the TGF-β/BMP and Wnt signal axis may contribute to the structure-sensing of BMSCs and the subsequent osteogenesis as shown in Figure 7 [42].

8.2. Vascularization

For bone regeneration to be successful, adequate and rapid vascularization is necessary. In addition, vascularization plays an important role in the viability of seeded cells within the CaP implants. many studies have found that materials can influence the angiogenesis of CaP in bone formation.

Firstly, Ca^{2+} plays an important role in angiogenesis by mediating the angiogenesis-related signaling pathway [298]. In addition, extracellular calcium has been suggested to be a key factor in bone marrow progenitor cell angiogenic responses [299]. Furthermore, the physical characteristics of the CaP material, such as porosity and pore size, are known to influence vascularization [300]. The size of the channel influences the behavior of vascularization in vivo. A large porosity (>50 μm) is necessary for cellular infiltration of bone ingrowth [301]. In a study [300] examining the relationship between pore size and angiogenesis, a scaffold with a pore size of 150 μm displayed significantly greater vascularization than scaffolds with 100 μm and 120 μm pores. In addition, the authors demonstrated that scaffolds with a pore size of 150 μm enhanced the formation of new blood vessels through the PI3K/Akt pathway. The expression of the representative angiogenic factors HIF1α, PLGF, and the migration factor CXCR4 were increased in pores with a 250 μm diameter, and by increasing the diameter of the pores to 500 μm, VEGF expression was enhanced, which led to the development of large vessels [302].

On the other hand, it is known that the CaP can accelerate the BMP2 expression of BMSCs [303], which may be due to Ca ions subsequently activating the PKC, EK1/2, and ERK1/2 pathways, and then entering the nucleus to up-regulate BMP-2 expression through Fos expression and activator protein 1 (AP-1) formation [304]. As well as regulating osteogenesis, BMPs also play an important role in regulating angiogenesis [305]. Moreover, endothelial cells and pericytes are responsible for angiogenesis, and they may also be able to differentiate into osteoblasts when inflammation is present [303]. CaP biomaterials have recruiting ability for these cells, as demonstrated previously [282]. Furthermore, these cells are capable of expressing cytokines such as BMP-2 and BMP-7, which are markedly up-regulated in response to inflammatory stresses [306,307]. In this regard, angiogenesis induced by CaP materials may affect osteogenesis by a variety of mechanisms. One is by secreting cytokines, such as BMPs, and MSCs are recruited to undergo osteoblastic differentiation, and by providing more endothelial cells and pericytes to transform into MSCs, which are then induced to differentiate into osteoblasts under the influence of different materials or cytokines [303].

9. Conclusions and Outlook

In summary, calcium phosphate-based materials are commonly and widely used for bone defect repair based on their unique properties resulting from similarity with the main inorganic components of human bones and teeth. Numerous studies have been conducted and have confirmed the excellent characteristics of their physicochemical properties and biological properties, including mechanical properties, biodegradability, biocompatibility, bioactivity, osteoinductivity and osteoconductivity. With preparation technologies developing, the number of CaP materials with varying species, sizes, and morphologies has mushroomed. Different CaP materials have different properties, which determine the approaches of application in accelerating bone tissue regeneration. The CaP materials with different species, sizes and structures have various physico-chemical and biological properties, including mechanical properties, specific surface areas, roughness, porosity, bioactivity, biodegradability, and osteoinductivity. To develop better bone repair materials, it is critical to understand these properties and the biological effects of different species, sizes, and structures of CaP materials. Applying bone repair materials in different approaches in accordance with their advantages is effective for bone regeneration. However, the mechanisms of controlling accurately the species, sizes, and morphologies, as well as the specific relationships between the species, sizes, and shapes of the CaP materials and their biological responses, are not well understood currently. Therefore, determining these relationships will be of great assistance in fabricating artificial CaP bone grafts and in repairing bone defects.

Considering the existing knowledge and available limitations, future studies may focus on the following areas.

(1) CaP materials have a huge potential for bone repair application because of their unique properties similarity to natural minerals. Various CaP materials and composites have been studied and developed for bone regeneration using the advanced physiological properties of CaP such as biocompatibility, osteogenic and osteoinductivity. However, clinical orthopaedic products of CaPs are still rare. With better understanding of biomimetic principles, biomimetic CaP materials have received more attentions. Exploration of the mechanism of biomineralization is important and beneficial for the development and application of CaP-based biomaterials.

(2) Many strategies have been investigated for fabricating CaP materials with different sizes and morphologies. However, strategies for precisely controlling the size of CaPs are still limited. The microfluidic system and microemulsion technologies are novel approaches which may be useful in regulating CaP material sizes accurately.

(3) Many studies have investigated the microscale of the scaffolds/ceramic surface structure and how this affects cell behavior, and have suggested that sizes of the surface microstructure with similarity to cell dimensions might induce the guiding of cell alignment and adhesion. In the future, biological performance and mechanisms of cells responding to different surfaces at the nanoscale will be of interest to researchers.

(4) By understanding natural bone structure thoroughly, artificial bone grafts with biomimetic hierarchical nano-/microstructures have garnered much attention, with promising results for the repair of bone defects. However, biomimetic construction with properly oriented structures ranging from the nanoscale to the microscale remains a challenge. Development of new technologies of mineralization, such as utilizing novel directing biomolecules and precursors, 3D printing with a more precise printing systems and exploiting new bio-inks may be promising routes to achieve these goals.

(5) Up to now, the mechanisms by which the CaP grafts influence bone regeneration are not well understood, such as the influence of nano-/microstructured surfaces on cell behavior. Thus, there is still much work to be done on the mechanisms and relationships between the sizes/shapes of CaP materials and their biological abilities. Further research will result in beneficial developments for the design of CaP biomaterials and for accelerating bone regeneration.

(6) Strategies in improving the physicochemical properties and biological performance of CaP materials are still emerging. For instance, HAP has good biocompatibility but poor mechanical properties. The design and preparation of hybrid scaffolds of CaPs in combination with organic polymers are effective and simple methods. In addition, metal ions as additives can improve the biological activities of CaP-based biomaterials. The addition of Mg^{2+} into CaPs has been shown as an effective strategy to induce the proliferation of osteoblasts and inhibit the absorption of osteoclasts. Meanwhile, other ions, such as Cu^{2+}, Fe^{3+}, Ti^{2+}, and Mn^{2+}, may produce unexpected biological responses, which will require further study.

Author Contributions: Conceptualization, X.H., Z.Z., F.C. and L.Z. (Longpo Zheng); software, X.H., L.Z. (Lei Zhang), X.L. and T.W.; writing—original draft preparation, X.H. and F.C.; writing—review and editing, X.H., X.Z., B.L., F.C. and L.Z. (Lei Zhang); supervision, Z.Z., L.Z. (Longpo Zheng) and F.C.; funding acquisition, Z.Z., F.C. and L.Z. (Longpo Zheng). All authors have read and agreed to the published version of the manuscript.

Funding: This research was funded by the National natural science foundation of China (81901892, 31771081), the Science and Technology Commission of Shanghai Municipality (21ZR1450000, 19YF1438200), S&T Innovation 2025 Major Special Programme of Ningbo (2018B10040), and Orthopedic Intelligent Minimally Invasive Diagnosis & Treatment Center, Shanghai Tenth People's Hospital (049918006).

Conflicts of Interest: The authors declare no conflict of interest.

References

1. Hellwinkel, J.E.; Working, Z.M.; Certain, L.; García, A.J.; Wenke, J.C.; Bahney, C.S. The intersection of fracture healing and infection: Orthopaedics research society workshop 2021. *J. Orthop. Res.* **2022**, *40*, 541–552. [CrossRef] [PubMed]
2. Nauth, A.; Schemitsch, E.; Norris, B.; Nollin, Z.; Watson, J.T. Critical-Size Bone Defects: Is There a Consensus for Diagnosis and Treatment? *J. Orthop. Trauma* **2018**, *32* (Suppl. 1), S7–S11. [CrossRef] [PubMed]
3. Baldwin, P.; Li, D.J.; Auston, D.A.; Mir, H.S.; Yoon, R.S.; Koval, K.J. Autograft, Allograft, and Bone Graft Substitutes: Clinical Evidence and Indications for Use in the Setting of Orthopaedic Trauma Surgery. *J. Orthop. Trauma* **2019**, *33*, 203–213. [CrossRef] [PubMed]
4. Wang, W.; Yeung, K.W.K. Bone grafts and biomaterials substitutes for bone defect repair: A review. *Bioact. Mater.* **2017**, *2*, 224–247 [CrossRef]
5. Aktuglu, K.; Erol, K.; Vahabi, A. Ilizarov bone transport and treatment of critical-sized tibial bone defects: A narrative review *J. Orthop. Traumatol.* **2019**, *20*, 22. [CrossRef]
6. Roddy, E.; DeBaun, M.R.; Daoud-Gray, A.; Yang, Y.P.; Gardner, M.J. Treatment of critical-sized bone defects: Clinical and tissue engineering perspectives. *Eur. J. Orthop. Surg. Traumatol.* **2018**, *28*, 351–362. [CrossRef]
7. Li, R.; Nauth, A.; Li, C.; Qamirani, E.; Atesok, K.; Schemitsch, E.H. Expression of VEGF gene isoforms in a rat segmental bone defect model treated with EPCs. *J. Orthop. Trauma* **2012**, *26*, 689–692. [CrossRef]
8. Manzini, B.M.; Machado, L.M.R.; Noritomi, P.Y.; da Silva, J.V.L. Advances in Bone tissue engineering: A fundamental review *J. Biosci.* **2021**, *46*, 1–18. [CrossRef]
9. Govoni, M.; Vivarelli, L.; Mazzotta, A.; Stagni, C.; Maso, A.; Dallari, D. Commercial Bone Grafts Claimed as an Alternative to Autografts: Current Trends for Clinical Applications in Orthopaedics. *Materials* **2021**, *14*, 3290. [CrossRef]
10. Vangsness, C.T.; Wagner, P.P.; Moore, T.M.; Roberts, M.R. Overview of safety issues concerning the preparation and processing of soft-tissue allografts. *Arthroscopy* **2006**, *22*, 1351–1358. [CrossRef]
11. Perić Kačarević, Ž.; Rider, P.; Alkildani, S.; Retnasingh, S.; Pejakić, M.; Schnettler, R.; Gosau, M.; Smeets, R.; Jung, O.; Barbeck, M An introduction to bone tissue engineering. *Int. J. Artif. Organs* **2020**, *43*, 69–86. [CrossRef] [PubMed]
12. Khademhosseini, A.; Langer, R. A decade of progress in tissue engineering. *Nat. Protoc.* **2016**, *11*, 1775–1781. [CrossRef] [PubMed]
13. Whitely, M.; Cereceres, S.; Dhavalikar, P.; Salhadar, K.; Wilems, T.; Smith, B.; Mikos, A.; Cosgriff-Hernandez, E. Improved in situ seeding of 3D printed scaffolds using cell-releasing hydrogels. *Biomaterials* **2018**, *185*, 194–204. [CrossRef]
14. Cipitria, A.; Boettcher, K.; Schoenhals, S.; Garske, D.S.; Schmidt-Bleek, K.; Ellinghaus, A.; Dienelt, A.; Peters, A.; Mehta, M.; Madl, C.M.; et al. In-situ tissue regeneration through SDF-1alpha driven cell recruitment and stiffness-mediated bone regeneration in a critical-sized segmental femoral defect. *Acta Biomater.* **2017**, *60*, 50–63. [CrossRef] [PubMed]
15. Li, Q.; Ma, L.; Gao, C. Biomaterials for in situ tissue regeneration: Development and perspectives. *J. Mater. Chem. B* **2015**, *3*, 8921–8938. [CrossRef] [PubMed]
16. Roycroft, P.; Cuypers, M. The Etymology of The Mineral Name 'Apatite': A Clarification. *Ir. J. Earth Sci.* **2015**, *33*, 71. [CrossRef]
17. de Jong, W.F. La Substance Minérale Dans les Os. *Recl. Des Trav. Chim. Des Pays-Bas* **1926**, *45*, 445–448. [CrossRef]
18. Albee, F.H. Studies in Bone Growth: Triple Calcium Phosphate as a Stimulus to Osteogenesis. *Ann. Surg.* **1920**, *71*, 32–39. [CrossRef]
19. Monroe, E.; Bass, D.; Votava, W.; Mc Mullen, J. New calcium phosphate ceramic material for bone and tooth implants. *J. Dent Res.* **1971**, *50*, 860–861. [CrossRef]
20. Ballouze, R.; Marahat, M.H.; Mohamad, S.; Saidin, N.A.; Kasim, S.R.; Ooi, J.P. Biocompatible magnesium-doped biphasic calcium phosphate for bone regeneration. *J. Biomed. Mater. Res. Part B Appl. Biomater.* **2021**, *109*, 1426–1435. [CrossRef]
21. Wang, G.C.; Lu, Z.F.; Zreiqat, H. 8—Bioceramics for skeletal bone regeneration. In *Bone Substitute Biomaterials*; Mallick, K., Ed.; Woodhead Publishing: Cambridge, UK, 2014; pp. 180–216.
22. Cama, G. 1—Calcium phosphate cements for bone regeneration. In *Bone Substitute Biomaterials*; Mallick, K., Ed.; Woodhead Publishing: Cambridge, UK, 2014; pp. 3–25.
23. Weiner, S.; Traub, W. Bone structure: From angstroms to microns. *FASEB J.* **1992**, *6*, 879–885. [CrossRef]
24. Guo, B.; Lei, B.; Li, P.; Ma, P.X. Functionalized scaffolds to enhance tissue regeneration. *Regen. Biomater.* **2015**, *2*, 47–57. [CrossRef] [PubMed]
25. Ma, J.; Wang, J.; Ai, X.; Zhang, S. Biomimetic self-assembly of apatite hybrid materials: From a single molecular template to bi-/multi-molecular templates. *Biotechnol. Adv.* **2014**, *32*, 744–760. [CrossRef] [PubMed]
26. Zhang, W.; Liao, S.S.; Cui, F.Z. Hierarchical Self-Assembly of Nano-Fibrils in Mineralized Collagen. *Chem. Mater.* **2003**, *15*, 3221–3226. [CrossRef]
27. Qiu, Z.-Y.; Cui, Y.; Tao, C.-S.; Zhang, Z.-Q.; Tang, P.-F.; Mao, K.-Y.; Wang, X.-M.; Cui, F.-Z. Mineralized Collagen: Rationale, Current Status, and Clinical Applications. *Materials* **2015**, *8*, 4733–4750. [CrossRef]
28. Khalid, H.; Chaudhry, A.A. 4—Basics of hydroxyapatite—structure, synthesis, properties, and clinical applications. In *Handbook of Ionic Substituted Hydroxyapatites*; Khan, A.S., Chaudhry, A.A., Eds.; Woodhead Publishing: Cambridge, UK, 2020; pp. 85–115.
29. Lin, K.L.; Chen, L.; Chang, J. Fabrication of Dense Hydroxyapatite Nanobioceramics with Enhanced Mechanical Properties via Two-Step Sintering Process. *Int. J. Appl. Ceram. Technol.* **2012**, *9*, 479–485. [CrossRef]
30. Chen, J.; Yang, Y.; Etim, I.P.; Tan, L.; Yang, K.; Misra, R.D.K.; Wang, J.; Su, X. Recent Advances on Development of Hydroxyapatite Coating on Biodegradable Magnesium Alloys: A Review. *Materials* **2021**, *14*, 5550. [CrossRef]

31. Jeong, J.; Kim, J.H.; Shim, J.H.; Hwang, N.S.; Heo, C.Y. Bioactive calcium phosphate materials and applications in bone regeneration. *Biomater. Res.* **2019**, *23*, 4. [CrossRef]
32. Guo, X.; Lei, L.; Xiao, H.; Zheng, J. Effect of remineralisation on the mechanical properties and tribological behaviour of human tooth dentine. *Biosurf. Biotribol.* **2020**, *6*, 92–95. [CrossRef]
33. Eliaz, N.; Metoki, N. Calcium Phosphate Bioceramics: A Review of Their History, Structure, Properties, Coating Technologies and Biomedical Applications. *Materials* **2017**, *10*, 334. [CrossRef]
34. Rh Owen, G.; Dard, M.; Larjava, H. Hydoxyapatite/beta-tricalcium phosphate biphasic ceramics as regenerative material for the repair of complex bone defects. *J. Biomed. Mater. Res. B Appl. Biomater.* **2018**, *106*, 2493–2512. [CrossRef] [PubMed]
35. Sun, R.; Åhlén, M.; Tai, C.-W.; Bajnóczi, G.; de Kleijne, F.; Ferraz, N.; Persson, I.; Strømme, M.; Cheung, O. Highly Porous Amorphous Calcium Phosphate for Drug Delivery and Bio-Medical Applications. *Nanomaterials* **2019**, *10*, 20. [CrossRef] [PubMed]
36. Eden, M. Structure and formation of amorphous calcium phosphate and its role as surface layer of nanocrystalline apatite: Implications for bone mineralization. *Materialia* **2021**, *17*, 101107. [CrossRef]
37. Kovrlija, I.; Locs, J.; Loca, D. Octacalcium phosphate: Innovative vehicle for the local biologically active substance delivery in bone regeneration. *Acta Biomater.* **2021**, *135*, 27–47. [CrossRef] [PubMed]
38. Dorozhkin, S.V. Calcium Orthophosphates in Nature, Biology and Medicine. *Materials* **2009**, *2*, 399–498. [CrossRef]
39. LeGeros, R.Z. Calcium phosphates in oral biology and medicine. *Monogr. Oral Sci.* **1991**, *15*, 1–201.
40. Moseke, C.; Gbureck, U. Tetracalcium phosphate: Synthesis, properties and biomedical applications. *Acta Biomater.* **2010**, *6*, 3815–3823. [CrossRef]
41. Shi, Z.; Huang, X.; Cai, Y.; Tang, R.; Yang, D. Size effect of hydroxyapatite nanoparticles on proliferation and apoptosis of osteoblast-like cells. *Acta Biomater.* **2009**, *5*, 338–345. [CrossRef]
42. Xiao, D.; Zhang, J.; Zhang, C.; Barbieri, D.; Yuan, H.; Moroni, L.; Feng, G. The role of calcium phosphate surface structure in osteogenesis and the mechanisms involved. *Acta Biomater.* **2020**, *106*, 22–33. [CrossRef]
43. Wang, J.; Chen, X.; Guo, B.; Yang, X.; Zhou, Y.; Zhu, X.; Zhang, K.; Fan, Y.; Tu, C.; Zhang, X. A serum protein adsorption profile on BCP ceramics and influence of the elevated adsorption of adhesive proteins on the behaviour of MSCs. *J. Mater. Chem. B* **2018**, *6*, 7383–7395. [CrossRef]
44. BenNissan, B. (Ed.) *Advances in Calcium Phosphate Biomaterials*; Springer: Berlin/Heidelberg, Germany, 2014; Volume 2, pp. 1–547.
45. Lodoso-Torrecilla, I.; van den Beucken, J.; Jansen, J.A. Calcium phosphate cements: Optimization toward biodegradability. *Acta Biomater.* **2021**, *119*, 6985. [CrossRef] [PubMed]
46. Liu, W.; Wang, T.; Yang, C.; Darvell, B.W.; Wu, J.; Lin, K.; Chang, J.; Pan, H.; Lu, W.W. Alkaline biodegradable implants for osteoporotic bone defects—Importance of microenvironment pH. *Osteoporos. Int.* **2016**, *27*, 93–104. [CrossRef] [PubMed]
47. Liu, W.; Dan, X.; Lu, W.W.; Zhao, X.; Ruan, C.; Wang, T.; Cui, X.; Zhai, X.; Ma, Y.; Wang, D.; et al. Spatial Distribution of Biomaterial Microenvironment pH and Its Modulatory Effect on Osteoclasts at the Early Stage of Bone Defect Regeneration. *ACS Appl. Mater. Interfaces* **2019**, *11*, 9557–9572. [CrossRef] [PubMed]
48. Surmenev, R.A.; Surmeneva, M.A.; Ivanova, A.A. Significance of calcium phosphate coatings for the enhancement of new bone osteogenesis—A review. *Acta Biomater.* **2014**, *10*, 557–579. [CrossRef]
49. Peacock, M. Calcium metabolism in health and disease. *Clin. J. Am. Soc. Nephrol.* **2010**, *5* (Suppl. 1), S23–S30. [CrossRef]
50. Zhang, J.; Wu, Q.; Yin, C.; Jia, X.; Zhao, Z.; Zhang, X.; Yuan, G.; Hu, H.; Zhao, Q. Sustained calcium ion release from bioceramics promotes CaSR-mediated M2 macrophage polarization for osteoinduction. *J. Leukoc. Biol.* **2021**, *110*, 485–496. [CrossRef]
51. Xu, T.; Yang, Y.; Suo, D.; Bei, H.P.; Xu, X.; Zhao, X. Electrosprayed Regeneration-Enhancer-Element Microspheres Power Osteogenesis and Angiogenesis Coupling. *Small* **2022**, *18*, e2200314.
52. Liu, D.; Genetos, D.C.; Shao, Y.; Geist, D.J.; Li, J.; Ke, H.Z.; Duncan, R.L. Activation of extracellular-signal regulated kinase (ERK1/2) by fluid shear is Ca(2+)- and ATP-dependent in MC3T3-E1 osteoblasts. *Bone* **2008**, *42*, 644–652. [CrossRef]
53. Danciu, T.E.; Adam, R.M.; Naruse, K.; Freeman, M.R.; Hauschka, P.V. Calcium regulates the PI3K-Akt pathway in stretched osteoblasts. *FEBS Lett.* **2003**, *536*, 193–197. [CrossRef]
54. Goretti Penido, M.; Alon, U.S. Phosphate homeostasis and its role in bone health. *Pediatr. Nephrol.* **2012**, *27*, 2039–2048. [CrossRef]
55. Khoshniat, S.; Bourgine, A.; Julien, M.; Weiss, P.; Guicheux, J.; Beck, L. The emergence of phosphate as a specific signaling molecule in bone and other cell types in mammals. *Cell. Mol. Life Sci.* **2011**, *68*, 205–218. [CrossRef]
56. Julien, M.; Khoshniat, S.; Lacreusette, A.; Gatius, M.; Bozec, A.; Wagner, E.F.; Wittrant, Y.; Masson, M.; Weiss, P.; Beck, L.; et al. Phosphate-dependent regulation of MGP in osteoblasts: Role of ERK1/2 and Fra-1. *J. Bone Miner. Res.* **2009**, *24*, 1856–1868. [CrossRef] [PubMed]
57. Tada, H.; Nemoto, E.; Foster, B.L.; Somerman, M.J.; Shimauchi, H. Phosphate increases bone morphogenetic protein-2 expression through cAMP-dependent protein kinase and ERK1/2 pathways in human dental pulp cells. *Bone* **2011**, *48*, 1409–1416. [CrossRef]
58. Brown, W.E.; Lehr, J.R.; Smith, J.P.; Frazier, A.W. Crystallography of Octacalcium Phosphate. *J. Am. Chem. Soc.* **1957**, *79*, 5318–5319. [CrossRef]
59. Lu, B.-Q.; Willhammar, T.; Sun, B.-B.; Hedin, N.; Gale, J.D.; Gebauer, D. Introducing the crystalline phase of dicalcium phosphate monohydrate. *Nat. Commun.* **2020**, *11*, 1546. [CrossRef] [PubMed]
60. Sommerfeldt, D.W.; Rubin, C.T. Biology of bone and how it orchestrates the form and function of the skeleton. *Eur. Spine J.* **2001**, *10* (Suppl. 2), S86–S95.

61. Dorozhkin, S.V.; Epple, M. Biological and medical significance of calcium phosphates. *Angew. Chem. Int. Ed.* **2002**, *41*, 3130–3146. [CrossRef]
62. White, T.J.; ZhiLi, D. Structural derivation and crystal chemistry of apatites. *Acta Crystallogr. B* **2003**, *59*, 1–16. [CrossRef] [PubMed]
63. Manjubala, I.; Sastry, T.P.; Kumar, R.V. Bone in-growth induced by biphasic calcium phosphate ceramic in femoral defect of dogs. *J. Biomater. Appl.* **2005**, *19*, 341–360. [CrossRef]
64. Suchanek, W.; Yoshimura, M. Processing and properties of hydroxyapatite-based biomaterials for use as hard tissue replacement implants. *J. Mater. Res.* **1998**, *13*, 94–117. [CrossRef]
65. Chen, S.; Shi, Y.; Zhang, X.; Ma, J. Biomimetic synthesis of Mg-substituted hydroxyapatite nanocomposites and three-dimensional printing of composite scaffolds for bone regeneration. *J. Biomed. Mater. Res. A* **2019**, *107*, 2512–2521. [CrossRef] [PubMed]
66. Arcos, D.; Vallet-Regi, M. Substituted hydroxyapatite coatings of bone implants. *J. Mater. Chem. B* **2020**, *8*, 1781–1800. [CrossRef] [PubMed]
67. Castiglioni, S.; Cazzaniga, A.; Albisetti, W.; Maier, J.A.M. Magnesium and osteoporosis: Current state of knowledge and future research directions. *Nutrients* **2013**, *5*, 3022–3033. [CrossRef] [PubMed]
68. He, L.; Zhang, X.; Liu, B.; Tian, Y.; Ma, W. Effect of magnesium ion on human osteoblast activity. *Braz. J. Med. Biol. Res.* **2016**, *49*, e5257. [CrossRef]
69. Santos, G.G.; Nunes, V.L.C.; Marinho, S.M.O.C.; Santos, S.R.A.; Rossi, A.M.; Miguel, F.B. Biological behavior of magnesium substituted hydroxyapatite during bone repair. *Braz. J. Biol.* **2021**, *81*, 53–61. [CrossRef] [PubMed]
70. de Lima, I.R.; Alves, G.G.; Soriano, C.A.; Campaneli, A.P.; Gasparoto, T.H.; Ramos, E.S.; de Sena, L.; Rossi, A.M.; Granjeiro, J.M. Understanding the impact of divalent cation substitution on hydroxyapatite: An in vitro multiparametric study on biocompatibility. *J. Biomed. Mater. Res. A* **2011**, *98*, 351–358. [CrossRef] [PubMed]
71. Wang, W.; Wan, P.; Liu, C.; Tan, L.; Li, W.; Li, L.; Yang, K. Degradation and biological properties of Ca-P contained micro-arc oxidation self-sealing coating on pure magnesium for bone fixation. *Regen. Biomater.* **2015**, *2*, 107–118. [CrossRef]
72. Nygren, H.; Bigdeli, N.; Ilver, L.; Malmberg, P. Mg-corrosion, hydroxyapatite, and bone healing. *Biointerphases* **2017**, *12*, 02C407. [CrossRef]
73. Dong, J.; Tumer, N.; Putra, N.E.; Zhu, J.; Li, Y.; Leeflang, M.A.; Taheri, P.; Fratila-Apachitei, L.E.; Mol, J.M.C.; Zadpoor, A.A.; et al. Extrusion-based 3D printed magnesium scaffolds with multifunctional MgF2 and MgF2-CaP coatings. *Biomater. Sci.* **2021**, *9*, 7159–7182. [CrossRef] [PubMed]
74. Weiner, S. Transient precursor strategy in mineral formation of bone. *Bone* **2006**, *39*, 431–433. [CrossRef]
75. Samavedi, S.; Whittington, A.R.; Goldstein, A.S. Calcium phosphate ceramics in bone tissue engineering: A review of properties and their influence on cell behavior. *Acta Biomater.* **2013**, *9*, 8037–8045. [CrossRef] [PubMed]
76. Cheah, C.W.; Al-Namnam, N.M.; Lau, M.N.; Lim, G.S.; Raman, R.; Fairbairn, P.; Ngeow, W.C. Synthetic Material for Bone, Periodontal, and Dental Tissue Regeneration: Where Are We Now, and Where Are We Heading Next? *Materials* **2021**, *14*, 6123. [CrossRef]
77. Besinis, A.; van Noort, R.; Martin, N. Infiltration of demineralized dentin with silica and hydroxyapatite nanoparticles. *Dent. Mater.* **2012**, *28*, 1012–1023. [CrossRef] [PubMed]
78. Ching, W.Y.; Rulis, P.; Misra, A. Ab initio elastic properties and tensile strength of crystalline hydroxyapatite. *Acta Biomater.* **2009**, *5*, 3067–3075. [CrossRef] [PubMed]
79. Mukherjee, S.; Kundu, B.; Sen, S.; Chanda, A. Improved properties of hydroxyapatite-carbon nanotube biocomposite: Mechanical, in vitro bioactivity and biological studies. *Ceram. Int.* **2014**, *40*, 5635–5643. [CrossRef]
80. Bose, S.; Banerjee, A.; Dasgupta, S.; Bandyopadhyay, A. Synthesis, Processing, Mechanical, and Biological Property Characterization of Hydroxyapatite Whisker-Reinforced Hydroxyapatite Composites. *J. Am. Ceram. Soc.* **2009**, *92*, 323–330. [CrossRef]
81. Sheikh, Z.; Najeeb, S.; Khurshid, Z.; Verma, V.; Rashid, H.; Glogauer, M. Biodegradable Materials for Bone Repair and Tissue Engineering Applications. *Materials* **2015**, *8*, 5744–5794. [CrossRef]
82. Iqbal, H.; Ali, M.; Zeeshan, R.; Mutahir, Z.; Iqbal, F.; Nawaz, M.A.H.; Shahzadi, L.; Chaudhry, A.A.; Yar, M.; Luan, S.; et al. Chitosan/hydroxyapatite (HA)/hydroxypropylmethyl cellulose (HPMC) spongy scaffolds-synthesis and evaluation as potential alveolar bone substitutes. *Colloid Surf. B* **2017**, *160*, 553–563. [CrossRef]
83. He, M.; Hou, Y.; Zhu, C.; He, M.; Jiang, Y.; Feng, G.; Liu, L.; Li, Y.; Chen, C.; Zhang, L. 3D-Printing Biodegradable PU/PAAM/Gel Hydrogel Scaffold with High Flexibility and Self-Adaptibility to Irregular Defects for Nonload-Bearing Bone Regeneration. *Bioconjug. Chem.* **2021**, *32*, 1915–1925. [CrossRef] [PubMed]
84. Toledano, M.; Vallecillo-Rivas, M.; Osorio, M.; Muñoz-Soto, E.; Toledano-Osorio, M.; Vallecillo, C.; Toledano, R.; Lynch, C.; Serrera-Figallo, M.-A.; Osorio, R. Zn-Containing Membranes for Guided Bone Regeneration in Dentistry. *Polymers* **2021**, *13*, 1797. [CrossRef]
85. Ellinger, R.F.; Nery, E.B.; Lynch, K.L. Histological assessment of periodontal osseous defects following implantation of hydroxyapatite and biphasic calcium phosphate ceramics: A case report. *Int. J. Periodontics Restor. Dent.* **1986**, *6*, 22–33.
86. Bouler, J.; Pilet, P.; Gauthier, O.; Verron, E. Biphasic calcium phosphate ceramics for bone reconstruction: A review of biological response. *Acta Biomater.* **2017**, *53*, 1–12. [CrossRef]
87. Dorozhkin, S.V. Biphasic, triphasic and multiphasic calcium orthophosphates. *Acta Biomater.* **2012**, *8*, 963–977. [CrossRef]

88. Zhu, Y.; Zhang, K.; Zhao, R.; Ye, X.; Chen, X.; Xiao, Z.; Yang, X.; Zhu, X.; Zhang, K.; Fan, Y.; et al. Bone regeneration with micro/nano hybrid-structured biphasic calcium phosphate bioceramics at segmental bone defect and the induced immunoregulation of MSCs. *Biomaterials* **2017**, *147*, 133–144. [CrossRef]
89. Dorozhkin, S. *Calcium Orthophosphates: Applications in Nature, Biology, and Medicine*; Jenny Stanford Publishing: New York, NY, USA, 2012; pp. 1–152.
90. Bohner, M.; Santoni, B.L.G.; Dobelin, N. beta-tricalcium phosphate for bone substitution: Synthesis and properties. *Acta Biomater.* **2020**, *113*, 23–41. [CrossRef]
91. Kjellin, P.; Rajasekharan, A.K.; Currie, F.; Handa, P. Investigation of calcium phosphate formation from calcium propionate and triethyl phosphate. *Ceram. Int.* **2016**, *42*, 14061–14065. [CrossRef]
92. Stahli, C.; Thuring, J.; Galea, L.; Tadier, S.; Bohner, M.; Döbelin, N. Hydrogen-substituted beta-tricalcium phosphate synthesized in organic media. *Acta Cryst. B Struct. Sci. Cryst. Eng. Mater.* **2016**, *72*, 875–884. [CrossRef]
93. Makarov, C.; Gotman, I.; Jiang, X.; Fuchs, S.; Kirkpatrick, C.J.; Gutmanas, E.Y. In situ synthesis of calcium phosphate-polycaprolactone nanocomposites with high ceramic volume fractions. *J. Mater. Sci. Mater. Med.* **2010**, *21*, 1771–1779. [CrossRef]
94. Tao, J.; Pan, H.; Zhai, H.; Wang, J.; Li, L.; Wu, J.; Jiang, W.; Xu, X.; Tang, R. Controls of Tricalcium Phosphate Single-Crystal Formation from Its Amorphous Precursor by Interfacial Energy. *Cryst. Growth Des.* **2009**, *9*, 3154–3160. [CrossRef]
95. Yin, X.L.; Stott, M.J.; Rubio, A. alpha- and beta-tricalcium phosphate: A density functional study. *Phys. Rev. B* **2003**, *68*, 205205. [CrossRef]
96. Bi, L.; Cheng, W.; Fan, H.; Pei, G. Reconstruction of goat tibial defects using an injectable tricalcium phosphate/chitosan in combination with autologous platelet-rich plasma. *Biomaterials* **2010**, *31*, 3201–3211. [CrossRef]
97. Mestres, G.; Le Van, C.; Ginebra, M.P. Silicon-stabilized alpha-tricalcium phosphate and its use in a calcium phosphate cement: Characterization and cell response. *Acta Biomater.* **2012**, *8*, 1169–1179. [CrossRef] [PubMed]
98. Carrodeguas, R.G.; De Aza, S. alpha-Tricalcium phosphate: Synthesis, properties and biomedical applications. *Acta Biomater.* **2011**, *7*, 3536–3546. [CrossRef]
99. Diez-Escudero, A.; Espanol, M.; Beats, S.; Ginebra, M.-P. In vitro degradation of calcium phosphates: Effect of multiscale porosity, textural properties and composition. *Acta Biomater.* **2017**, *60*, 81–92. [CrossRef] [PubMed]
100. Eanes, E.D.; Gillessen, I.H.; Posner, A.S. Intermediate states in the precipitation of hydroxyapatite. *Nature* **1965**, *208*, 365–367. [CrossRef]
101. Brecevic, L.; Furedi-Milhofer, H. Precipitation of calcium phosphates from electrolyte solutions. II. The formation and transformation of the precipitates. *Calcif. Tissue Res.* **1972**, *10*, 82–90. [CrossRef]
102. Glimcher, M.J. Recent Studies of the Mineral Phase in Bone and Its Possible Linkage to the Organic Matrix by Protein-Bound Phosphate Bonds. *Philos. Trans. R. Soc. Lond. B Biol. Sci.* **1984**, *304*, 479–508.
103. O'Neill, W.C. The fallacy of the calcium-phosphorus product. *Kidney Int.* **2007**, *72*, 792–796. [CrossRef]
104. LeGeros, R.Z. Formation and transformation of calcium phosphates: Relevance to vascular calcification. *Z Kardiol.* **2001**, *90* (Suppl. 3), 116–124. [CrossRef]
105. Iafisco, M.; Degli Esposti, L.; Rodriguez, G.B.R.; Carella, F.; Gomez-Morales, J.; Ionescu, A.C.; Brambilla, E.; Tampieri, A.; Delgado-López, J.M. Fluoride-doped amorphous calcium phosphate nanoparticles as a promising biomimetic material for dental remineralization. *Sci. Rep.* **2018**, *8*, 17016. [CrossRef]
106. Zhao, J.; Liu, Y.; Sun, W.-B.; Zhang, H. Amorphous calcium phosphate and its application in dentistry. *Chem. Cent. J.* **2011**, *5*, 40. [CrossRef]
107. Sears, L.M.; Wu, L.; Morrow, B.R.; Hollis, W.; Cagna, D.R.; Hong, L. Effects of NanoAg-ACP Microparticles as Bioactive Fillers on the Mechanical and Remineralization Properties of Dental Resin Cement. *J. Prosthodont.* **2021**, 1–9. [CrossRef] [PubMed]
108. Santhakumar, S.; Oyane, A.; Nakamura, M.; Yoshino, Y.; Alruwaili, M.K.; Miyaji, H. Bone Tissue Regeneration by Collagen Scaffolds with Different Calcium Phosphate Coatings: Amorphous Calcium Phosphate and Low-Crystalline Apatite. *Materials* **2021**, *14*, 5860. [CrossRef] [PubMed]
109. Jiang, Y.Y.; Zhou, Z.F.; Zhu, Y.J.; Chen, F.F.; Lu, B.Q.; Cao, W.T.; Zhang, Y.G.; Cai, Z.D.; Chen, F. Enzymatic Reaction Generates Biomimic Nanominerals with Superior Bioactivity. *Small* **2018**, *14*, e1804321. [CrossRef]
110. Chen, X.; Yan, J.; Jiang, Y.; Fan, Y.; Ying, Z.; Tan, S.; Zhou, Z.; Liu, J.; Chen, F.; He, S. Platelet-Activating Biominerals Enhanced Injectable Hydrogels with Superior Bioactivity for Bone Regeneration. *Front. Bioeng. Biotechnol.* **2022**, *10*, 826855. [CrossRef] [PubMed]
111. Weiner, S.; Sagi, I.; Addadi, L. Structural biology. Choosing the crystallization path less traveled. *Science* **2005**, *309*, 1027–1028. [CrossRef] [PubMed]
112. Crane, N.J.; Popescu, V.; Morris, M.D.; Steenhuis, P.; Ignelzi, M.A. Raman spectroscopic evidence for octacalcium phosphate and other transient mineral species deposited during intramembranous mineralization. *Bone* **2006**, *39*, 434–442. [CrossRef]
113. Gelli, R.; Ridi, F.; Baglioni, P. The importance of being amorphous: Calcium and magnesium phosphates in the human body. *Adv. Colloid Interface Sci.* **2019**, *269*, 219–235. [CrossRef]
114. Golub, E.E. Role of matrix vesicles in biomineralization. *Biochim. Biophys. Acta* **2009**, *1790*, 1592–1598. [CrossRef]
115. Addadi, L.; Gal, A.; Faivre, D.; Scheffel, A.; Weiner, S. Control of Biogenic Nanocrystal Formation in Biomineralization. *Isr. J. Chem.* **2016**, *56*, 227–241. [CrossRef]

116. Tertuliano, O.A.; Greer, J.R. The nanocomposite nature of bone drives its strength and damage resistance. *Nat. Mater.* **2016**, *15*, 1195–1202. [CrossRef]
117. Wang, Y.; Azaïs, T.; Robin, M.; Vallée, A.; Catania, C.; Legriel, P.; Pehau-Arnaudet, G.; Babonneau, F.; Giraud-Guille, M.-M.; Nassif, N. The predominant role of collagen in the nucleation, growth, structure and orientation of bone apatite. *Nat. Mater.* **2012**, *11*, 724–733. [CrossRef] [PubMed]
118. Thein-Han, W.; Liu, J.; Xu, H.H. Calcium phosphate cement with biofunctional agents and stem cell seeding for dental and craniofacial bone repair. *Dent. Mater.* **2012**, *28*, 1059–1070. [CrossRef]
119. Xu, H.H.; Weir, M.D.; Sun, L. Calcium and phosphate ion releasing composite: Effect of pH on release and mechanical properties. *Dent. Mater.* **2009**, *25*, 535–542. [CrossRef] [PubMed]
120. Yoshikawa, M.; Hayami, S.; Tsuji, I.; Toda, T. Histopathological study of a newly developed root canal sealer containing tetracalcium-dicalcium phosphates and 1.0% chondroitin sulfate. *J. Endod.* **1997**, *23*, 162–166. [CrossRef]
121. Liu, C.; Shao, H.; Chen, F.; Zheng, H. Effects of the granularity of raw materials on the hydration and hardening process of calcium phosphate cement. *Biomaterials* **2003**, *24*, 4103–4113. [CrossRef]
122. Qin, T.; Li, X.; Long, H.; Bin, S.; Xu, Y. Bioactive Tetracalcium Phosphate Scaffolds Fabricated by Selective Laser Sintering for Bone Regeneration Applications. *Materials* **2020**, *13*, 2268. [CrossRef]
123. Yoshimine, Y.; Maeda, K. Histologic evaluation of tetracalcium phosphate-based cement as a direct pulp-capping agent. *Oral Surg. Oral Med. Oral Pathol. Oral Radiol. Endod.* **1995**, *79*, 351–358. [CrossRef]
124. Tsai, C.-H.; Lin, R.-M.; Ju, C.-P.; Lin, J.-H.C. Bioresorption behavior of tetracalcium phosphate-derived calcium phosphate cement implanted in femur of rabbits. *Biomaterials* **2008**, *29*, 984–993. [CrossRef]
125. Suzuki, O. 1—Evolution of octacalcium phosphate biomaterials. In *Octacalcium Phosphate Biomaterials*; Suzuki, O., Insley, G., Eds.; Woodhead Publishing: Cambridge, UK, 2020; pp. 1–15.
126. LeGeros, R.Z.; Daculsi, G.; Orly, I.; Abergas, T.; Torres, W. Solution-mediated transformation of octacalcium phosphate (OCP) to apatite. *Scanning Microsc.* **1989**, *3*, 129–138.
127. Tseng, Y.H.; Mou, C.Y.; Chan, J.C. Solid-state NMR study of the transformation of octacalcium phosphate to hydroxyapatite: A mechanistic model for central dark line formation. *J. Am. Chem. Soc.* **2006**, *128*, 6909–6918. [CrossRef]
128. Yokoi, T.; Goto, T.; Kato, T.; Takahashi, S.; Nakamura, J.; Sekino, T.; Ohtsuki, C.; Kawashita, M. Hydroxyapatite Formation from Octacalcium Phosphate and Its Related Compounds: A Discussion of the Transformation Mechanism. *Bull. Chem. Soc. Jpn.* **2020**, *93*, 701–707. [CrossRef]
129. Shiwaku, Y.; Anada, T.; Yamazaki, H.; Honda, Y.; Morimoto, S.; Sasaki, K.; Suzuki, O. Structural, morphological and surface characteristics of two types of octacalcium phosphate-derived fluoride-containing apatitic calcium phosphates. *Acta Biomater.* **2012**, *8*, 4417–4425. [CrossRef] [PubMed]
130. Arellano-Jimenez, M.J.; Garcia-Garcia, R.; Reyes-Gasga, J. Synthesis and hydrolysis of octacalcium phosphate and its characterization by electron microscopy and X-ray diffraction. *J. Phys. Chem. Solids* **2009**, *70*, 390–395. [CrossRef]
131. Suzuki, O.; Yagishita, H.; Amano, T.; Aoba, T. Reversible structural changes of octacalcium phosphate and labile acid phosphate. *J. Dent. Res.* **1995**, *74*, 1764–1769. [CrossRef]
132. Suzuki, O.; Kamakura, S.; Katagiri, T.; Nakamura, M.; Zhao, B.; Honda, Y.; Kamijo, R. Bone formation enhanced by implanted octacalcium phosphate involving conversion into Ca-deficient hydroxyapatite. *Biomaterials* **2006**, *27*, 2671–2681. [CrossRef] [PubMed]
133. Yokoi, T.; Kim, I.Y.; Ohtsuki, C. Mineralization of Calcium Phosphate on Octacalcium Phosphate in a Solution Mimicking In Vivo Conditions. *Phosphorus Res. Bull.* **2012**, *26*, 71–76. [CrossRef]
134. Suzuki, O. Octacalcium phosphate: Osteoconductivity and crystal chemistry. *Acta Biomater.* **2010**, *6*, 3379–3387. [CrossRef] [PubMed]
135. Sugiura, Y.; Munar, M.L.; Ishikawa, K. Fabrication of octacalcium phosphate block through a dissolution-precipitation reaction using a calcium sulphate hemihydrate block as a precursor. *J. Mater. Sci. Mater. Med.* **2018**, *29*, 151. [CrossRef]
136. Imamura, Y.; Tanaka, Y.; Nagai, A.; Yamashita, K.; Takagi, Y. Self-sealing ability of OCP-mediated cement as a deciduous root canal filling materia. *Dent. Mater. J.* **2010**, *29*, 582–588. [CrossRef] [PubMed]
137. Yang, Y.; Wang, H.; Yan, F.-Y.; Qi, Y.; Lai, Y.-K.; Zeng, D.-M.; Chen, G.; Zhang, K.-Q. Bioinspired porous octacalcium phosphate/silk fibroin composite coating materials prepared by electrochemical deposition. *ACS Appl. Mater. Interfaces* **2015**, *7*, 5634–5642. [CrossRef]
138. Kamakura, S.; Sasaki, K.; Honda, Y.; Anada, T.; Suzuki, O. Octacalcium phosphate combined with collagen orthotopically enhances bone regeneration. *J. Biomed. Mater. Res. B Appl. Biomater.* **2006**, *79*, 210–217. [CrossRef] [PubMed]
139. Kawai, T.; Anada, T.; Honda, Y.; Kamakura, S.; Matsui, K.; Matsui, A.; Sasaki, K.; Morimoto, S.; Echigo, S.; Suzuki, O. Synthetic octacalcium phosphate augments bone regeneration correlated with its content in collagen scaffold. *Tissue Eng. Part A* **2009**, *15*, 23–32. [CrossRef]
140. Suzuki, O.; Nakamura, M.; Miyasaka, Y.; Kagayama, M.; Sakurai, M. Bone formation on synthetic precursors of hydroxyapatite. *Tohoku J. Exp. Med.* **1991**, *164*, 37–50. [CrossRef]
141. Suzuki, O.; Nakamura, M.; Miyasaka, Y.; Kagayama, M.; Sakurai, M. Maclura pomifera agglutinin-binding glycoconjugates on converted apatite from synthetic octacalcium phosphate implanted into subperiosteal region of mouse calvaria. *Bone Miner.* **1993**, *20*, 151–166. [CrossRef]

42. Kikawa, T.; Kashimoto, O.; Imaizumi, H.; Kokubun, S.; Suzuki, O. Intramembranous bone tissue response to biodegradable octacalcium phosphate implant. *Acta Biomater.* **2009**, *5*, 1756–1766. [CrossRef] [PubMed]
43. Miyatake, N.; Kishimoto, K.N.; Anada, T.; Imaizumi, H.; Itoi, E.; Suzuki, O. Effect of partial hydrolysis of octacalcium phosphate on its osteoconductive characteristics. *Biomaterials* **2009**, *30*, 1005–1014. [CrossRef]
44. Honda, Y.; Anada, T.; Kamakura, S.; Morimoto, S.; Kuriyagawa, T.; Suzuki, O. The effect of microstructure of octacalcium phosphate on the bone regenerative property. *Tissue Eng. Part A* **2009**, *15*, 1965–1973. [CrossRef] [PubMed]
45. Rey, C.; Combes, C.; Drouet, C.; Grossin, D. Bioactive Ceramics: Physical Chemistry. In *Comprehensive Biomaterials*; Ducheyne, P., Ed.; Elsevier: Amsterdam, The Netherlands, 2011; pp. 187–221.
46. Dorozhkin, S.V. Calcium orthophosphates in dentistry. *J. Mater. Sci. Mater. Med.* **2013**, *24*, 1335–1363. [CrossRef]
47. Yamamoto, H.; Niwa, S.; Hori, M.; Hattori, T.; Sawai, K.; Aoki, S.; Hirano, M.; Takeuchi, H. Mechanical strength of calcium phosphate cement in vivo and in vitro. *Biomaterials* **1998**, *19*, 1587–1591. [CrossRef]
48. Constantz, B.R.; Barr, B.M.; Ison, I.C.; Fulmer, M.K.; Baker, J.; McKinney, L.; Goodman, S.B.; Gunasekaren, S.; Delaney, D.C.; Ross, J.; et al. Histological, chemical, and crystallographic analysis of four calcium phosphate cements in different rabbit osseous sites. *J. Biomed. Mater. Res.* **1998**, *43*, 451–461. [CrossRef]
49. Uskokovic, V.; Uskokovic, D.P. Nanosized hydroxyapatite and other calcium phosphates: Chemistry of formation and application as drug and gene delivery agents. *J. Biomed. Mater. Res. B Appl. Biomater.* **2011**, *96*, 152–191. [CrossRef]
50. Rho, J.Y.; Kuhn-Spearing, L.; Zioupos, P. Mechanical properties and the hierarchical structure of bone. *Med. Eng. Phys.* **1998**, *20*, 92–102. [CrossRef]
51. Hassenkam, T.; Fantner, G.E.; Cutroni, J.A.; Weaver, J.C.; Morse, D.E.; Hansma, P.K. High-resolution AFM imaging of intact and fractured trabecular bone. *Bone* **2004**, *35*, 4–10. [CrossRef]
52. Lin, K.; Wu, C.; Chang, J. Advances in synthesis of calcium phosphate crystals with controlled size and shape. *Acta Biomater.* **2014**, *10*, 4071–4102. [CrossRef]
53. Loo, S.C.J.; Siew, Y.E.; Ho, S.; Boey, F.Y.C.; Ma, J. Synthesis and hydrothermal treatment of nanostructured hydroxyapatite of controllable sizes. *J. Mater. Sci. Mater. Med.* **2008**, *19*, 1389–1397. [CrossRef]
54. Ma, X.; Chen, Y.; Qian, J.; Yuan, Y.; Liu, C. Controllable synthesis of spherical hydroxyapatite nanoparticles using inverse microemulsion method. *Mater. Chem. Phys.* **2016**, *183*, 220–229. [CrossRef]
55. Waiyawat, J.; Kongsema, M.; Sinthuvanich, C.; Chienthavorn, O.; Teanchai, C.; Akkarachaneeyakorn, K. Fabrication of calcium phosphate composite polymer/SLS-stabilized emulsion-based bioactive gels and their application for dentine tubule occlusion. *J. Oral Biosci.* **2020**, *62*, 64–71. [CrossRef] [PubMed]
56. Sadat-Shojai, M.; Khorasani, M.-T.; Dinpanah-Khoshdargi, E.; Jamshidi, A. Synthesis methods for nanosized hydroxyapatite with diverse structures. *Acta Biomater.* **2013**, *9*, 7591–7621. [CrossRef]
57. Kandori, K.; Kuroda, T.; Togashi, S.; Katayama, E. Preparation of Calcium Hydroxyapatite Nanoparticles Using Microreactor and Their Characteristics of Protein Adsorption. *J. Phys. Chem. B* **2011**, *115*, 653–659. [CrossRef] [PubMed]
58. Zhang, C.; Yang, J.; Quan, Z.; Yang, P.; Li, C.; Hou, Z.; Lin, J. Hydroxyapatite Nano- and Microcrystals with Multiform Morphologies: Controllable Synthesis and Luminescence Properties. *Cryst. Growth Des.* **2009**, *9*, 2725–2733. [CrossRef]
59. Galea, L.; Bohner, M.; Thuering, J.; Doebelin, N.; Aneziris, C.G.; Graule, T. Control of the size, shape and composition of highly uniform, non-agglomerated, sub-micrometer beta-tricalcium phosphate and dicalcium phosphate platelets. *Biomaterials* **2013**, *34*, 6388–6401. [CrossRef] [PubMed]
60. Abels, M.; Alkildani, S.; Pröhl, A.; Xiong, X.; Krastev, R.; Korzinskas, T.; Stojanovic, S.; Jung, O.; Najman, S.; Barbeck, M. The Granule Size Mediates the In Vivo Foreign Body Response and the Integration Behavior of Bone Substitutes. *Materials* **2021**, *14*, 7372. [CrossRef]
61. Li, X.; Zhou, Q.; Wu, Y.; Feng, C.; Yang, X.; Wang, L.; Xiao, Y.; Zhang, K.; Zhu, X.; Liu, L.; et al. Enhanced bone regenerative properties of calcium phosphate ceramic granules in rabbit posterolateral spinal fusion through a reduction of grain size. *Bioact. Mater.* **2022**, *11*, 90–106. [CrossRef]
62. Damm, D.; Rojas-Sánchez, L.; Theobald, H.; Sokolova, V.; Wyatt, R.T.; Überla, K.; Epple, M.; Temchura, V. Calcium Phosphate Nanoparticle-Based Vaccines as a Platform for Improvement of HIV-1 Env Antibody Responses by Intrastructural Help. *Nanomaterials* **2019**, *9*, 1389. [CrossRef]
63. Munir, M.U.; Salman, S.; Javed, I.; Bukhari, S.N.A.; Ahmad, N.; Shad, N.A.; Aziz, F. Nano-hydroxyapatite as a delivery system: Overview and advancements. *Artif. Cells Nanomed. Biotechnol.* **2021**, *49*, 717–727. [CrossRef]
64. Sun, B.; Zhao, X.; Gu, W.; Cao, P.; Movahedi, F.; Wu, Y.; Xu, Z.P.; Gu, W. ATP stabilised and sensitised calcium phosphate nanoparticles as effective adjuvants for a DNA vaccine against cancer. *J. Mater. Chem. B* **2021**, *9*, 7435–7446. [CrossRef] [PubMed]
65. Wu, Y.; Gu, W.; Li, L.; Chen, C.; Xu, Z.P. Enhancing PD-1 Gene Silence in, T.; Lymphocytes by Comparing the Delivery Performance of Two Inorganic Nanoparticle Platforms. *Nanomaterials* **2019**, *9*, 159. [CrossRef]
66. Sokolova, V.; Kostka, K.; Shalumon, K.T.; Prymak, O.; Chen, J.-P.; Epple, M. Synthesis and characterization of PLGA/HAP scaffolds with DNA-functionalised calcium phosphate nanoparticles for bone tissue engineering. *J. Mater. Sci. Mater. Med.* **2020**, *31*, 102. [CrossRef]
67. Chen, X.-R.; Bai, J.; Yuan, S.-J.; Yu, C.-X.; Huang, J.; Zhang, T.-L.; Wang, K. Calcium phosphate nanoparticles are associated with inorganic phosphate-induced osteogenic differentiation of rat bone marrow stromal cells. *Chem. Biol. Interact.* **2015**, *238*, 111–117. [CrossRef]

168. Kim, Y.H.; Oreffo, R.O.C.; Dawson, J.I. From hurdle to springboard: The macrophage as target in biomaterial-based bone regeneration strategies. *Bone* **2022**, *159*, 116389. [CrossRef]
169. Linares, J.; Fernández, A.B.; Feito, M.J.; Matesanz, M.C.; Sánchez-Salcedo, S.; Arcos, D.; Vallet-Regí, M.; Rojo, J.M.; Portolés, M.T. Effects of nanocrystalline hydroxyapatites on macrophage polarization. *J. Mater. Chem. B* **2016**, *4*, 1951–1959. [CrossRef] [PubMed]
170. Mahon, O.R.; O'Hanlon, S.; Cunningham, C.C.; McCarthy, G.M.; Hobbs, C.; Nicolosi, V.; Kelly, D.J.; Dunne, A. Orthopaedic implant materials drive M1 macrophage polarization in a spleen tyrosine kinase- and mitogen-activated protein kinase-dependent manner. *Acta Biomater.* **2018**, *65*, 426–435. [CrossRef] [PubMed]
171. Mahon, O.R.; Browe, D.C.; Gonzalez-Fernandez, T.; Pitacco, P.; Whelan, I.T.; Von Euw, S.; Hobbs, C.; Nicolosi, V.; Cunningham, K.T.; Mills, K.H.; et al. Nano-particle mediated M2 macrophage polarization enhances bone formation and MSC osteogenesis in an IL-10 dependent manner. *Biomaterials* **2020**, *239*, 119833. [CrossRef]
172. Ji, C.; Qiu, M.; Ruan, H.; Li, C.; Cheng, L.; Wang, J.; Li, C.; Qi, J.; Cui, W.; Deng, L. Transcriptome Analysis Revealed the Symbiosis Niche of 3D Scaffolds to Accelerate Bone Defect Healing. *Adv. Sci.* **2022**, *9*, e2105194. [CrossRef]
173. Raymond, Y.; Bonany, M.; Lehmann, C.; Thorel, E.; Benítez, R.; Franch, J.; Espanol, M.; Solé-Martí, X.; Manzanares, M.-C.; Canal, C.; et al. Hydrothermal processing of 3D-printed calcium phosphate scaffolds enhances bone formation in vivo: A comparison with biomimetic treatment. *Acta Biomater.* **2021**, *135*, 671–688. [CrossRef]
174. Zhou, Z.; Fan, Y.; Jiang, Y.; Shi, S.; Xue, C.; Zhao, X.; Tan, S.; Chen, X.; Feng, C.; Zhu, Y.; et al. Mineralized Enzyme-Based Biomaterials with Superior Bioactivities for Bone Regeneration. *ACS Appl Mater. Interfaces* **2022**, *14*, 36315–36330. [CrossRef]
175. Kurobane, T.; Shiwaku, Y.; Anada, T.; Hamai, R.; Tsuchiya, K.; Baba, K.; Iikubo, M.; Takahashi, T.; Suzuki, O. Angiogenesis involvement by octacalcium phosphate-gelatin composite-driven bone regeneration in rat calvaria critical-sized defect. *Acta Biomater.* **2019**, *88*, 514–526. [CrossRef]
176. Sheikh, Z.; Zhang, Y.L.; Tamimi, F.; Barralet, J. Effect of processing conditions of dicalcium phosphate cements on graft resorption and bone formation. *Acta Biomater.* **2017**, *53*, 526–535. [CrossRef]
177. Ko, C.-L.; Chen, J.-C.; Tien, Y.-C.; Hung, C.-C.; Wang, J.-C.; Chen, W.-C. Osteoregenerative capacities of dicalcium phosphate-rich calcium phosphate bone cement. *J. Biomed. Mater. Res. A* **2015**, *103*, 203–210. [CrossRef]
178. Li, X.; Song, T.; Chen, X.; Wang, M.; Yang, X.; Xiao, Y.; Zhang, X. Osteoinductivity of Porous Biphasic Calcium Phosphate Ceramic Spheres with Nanocrystalline and Their Efficacy in Guiding Bone Regeneration. *ACS Appl. Mater. Interfaces* **2019**, *11*, 3722–3736. [CrossRef] [PubMed]
179. Karageorgiou, V.; Kaplan, D. Porosity of 3D biomaterial scaffolds and osteogenesis. *Biomaterials* **2005**, *26*, 5474–5491. [CrossRef] [PubMed]
180. Qian, G.; Fan, P.; He, F.; Ye, J. Novel Strategy to Accelerate Bone Regeneration of Calcium Phosphate Cement by Incorporating 3D Plotted Poly(lactic-co-glycolic acid) Network and Bioactive Wollastonite. *Adv. Healthc. Mater.* **2019**, *8*, e1801325. [CrossRef]
181. He, F.; Lu, T.; Fang, X.; Feng, S.; Feng, S.; Tian, Y.; Li, Y.; Zuo, F.; Deng, X.; Ye, J. Novel Extrusion-Microdrilling Approach to Fabricate Calcium Phosphate-Based Bioceramic Scaffolds Enabling Fast Bone Regeneration. *ACS Appl. Mater. Interfaces* **2020**, *12*, 32340–32351. [CrossRef]
182. Murphy, C.M.; Haugh, M.G.; O'Brien, F.J. The effect of mean pore size on cell attachment, proliferation and migration in collagen-glycosaminoglycan scaffolds for bone tissue engineering. *Biomaterials* **2010**, *31*, 461–466. [CrossRef]
183. Lecuit, T.; Lenne, P.F. Cell surface mechanics and the control of cell shape, tissue patterns and morphogenesis. *Nat. Rev. Mol. Cell Biol.* **2007**, *8*, 633–644. [CrossRef]
184. Duan, R.; Barbieri, D.; Luo, X.; Weng, J.; Bao, C.; de Bruijn, J.D.; Yuan, H. Variation of the bone forming ability with the physicochemical properties of calcium phosphate bone substitutes. *Biomater. Sci.* **2017**, *6*, 136–145. [CrossRef]
185. Bohner, M.; Baroud, G.; Bernstein, A.; Döbelin, N.; Galea, L.; Hesse, B.; Heuberger, R.; Meille, S.; Michel, P.; von Rechenberg, B.; et al. Characterization and distribution of mechanically competent mineralized tissue in micropores of β-tricalcium phosphate bone substitutes. *Mater. Today* **2017**, *20*, 106–115. [CrossRef]
186. Rustom, L.E.; Poellmann, M.J.; Wagoner Johnson, A.J. Mineralization in micropores of calcium phosphate scaffolds. *Acta Biomater.* **2019**, *83*, 435–455. [CrossRef] [PubMed]
187. Rustom, L.E.; Boudou, T.; Lou, S.; Pignot-Paintrand, I.; Nemke, B.W.; Lu, Y.; Markel, M.D.; Picart, C.; Johnson, A.J.W. Micropore-induced capillarity enhances bone distribution in vivo in biphasic calcium phosphate scaffolds. *Acta Biomater.* **2016**, *44*, 144–154. [CrossRef] [PubMed]
188. Yamasaki, H.; Sakai, H. Osteogenic Response to Porous Hydroxyapatite Ceramics under the Skin of Dogs. *Biomaterials* **1992**, *13*, 308–312. [CrossRef]
189. Zhang, J.; Luo, X.; Barbieri, D.; Barradas, A.M.; de Bruijn, J.D.; van Blitterswijk, C.A.; Yuan, H. The size of surface microstructures as an osteogenic factor in calcium phosphate ceramics. *Acta Biomater.* **2014**, *10*, 3254–3263. [CrossRef] [PubMed]
190. Li, J.; Li, G.; Zhang, K.; Liao, Y.; Yang, P.; Maitz, M.F.; Huang, N. Co-culture of vascular endothelial cells and smooth muscle cells by hyaluronic acid micro-pattern on titanium surface. *Appl. Surf. Sci.* **2013**, *273*, 24–31. [CrossRef]
191. Holthaus, M.; Stolle, J.; Treccani, L.; Rezwan, K. Orientation of human osteoblasts on hydroxyapatite-based microchannels. *Acta Biomater.* **2012**, *8*, 394–403. [CrossRef]
192. Lu, X.; Leng, Y. Comparison of the Osteoblast and Myoblast Behavior on Hydroxyapatite Microgrooves. *J. Biomed. Mater. Res. B* **2009**, *90*, 438–445. [CrossRef]

193. Yang, S.-P.; Yang, C.-Y.; Lee, T.-M.; Lui, T.-S. Effects of calcium-phosphate topography on osteoblast mechanobiology determined using a cytodetacher. *Mat. Sci. Eng. C Mater.* **2012**, *32*, 254–262. [CrossRef]
194. Pang, S.; He, Y.; He, P.; Luo, X.; Guo, Z.; Li, H. Fabrication of two distinct hydroxyapatite coatings and their effects on MC3T3-E1 cell behavior. *Colloids Surf. B Biointerfaces* **2018**, *171*, 40–48. [CrossRef]
195. Xiao, D.; Guo, T.; Yang, F.; Feng, G.; Shi, F.; Li, J.; Wang, D.; Duan, K.; Weng, J. In situ formation of nanostructured calcium phosphate coatings on porous hydroxyapatite scaffolds using a hydrothermal method and the effect on mesenchymal stem cell behavior. *Ceram. Int.* **2017**, *43*, 1588–1596. [CrossRef]
196. Zhang, J.; Dalbay, M.T.; Luo, X.; Vrij, E.; Barbieri, D.; Moroni, L.; de Bruijn, J.D.; van Blitterswijk, C.A.; Chapple, J.P.; Knight, M.M.; et al. Topography of calcium phosphate ceramics regulates primary cilia length and TGF receptor recruitment associated with osteogenesis. *Acta Biomater.* **2017**, *57*, 487–497. [CrossRef]
197. Haston, S.; Pozzi, S.; Carreno, G.; Manshaei, S.; Panousopoulos, L.; Gonzalez-Meljem, J.M.; Apps, J.R.; Virasami, A.; Thavaraj, S.; Gutteridge, A.; et al. MAPK pathway control of stem cell proliferation and differentiation in the embryonic pituitary provides insights into the pathogenesis of papillary craniopharyngioma. *Development* **2017**, *144*, 2141–2152.
198. Zhao, C.; Wang, X.; Gao, L.; Jing, L.; Zhou, Q.; Chang, J. The role of the micro-pattern and nano-topography of hydroxyapatite bioceramics on stimulating osteogenic differentiation of mesenchymal stem cells. *Acta Biomater.* **2018**, *73*, 509–521. [CrossRef]
199. Murshed, M. Mechanism of Bone Mineralization. *Cold Spring Harb. Perspect. Med.* **2018**, *8*, a031229. [CrossRef]
200. Barakat, N.; Khalil, K.; Sheikh, F.A.; Omran, A.; Gaihre, B.; Khil, S.M.; Kim, H.Y. Physiochemical characterizations of hydroxyapatite extracted from bovine bones by three different methods: Extraction of biologically desirable HAp. *Mater. Sci. Eng. C* **2008**, *28*, 1381–1387. [CrossRef]
201. Xu, J.; Khor, K.; Sui, J.; Zhang, J.; Chen, W. Protein expression profiles in osteoblasts in response to differentially shaped hydroxyapatite nanoparticles. *Biomaterials* **2009**, *30*, 5385–5391. [CrossRef]
202. Li, B.; Chen, X.; Guo, B.; Wang, X.; Fan, H.; Zhang, X. Fabrication and cellular biocompatibility of porous carbonated biphasic calcium phosphate ceramics with a nanostructure. *Acta Biomater.* **2009**, *5*, 134–143. [CrossRef]
203. Zhou, W.Y.; Wang, M.; Cheung, W.L.; Guo, B.C.; Jia, D.M. Synthesis of carbonated hydroxyapatite nanospheres through nanoemulsion. *J. Mater. Sci. Mater. Med.* **2008**, *19*, 103–110. [CrossRef] [PubMed]
204. Durucan, C.; Brown, P.W. alpha-Tricalcium phosphate hydrolysis to hydroxyapatite at and near physiological temperature. *J. Mater. Sci. Mater. Med.* **2000**, *11*, 365–371. [CrossRef]
205. Rajabi-Zamani, A.H.; Behnamghader, A.; Kazemzadeh, A. Synthesis of nanocrystalline carbonated hydroxyapatite powder via nonalkoxide sol-gel method. *Mater. Sci. Eng. C* **2008**, *28*, 1326–1329. [CrossRef]
206. Lemos, A.; Rocha, J.; Quaresma, S.; Kannan, S.; Oktar, F.; Agathopoulos, S.; Ferreira, J. Hydroxyapatite nano-powders produced hydrothermally from nacreous material. *J. Eur. Ceram. Soc.* **2006**, *26*, 3639–3646. [CrossRef]
207. Zhao, J.; Zhao, J.; Qian, Y.; Zhang, X.; Zhou, F.; Zhang, H.; Lu, H.; Chen, J.; Wang, X.; Yu, W. Solution combustion synthesis of calcium phosphate particles for controlled release of bovine serum albumin. *Mater. Sci. Eng. C* **2015**, *50*, 194–200. [CrossRef]
208. Aizawa, M.; Porter, A.E.; Best, S.M.; Bonfield, W. Ultrastructural observation of single-crystal apatite fibres. *Biomaterials* **2005**, *26*, 3427–3433. [CrossRef] [PubMed]
209. Cai, Y.; Liu, Y.; Yan, W.; Hu, Q.; Tao, J.; Zhang, M.; Shi, Z.; Tang, R. Role of hydroxyapatite nanoparticle size in bone cell proliferation. *J. Mater. Chem.* **2007**, *17*, 3780–3787. [CrossRef]
210. Perkin, K.K.; Turner, J.L.; Wooley, K.; Mann, S. Fabrication of hybrid nanocapsules by calcium phosphate mineralization of shell cross-linked polymer micelles and nanocages. *Nano Lett.* **2005**, *5*, 1457–1461. [CrossRef] [PubMed]
211. Shchukin, D.G.; Sukhorukov, G.B.; Mohwald, H. Biomimetic fabrication of nanoengineered hydroxyapatite/polyelectrolyte composite shell. *Chem. Mater.* **2003**, *15*, 3947–3950. [CrossRef]
212. Wang, A.; Liu, D.; Yin, H.; Wu, H.; Wada, Y.; Ren, M.; Jiang, T.; Cheng, X.; Xu, Y. Size-controlled synthesis of hydroxyapatite nanorods by chemical precipitation in the presence of organic modifiers. *Mater. Sci. Eng. C* **2007**, *27*, 865–869. [CrossRef]
213. Huang, F.; Shen, Y.; Xie, A.; Zhu, J.; Zhang, C.; Li, S.; Zhu, J. Study on synthesis and properties of hydroxyapatite nanorods and its complex containing biopolymer. *J. Mater. Sci.* **2007**, *42*, 8599–8605. [CrossRef]
214. Stupp, S.I.; Braun, P.V. Molecular manipulation of microstructures: Biomaterials, ceramics, and semiconductors. *Science* **1997**, *277*, 1242–1248. [CrossRef] [PubMed]
215. Chung, S.-Y.; Kim, Y.-M.; Kim, J.-G.; Kim, Y.-J. Multiphase transformation and Ostwald's rule of stages during crystallization of a metal phosphate. *Nat. Phys.* **2009**, *5*, 68–73. [CrossRef]
216. Zhang, Y.-G.; Zhu, Y.-J.; Chen, F.; Sun, T.-W.; Jiang, Y.-Y. Ultralong hydroxyapatite microtubes: Solvothermal synthesis and application in drug loading and sustained drug release. *CrystEngComm* **2017**, *19*, 1965–1973. [CrossRef]
217. Mohd Pu'Ad, N.A.S.; Koshy, P.; Abdullah, H.Z.; Idris, M.I.; Lee, T.C. Syntheses of hydroxyapatite from natural sources. *Heliyon* **2019**, *5*, e01588. [CrossRef]
218. Nasiri-Tabrizi, B.; Honarmandi, P.; Ebrahimi-Kahrizsangi, R. Synthesis of nanosize single-crystal hydroxyapatite via mechanochemical method. *Mater. Lett.* **2009**, *63*, 543–546. [CrossRef]
219. Wang, F.; Li, M.S.; Lu, Y.P.; Qi, Y.X. A simple sol-gel technique for preparing hydroxyapatite nanopowders. *Mater. Lett.* **2005**, *59*, 916–919.
220. Lai, C.; Tang, S.; Wang, Y.; Wei, K. Formation of calcium phosphate nanoparticles in reverse microemulsions. *Mater. Lett.* **2005**, *59*, 210–214. [CrossRef]

221. Morgan, T.T.; Muddana, H.S.; Altinoğlu, E.I.; Rouse, S.M.; Tabaković, A.; Tabouillot, T.; Russin, T.J.; Shanmugavelandy, S.S.; Butler, P.J.; Eklund, P.C.; et al. Encapsulation of Organic Molecules in Calcium Phosphate Nanocomposite Particles for Intracellular Imaging and Drug Delivery. *Nano Lett.* **2008**, *8*, 4108–4115. [CrossRef]
222. Shen, Y.; Liu, J.; Lin, K.; Zhang, W. Synthesis of strontium substituted hydroxyapatite whiskers used as bioactive and mechanical reinforcement material. *Mater. Lett.* **2012**, *70*, 76–79. [CrossRef]
223. Xin, R.L.; Ren, F.Z.; Leng, Y. Synthesis and characterization of nano-crystalline calcium phosphates with EDTA-assisted hydrothermal method. *Mater. Des.* **2010**, *31*, 1691–1694. [CrossRef]
224. Lin, K.; Chang, J.; Cheng, R.; Ruan, M. Hydrothermal microemulsion synthesis of stoichiometric single crystal hydroxyapatite nanorods with mono-dispersion and narrow-size distribution. *Mater. Lett.* **2007**, *61*, 1683–1687. [CrossRef]
225. Wang, X.; Zhuang, J.; Peng, Q.; Li, Y.D. Liquid-solid-solution synthesis of biomedical hydroxyapatite nanorods. *Adv. Mater.* **2006**, *18*, 2031–2034. [CrossRef]
226. Lin, K.; Chang, J.; Liu, X.; Chen, L.; Zhou, Y. Synthesis of element-substituted hydroxyapatite with controllable morphology and chemical composition using calcium silicate as precursor. *CrystEngComm* **2011**, *13*, 4850–4855. [CrossRef]
227. Xiao, J.; Zhu, Y.; Ruan, Q.; Liu, Y.; Zeng, Y.; Xu, F.; Zhang, L. Biomacromolecule and Surfactant Complex Matrix for Oriented Stack of 2-Dimensional Carbonated Hydroxyapatite Nanosheets as Alignment in Calcified Tissues. *Cryst. Growth Des.* **2010**, *10*, 1492–1499. [CrossRef]
228. Viswanath, B.; Ravishankar, N. Controlled synthesis of plate-shaped hydroxyapatite and implications for the morphology of the apatite phase in bone. *Biomaterials* **2008**, *29*, 4855–4863. [CrossRef] [PubMed]
229. Zhuang, Z.; Yamamoto, H.; Aizawa, M. Synthesis of plate-shaped hydroxyapatite via an enzyme reaction of urea with urease and its characterization. *Powder Technol.* **2012**, *222*, 193–200. [CrossRef]
230. Zou, Z.; Liu, X.; Chen, L.; Lin, K.; Chang, J. Dental enamel-like hydroxyapatite transformed directly from monetite. *J. Mater. Chem.* **2012**, *22*, 22637–22641. [CrossRef]
231. Qi, C.; Zhu, Y.-J.; Lu, B.-Q.; Zhao, X.-Y.; Zhao, J.; Chen, F. Hydroxyapatite nanosheet-assembled porous hollow microspheres: DNA-templated hydrothermal synthesis, drug delivery and protein adsorption. *J. Mater. Chem.* **2012**, *22*, 22642–22650. [CrossRef]
232. Ma, M.G.; Zhu, J.F. Solvothermal Synthesis and Characterization of Hierarchically Nanostructured Hydroxyapatite Hollow Spheres. *Eur. J. Inorg. Chem.* **2009**, *2009*, 5522–5526. [CrossRef]
233. Nathanael, A.J.; Hong, S.I.; Mangalaraj, D.; Ponpandian, N.; Chen, P.C. Template-Free Growth of Novel Hydroxyapatite Nanorings: Formation Mechanism and Their Enhanced Functional Properties. *Cryst. Growth Des.* **2012**, *12*, 3565–3574. [CrossRef]
234. Chen, F.; Zhu, Y.-J.; Wang, K.-W.; Zhao, K.-L. Surfactant-free solvothermal synthesis of hydroxyapatite nanowire/nanotube ordered arrays with biomimetic structures. *CrystEngComm* **2011**, *13*, 1858–1863. [CrossRef]
235. Ebrahimi, M.; Botelho, M.; Lu, W.; Monmaturapoj, N. Synthesis and characterization of biomimetic bioceramic nanoparticles with optimized physicochemical properties for bone tissue engineering. *J. Biomed. Mater. Res. A* **2019**, *107*, 1654–1666. [CrossRef]
236. Oliveira, T.M.; Berti, F.C.B.; Gasoto, S.C.; Schneider, B.J.; Stimamiglio, M.A.; Berti, L.F. Calcium Phosphate-Based Bioceramics in the Treatment of Osteosarcoma: Drug Delivery Composites and Magnetic Hyperthermia Agents. *Front. Med. Technol.* **2021**, *3*, 700266. [CrossRef]
237. Chen, J.; Ashames, A.; Buabeid, M.A.; Fahelelbom, K.; Ijaz, M.; Murtaza, G. Nanocomposites drug delivery systems for the healing of bone fractures. *Int. J. Pharm.* **2020**, *585*, 119477. [CrossRef]
238. Chindamo, G.; Sapino, S.; Peira, E.; Chirio, D.; Gonzalez, M.C.; Gallarate, M. Bone Diseases: Current Approach and Future Perspectives in Drug Delivery Systems for Bone Targeted Therapeutics. *Nanomaterials* **2020**, *10*, 875. [CrossRef]
239. Lezaja, M.; Veljovic, D.N.; Jokic, B.M.; Cvijovic-Alagic, I.; Zrilic, M.M.; Miletic, V. Effect of hydroxyapatite spheres, whiskers, and nanoparticles on mechanical properties of a model BisGMA/TEGDMA composite initially and after storage. *J. Biomed. Mater. Res. Part B Appl. Biomater.* **2013**, *101*, 1469–1476. [CrossRef] [PubMed]
240. Dou, Y.; Lin, K.; Chang, J. Polymer nanocomposites with controllable distribution and arrangement of inorganic nanocomponents. *Nanoscale* **2011**, *3*, 1508–1511. [CrossRef] [PubMed]
241. Neira, I.S.; Kolen'Ko, Y.V.; Kommareddy, K.P.; Manjubala, I.; Yoshimura, M.; Guitián, F. Reinforcing of a calcium phosphate cement with hydroxyapatite crystals of various morphologies. *ACS Appl. Mater. Interfaces* **2010**, *2*, 3276–3284. [CrossRef] [PubMed]
242. Oyane, A.; Araki, H.; Nakamura, M.; Shimizu, Y.; Shubhra, Q.T.H.; Ito, A.; Tsurushima, H. Controlled superficial assembly of DNA-amorphous calcium phosphate nanocomposite spheres for surface-mediated gene delivery. *Colloids Surf. B Biointerfaces* **2016**, *141*, 519–527. [CrossRef]
243. Lester, E.; Tang, S.V.Y.; Khlobystov, A.; Rose, V.L.; Buttery, L.; Roberts, C.J. Producing nanotubes of biocompatible hydroxyapatite by continuous hydrothermal synthesis. *CrystEngComm* **2013**, *15*, 3256–3260. [CrossRef]
244. Wang, J.; Qin, W.; Liu, X.; Liu, H. Synthesis and characterization of hydroxyapatite on hydrolyzed polyacrylonitrile nanofiber templates. *RSC Adv.* **2013**, *3*, 11132–11139. [CrossRef]
245. Ruffini, A.; Sprio, S.; Tampieri, A. Study of the hydrothermal transformation of wood-derived calcium carbonate into 3D hierarchically organized hydroxyapatite. *Chem. Eng. J.* **2013**, *217*, 150–158. [CrossRef]
246. Lin, K.; Chang, J.; Zhu, Y.; Wu, W.; Cheng, G.; Zeng, Y.; Ruan, M. A Facile One-Step Surfactant-Free and Low-Temperature Hydrothermal Method to Prepare Uniform 3D Structured Carbonated Apatite Flowers. *Cryst. Growth Des.* **2009**, *9*, 177–181. [CrossRef]

247. Zhang, B.; Hou, X.; Dang, L.; Wei, H. Selective polymorphic crystal growth on self-assembled monolayer using molecular modeling as an assistant method. *J. Cryst. Growth* **2019**, *518*, 81–88. [CrossRef]
248. Bleek, K.; Taubert, A. New developments in polymer-controlled, bioinspired calcium phosphate mineralization from aqueous solution. *Acta Biomater.* **2013**, *9*, 6283–6321. [CrossRef]
249. LeGeros, R.Z. Calcium Phosphate-Based Osteoinductive Materials. *Chem. Rev.* **2008**, *108*, 4742–4753. [CrossRef] [PubMed]
250. Chen, H.; Clarkson, B.H.; Sun, K.; Mansfield, J.F. Self-assembly of synthetic hydroxyapatite nanorods into an enamel prism-like structure. *J. Colloid Interf. Sci.* **2005**, *288*, 97–103. [CrossRef]
251. Xia, L.; Lin, K.; Jiang, X.; Xu, Y.; Zhang, M.; Chang, J.; Zhang, Z. Enhanced osteogenesis through nano-structured surface design of macroporous hydroxyapatite bioceramic scaffolds via activation of ERK and p38 MAPK signaling pathways. *J. Mater. Chem. B* **2013**, *1*, 5403–5416. [CrossRef] [PubMed]
252. Lin, K.; Xia, L.; Gan, J.; Zhang, Z.; Chen, H.; Jiang, X.; Chang, J. Tailoring the Nanostructured Surfaces of Hydroxyapatite Bioceramics to Promote Protein Adsorption, Osteoblast Growth, and Osteogenic Differentiation. *ACS Appl. Mater. Interfaces* **2013**, *5*, 8008–8017. [CrossRef] [PubMed]
253. Liu, X.; Miao, Y.; Liang, H.; Diao, J.; Hao, L.; Shi, Z.; Zhao, N.; Wang, Y. 3D-printed bioactive ceramic scaffolds with biomimetic micro/nano-HAp surfaces mediated cell fate and promoted bone augmentation of the bone-implant interface in vivo. *Bioact. Mater.* **2022**, *12*, 120–132. [CrossRef]
254. Wang, W.; Oaki, Y.; Ohtsuki, C.; Nakano, T.; Imai, H. Formation of c-axis-oriented columnar structures through controlled epitaxial growth of hydroxyapatite. *J. Asian Ceram. Soc.* **2013**, *1*, 143–148. [CrossRef]
255. Zhang, C.-Y.; Zhang, W.; Yao, H.-B.; Zhu, H.-Z.; Mao, L.-B.; Yu, S.-H. Bioinspired Crystallization of Continuous Calcium Phosphate Films on a Langmuir Monolayer of Zein Protein: Their Mechanical Performance, Hydrophilicity, and Biocompatibility. *Cryst. Growth Des.* **2013**, *13*, 3505–3513. [CrossRef]
256. Ryu, J.; Ku, S.H.; Lee, M.; Park, C.B. Bone-like peptide/hydroxyapatite nanocomposites assembled with multi-level hierarchical structures. *Soft Matter* **2011**, *7*, 7201–7206. [CrossRef]
257. Liu, S.Z.; Yin, Y.J.; Chen, H.F. PEO-assisted precipitation of human enamel-like fluorapatite films for tooth whitening. *CrystEngComm* **2013**, *15*, 5853–5859. [CrossRef]
258. Chen, F.; Zhu, Y.-J.; Zhao, X.-Y.; Lu, B.-Q.; Wu, J. Solvothermal synthesis of oriented hydroxyapatite nanorod/nanosheet arrays using creatine phosphate as phosphorus source. *CrystEngComm* **2013**, *15*, 4527–4531. [CrossRef]
259. Zhang, B.; Sun, H.; Wu, L.; Ma, L.; Xing, F.; Kong, Q.; Fan, Y.; Zhou, C.; Zhang, X. 3D printing of calcium phosphate bioceramic with tailored biodegradation rate for skull bone tissue reconstruction. *Bio-Des. Manuf.* **2019**, *2*, 161–171. [CrossRef]
260. Wu, L.; Zhou, C.; Zhang, B.; Lei, H.; Wang, W.; Pu, X.; Liu, L.; Liang, J.; Fan, Y.; Zhang, X. Construction of Biomimetic Natural Wood Hierarchical Porous-Structure Bioceramic with Micro/Nanowhisker Coating to Modulate Cellular Behavior and Osteoinductive Activity. *ACS Appl. Mater. Interfaces* **2020**, *12*, 48395–48407. [CrossRef]
261. Kon, E.; Salamanna, F.; Filardo, G.; Di Matteo, B.; Shabshin, N.; Shani, J.; Fini, M.; Perdisa, F.; Parrilli, A.; Sprio, S.; et al. Bone Regeneration in Load-Bearing Segmental Defects, Guided by Biomorphic, Hierarchically Structured Apatitic Scaffold. *Front. Bioeng. Biotechnol.* **2021**, *9*, 734486. [CrossRef]
262. Li, J.; Chen, Y.-C.; Tseng, Y.-C.; Mozumdar, S.; Huang, L. Biodegradable calcium phosphate nanoparticle with lipid coating for systemic siRNA delivery. *J. Control. Release* **2010**, *142*, 416–421. [CrossRef]
263. Liu, Y.; Hunziker, E.; Randall, N.; de Groot, K.; Layrolle, P. Proteins incorporated into biomimetically prepared calcium phosphate coatings modulate their mechanical strength and dissolution rate. *Biomaterials* **2003**, *24*, 65–70. [CrossRef]
264. Chernousova, S.; Epple, M. Live-cell imaging to compare the transfection and gene silencing efficiency of calcium phosphate nanoparticles and a liposomal transfection agent. *Gene Ther.* **2017**, *24*, 282–289. [CrossRef] [PubMed]
265. Schumacher, M.; Reither, L.; Thomas, J.; Kampschulte, M.; Gbureck, U.; Lode, A.; Gelinsky, M. Calcium phosphate bone cement/mesoporous bioactive glass composites for controlled growth factor delivery. *Biomater. Sci.* **2017**, *5*, 578–588. [CrossRef]
266. Madhumathi, K.; Rubaiya, Y.; Doble, M.; Venkateswari, R.; Sampath Kumar, T.S. Antibacterial, anti-inflammatory, and bone-regenerative dual-drug-loaded calcium phosphate nanocarriers-in vitro and in vivo studies. *Drug Deliv. Transl. Res.* **2018**, *8*, 1066–1077. [CrossRef]
267. Levingstone, T.J.; Herbaj, S.; Dunne, N.J. Calcium Phosphate Nanoparticles for Therapeutic Applications in Bone Regeneration. *Nanomaterials* **2019**, *9*, 1570. [CrossRef]
268. Iafisco, M.; Palazzo, B.; Marchetti, M.; Margiotta, N.; Ostuni, R.; Natile, G.; Morpurgo, M.; Gandin, V.; Marzano, C.; Roveri, N. Smart delivery of antitumoral platinum complexes from biomimetic hydroxyapatite nanocrystals. *J. Mater. Chem.* **2009**, *19*, 8385–8392. [CrossRef]
269. Matsumoto, T.; Okazaki, M.; Inoue, M.; Yamaguchi, S.; Kusunose, T.; Toyonaga, T.; Hamada, Y.; Takahashi, J. Hydroxyapatite particles as a controlled release carrier of protein. *Biomaterials* **2004**, *25*, 3807–3812. [CrossRef]
270. Palazzo, B.; Iafisco, M.; Laforgia, M.; Margiotta, N.; Natile, G.; Bianchi, C.; Walsh, D.; Mann, S.; Roveri, N. Biomimetic Hydroxyapatite–Drug Nanocrystals as Potential Bone Substitutes with Antitumor Drug Delivery Properties. *Adv. Funct. Mater.* **2007**, *17*, 2180–2188. [CrossRef]
271. Uskoković, V.; Batarni, S.S.; Schweicher, J.; King, A.; Desai, T.A. Effect of calcium phosphate particle shape and size on their antibacterial and osteogenic activity in the delivery of antibiotics in vitro. *ACS Appl. Mater. Interfaces* **2013**, *5*, 2422–2431. [CrossRef]

272. Dorozhkin, S.V. Nanosized and nanocrystalline calcium orthophosphates. *Acta Biomater.* **2010**, 6, 715–734. [CrossRef]
273. Pountos, I.; Georgouli, T.; Bird, H.; Kontakis, G.; Giannoudis, P.V. The effect of antibiotics on bone healing: Current evidence. *Expert Opin. Drug Saf.* **2011**, 10, 935–945. [CrossRef]
274. Dimitriou, R.; Tsiridis, E.; Giannoudis, P.V. Current concepts of molecular aspects of bone healing. *Injury* **2005**, 36, 1392–1404. [CrossRef] [PubMed]
275. Dong, X.; Wang, Q.; Wu, T.; Pan, H. Understanding adsorption-desorption dynamics of BMP-2 on hydroxyapatite (001) surface. *Biophys. J.* **2007**, 93, 750–759. [CrossRef]
276. Xie, G.; Sun, J.; Zhong, G.; Liu, C.; Wei, J. Hydroxyapatite nanoparticles as a controlled-release carrier of BMP-2: Absorption and release kinetics in vitro. *J. Mater. Sci. Mater. Med.* **2010**, 21, 1875–1880. [CrossRef]
277. Qadir, A.; Gao, Y.; Suryaji, P.; Tian, Y.; Lin, X.; Dang, K.; Jiang, S.; Li, Y.; Miao, Z.; Qian, A. Non-Viral Delivery System and Targeted Bone Disease Therapy. *Int. J. Mol. Sci.* **2019**, 20, 565. [CrossRef]
278. Fu, X.; Liu, G.; Halim, A.; Ju, Y.; Luo, Q.; Song, G. Mesenchymal Stem Cell Migration and Tissue Repair. *Cells* **2019**, 8, 784. [CrossRef] [PubMed]
279. Chai, Y.; Carlier, A.; Bolander, J.; Roberts, S.; Geris, L.; Schrooten, J.; Van Oosterwyck, H.; Luyten, F. Current views on calcium phosphate osteogenicity and the translation into effective bone regeneration strategies. *Acta Biomater.* **2012**, 8, 3876–3887. [CrossRef] [PubMed]
280. Breitwieser, G.E. Extracellular calcium as an integrator of tissue function. *Int. J. Biochem. Cell Biol.* **2008**, 40, 1467–1480. [CrossRef] [PubMed]
281. Aguirre, A.; González, A.; Planell, J.A.; Engel, E. Extracellular calcium modulates in vitro bone marrow-derived Flk-1+ CD34+ progenitor cell chemotaxis and differentiation through a calcium-sensing receptor. *Biochem. Biophys. Res. Commun.* **2010**, 393, 156–161. [CrossRef]
282. Dvorak, M.M.; Riccardi, D. Ca2+ as an extracellular signal in bone. *Cell Calcium* **2004**, 35, 249–255. [CrossRef]
283. Godwin, S.L.; Soltoff, S.P. Extracellular calcium and platelet-derived growth factor promote receptor-mediated chemotaxis in osteoblasts through different signaling pathways. *J. Biol. Chem.* **1997**, 272, 11307–11312. [CrossRef]
284. Özkucur, N.; Perike, S.; Sharma, P.; Funk, R.H. Persistent directional cell migration requires ion transport proteins as direction sensors and membrane potential differences in order to maintain directedness. *BMC Cell Biol.* **2011**, 12, 4. [CrossRef]
285. Özkucur, N.; Monsees, T.K.; Perike, S.; Do, H.Q.; Funk, R.H.W. Local calcium elevation and cell elongation initiate guided motility in electrically stimulated osteoblast-like cells. *PLoS ONE* **2009**, 4, e6131. [CrossRef]
286. Zayzafoon, M. Calcium/calmodulin signaling controls osteoblast growth and differentiation. *J. Cell. Biochem.* **2006**, 97, 56–70. [CrossRef]
287. Dvorak, M.M.; Siddiqua, A.; Ward, D.T.; Carter, D.H.; Dallas, S.L.; Nemeth, E.F.; Riccardi, D. Physiological changes in extracellular calcium concentration directly control osteoblast function in the absence of calciotropic hormones. *Proc. Natl. Acad. Sci. USA* **2004**, 101, 5140–5145. [CrossRef]
288. Dvorak, M.M.; Chen, T.-H.; Orwoll, B.; Garvey, C.; Chang, W.; Bikle, D.D.; Shoback, D.M. Constitutive activity of the osteoblast Ca^{2+}-sensing receptor promotes loss of cancellous bone. *Endocrinology* **2007**, 148, 3156–3163. [CrossRef]
289. Ahlstrom, M.; Pekkinen, M.; Riehle, U.; Lamberg-Allardt, C. Extracellular calcium regulates parathyroid hormone-related peptide expression in osteoblasts and osteoblast progenitor cells. *Bone* **2008**, 42, 483–490. [CrossRef] [PubMed]
290. Nakade, O.; Takahashi, K.; Takuma, T.; Aoki, T.; Kaku, T. Effect of extracellular calcium on the gene expression of bone morphogenetic protein-2 and -4 of normal human bone cells. *J. Bone Miner. Metab.* **2001**, 19, 13–19. [CrossRef] [PubMed]
291. Park, J.W.; Park, K.B.; Suh, J.Y. Effects of calcium ion incorporation on bone healing of Ti6Al4V alloy implants in rabbit tibiae. *Biomaterials* **2007**, 28, 3306–3313. [CrossRef]
292. Liu, Y.; Cooper, P.; Barralet, J.; Shelton, R.M. Influence of calcium phosphate crystal assemblies on the proliferation and osteogenic gene expression of rat bone marrow stromal cells. *Biomaterials* **2007**, 28, 1393–1403. [CrossRef]
293. Nayab, S.N.; Jones, F.H.; Olsen, I. Modulation of the human bone cell cycle by calcium ion-implantation of titanium. *Biomaterials* **2007**, 28, 38–44. [CrossRef]
294. Piters, E.; Boudin, E.; Van Hul, W. Wnt signaling: A win for bone. *Arch. Biochem. Biophys.* **2008**, 473, 112–116. [CrossRef]
295. Beck, G.R., Jr.; Knecht, N. Osteopontin regulation by inorganic phosphate is ERK1/2-, protein kinase C-, and proteasome-dependent. *J. Biol. Chem.* **2003**, 278, 41921–41929. [CrossRef]
296. Murshed, M.; Harmey, D.; Millán, J.L.; McKee, M.D.; Karsenty, G. Unique coexpression in osteoblasts of broadly expressed genes accounts for the spatial restriction of ECM mineralization to bone. *Genes Dev.* **2005**, 19, 1093–1104. [CrossRef]
297. Suzuki, A.; Ghayor, C.; Guicheux, J.; Magne, D.; Quillard, S.; Kakita, A.; Ono, Y.; Miura, Y.; Oiso, Y.; Itoh, M.; et al. Enhanced expression of the inorganic phosphate transporter Pit-1 is involved in BMP-2-induced matrix mineralization in osteoblast-like cells. *J. Bone Miner. Res.* **2006**, 21, 674–683. [CrossRef]
298. Munaron, L. Intracellular calcium, endothelial cells and angiogenesis. *Recent Pat. Anticancer. Drug Discov.* **2006**, 1, 105–119. [CrossRef]
299. Aguirre, A.; González, A.; Navarro, M.; Castaño Linares, Ó.; Planell Estany, J.A.; Engel, E. Control of microenvironmental cues with a smart biomaterial composite promotes endothelial progenitor cell angiogenesis. *Eur. Cell Mater.* **2012**, 24, 90–106. [CrossRef]

300. Xiao, X.; Wang, W.; Liu, D.; Zhang, H.; Gao, P.; Geng, L.; Yuan, Y.; Lu, J.; Wang, Z. The promotion of angiogenesis induced by three-dimensional porous beta-tricalcium phosphate scaffold with different interconnection sizes via activation of PI3K/Akt pathways. *Sci. Rep.* **2015**, *5*, 9409. [CrossRef] [PubMed]
301. Kim, H.D.; Amirthalingam, S.; Kim, S.L.; Lee, S.S.; Rangasamy, J.; Hwang, N.S. Biomimetic Materials and Fabrication Approaches for Bone Tissue Engineering. *Adv. Healthc. Mater.* **2017**, *6*, 1700612. [CrossRef] [PubMed]
302. Yu, T.; Dong, C.; Shen, Z.; Chen, Y.; Yu, B.; Shi, H.; Zhou, C.; Ye, J. Vascularization of plastic calcium phosphate cement in vivo induced by in-situ-generated hollow channels. *Mater. Sci. Eng. C* **2016**, *68*, 153–162. [CrossRef]
303. Tang, Z.; Li, X.; Tan, Y.; Fan, H.; Zhang, X. The material and biological characteristics of osteoinductive calcium phosphate ceramics. *Regen. Biomater.* **2018**, *5*, 43–59. [CrossRef]
304. Barradas, A.M.; Fernandes, H.A.; Groena, N.; ChinChaibc, Y.; Schrootencd, J.; van de Peppel, J.; van Leeuwen, J.P.; van Blitterswijk, C.A.; de Boer, J. A calcium-induced signaling cascade leading to osteogenic differentiation of human bone marrow-derived mesenchymal stromal cells. *Biomaterials* **2012**, *33*, 3205–3215. [CrossRef] [PubMed]
305. de Jesus Perez, V.A.; Alastalo, T.P.; Wu, J.C.; Axelrod, J.D.; Cooke, J.P.; Amieva, M.; Rabinovitch, M. Bone morphogenetic protein 2 induces pulmonary angiogenesis via Wnt-beta-catenin and Wnt-RhoA-Rac1 pathways. *J. Cell Biol.* **2009**, *184*, 83–99. [CrossRef] [PubMed]
306. Sorescu, G.P.; Song, H.; Tressel, S.L.; Hwang, J.; Dikalov, S.; Smith, D.A.; Boyd, N.L.; Platt, M.O.; Lassègue, B.; Griendling, K.; et al. Bone morphogenic protein 4 produced in endothelial cells by oscillatory shear stress induces monocyte adhesion by stimulating reactive oxygen species production from a nox1-based NADPH oxidase. *Circ. Res.* **2004**, *95*, 773–779. [CrossRef]
307. Cola, C.; Almeida, M.; Li, D.; Romeo, F.; Mehta, J.L. Regulatory role of endothelium in the expression of genes affecting arterial calcification. *Biochem. Biophys. Res. Commun.* **2004**, *320*, 424–427. [CrossRef]

Journal of
Functional Biomaterials

MDPI

Article

Biodegradable Mg-Sc-Sr Alloy Improves Osteogenesis and Angiogenesis to Accelerate Bone Defect Restoration

Nadia Aboutalebianaraki [1,2], Craig J. Neal [1,2,3], Sudipta Seal [1,2,3] and Mehdi Razavi [1,2,*]

1 Biionix™ (Bionic Materials, Implants & Interfaces) Cluster, Department of Medicine, University of Central Florida College of Medicine, Orlando, FL 32827, USA
2 Department of Materials Science and Engineering, University of Central Florida, Orlando, FL 32816, USA
3 Advanced Materials Processing and Analysis Center, Department of Materials Science and Engineering, University of Central Florida, Orlando, FL 32826, USA
* Correspondence: mehdi.razavi@ucf.edu

Abstract: Magnesium (Mg) and its alloys are considered to be biodegradable metallic biomaterials for potential orthopedic implants. While the osteogenic properties of Mg alloys have been widely studied, few reports focused on developing a bifunctional Mg implant with osteogenic and angiogenic properties. Herein, a Mg-Sc-Sr alloy was developed, and this alloy's angiogenesis and osteogenesis effects were evaluated in vitro for the first time. X-ray Fluorescence (XRF), X-ray diffraction (XRD), and metallography images were used to evaluate the microstructure of the developed Mg-Sc-Sr alloy. Human umbilical vein/vascular endothelial cells (HUVECs) were used to evaluate the angiogenic character of the prepared Mg-Sc-Sr alloy. A mix of human bone-marrow-derived mesenchymal stromal cells (hBM-MSCs) and HUVEC cell cultures were used to assess the osteogenesis-stimulating effect of Mg-Sc-Sr alloy through alkaline phosphatase (ALP) and Von Kossa staining. Higher ALP activity and the number of calcified nodules (27% increase) were obtained for the Mg-Sc-Sr-treated groups compared to Mg-treated groups. In addition, higher VEGF expression (45.5% increase), tube length (80.8% increase), and number of meshes (37.9% increase) were observed. The Mg-Sc-Sr alloy showed significantly higher angiogenesis and osteogenic differentiation than pure Mg and the control group, suggesting such a composition as a promising candidate in bone implants.

Keywords: magnesium alloy; osteogenesis; angiogenesis; bone implants

1. Introduction

The growing number of musculoskeletal disorders caused by poorly healed fractures and diseases has become a serious global challenge in public health, significantly increasing overall healthcare costs for patients [1–5]. Recently, permanent metallic biomaterials, including titanium alloys, stainless steels, and cobalt-chromium alloys, have been used to fix bone fractures [5–7]. These metallic implants can only provide physical support at the fracture; they do not contribute to bone regeneration [5–7]. Further, concerns have been raised over the deleterious medical effects of mechanical property mismatches between such materials and cortical bones. Differences in elastic modulus between the implant and natural bone lead to unnatural, inhomogeneous mechanical loading, resulting in implant loosening, reduction in osteogenesis, implant failure, and removal surgery [1,7]. Meanwhile, the presence of these permanent implants in the fractured site after the complete healing of the bone can result in the release of toxic degradation-related particles and ions in the long term. This may cause chronic inflammation by triggering a host immune response, leading to tissue damage and implant failure [5,8].

Mg is a biodegradable metal that eliminates the need for removal surgery [9,10]. It has a Young's modulus (10–40 GPa) close to that of human cortical bone (3–20 GPa) [11,12]. More importantly, Mg^{2+}, which evolves from the implant over time post-implantation, is one of the most predominant elements in the human body, mainly in bones and teeth [12].

Mg ions can participate in numerous bone-related biological functions and processes related to cell metabolism [13,14]. Mg^{2+} can improve new bone formation, regulate the resorption process, upregulate the expression of osteogenic genes, increase the mineralization of Ca^{2+}, promote osteoblast adhesion to the implant surface, and regulate muscle contraction [14–16].

Angiogenesis is a critical factor in facilitating new bone formation by mediating the mass transport of sufficient oxygen and nutrients through new vessel formation in the early healing process and increasing the success rate of implantation [9,17,18]. Mg has also been found to enhance the extent and rate of angiogenesis at implant sites [18,19]. Additionally, Mg^{2+} induces nitric oxide production by endothelial cells, which upregulates the vascular endothelial growth factor (VEGF) expression and improves angiogenesis [9,20,21]. These characteristics indicate the excellent potential of Mg as a biomaterial in bone healing applications [5,9,22].

Although previous investigations have demonstrated the potential of Mg alloys in bone defect restoration, the high corrosion rate of Mg implants leads to the generation of substantial amounts of subcutaneous hydrogen (H_2) gas at the implantation site. The gas formation reduces implant mechanical integrity, resulting in damage to the surrounding tissues and implant failure [5,10,15]. Mg alloy's highly alkaline local pH value can also hinder the healing process since a high pH would damage cells [23]. Therefore, controlling the degradation rate and resultant H_2 generation of Mg implants is crucial for the successful bone healing process, limitation of tissue damage, and maintenance of a healthy tissue-implant microenvironment [8,9]. Recently, several alloying elements frequently used with Mg, such as zirconium or aluminum, were found to produce unfavorable effects on tissue health (i.e., toxicity induction) [24]. Suitable alloying components for bone implants should be biocompatible and enhance biomechanical support, osseointegration, osteogenesis, and angiogenesis [9,21,25].

Aiming to reduce the corrosion rate of Mg, we have used scandium (Sc) as an alloying element in Mg due to its biocompatibility, high maximum solubility limit (24.61 wt.%), and similar lattice parameters to Mg [2,26] (which should significantly decrease lattice distortion and, thereby, effectively hinder stress concentration). Incorporating bioactive elements such as Sr, F, and Zn in Mg alloys has gained interest recently [2,26,27]. Among these bioactive elements, Sr, due to its similar structure to Ca, can provide the additional advantage of inducing osteoprogenitor cell differentiation into bone-forming osteoblasts by activating the Wnt/β-catenin signaling pathway and calcium-sensing receptors (CaSR). This also reduces the bone resorption that, in conjunction with the previously detailed processes, leads to superior new bone formation [5,8,14,28,29]. Sr ions can also bind to the bone matrix, increase bone mineralization [10,29], and enhance angiogenic properties [27,28,30].

The Mg-Zr-Sr-Sc alloy was developed based on a previous report [2]. However, this is the first time that Mg-Sc-Sr was developed, and the mechanism of bone formation using ALP and Von Kossa staining and vascularization using wound healing were studied. Finally, for the first time, this study offers a tube formation assay conducted on the Mg-Sc-Sr alloy.

To integrate osteogenesis and angiogenesis, we developed an Mg alloy containing Sc and Sr (Mg-Sc-Sr alloy). We hypothesized that the Sc component would improve corrosion resistance, while Sr addition would enhance/induce bone-forming processes. XRD, XRF, and metallography were performed to analyze the Mg-Sc-Sr alloy microstructure. An evaluation of early angiogenesis stages in human umbilical vein/vascular endothelial cells (HUVECs) was performed by vascular endothelial growth factor induction using an enzyme-linked immunosorbent assay (VEGF ELISA), tube formation, and scratch tests.

Moreover, in this work, human bone marrow-derived mesenchymal stromal cells (hBM-MSCs) and HUVECs were co-cultured to study the in vitro osteogenesis effects of the Mg-Sc-Sr alloy implant in a similar environment to bone tissue. The osteogenesis character of Mg-Sc-Sr alloy was examined by analyzing the Alkaline phosphatase (ALP) activity and Von Kossa staining.

2. Materials and Methods

2.1. Mg Alloy Fabrication

The Mg (as control) and Mg-Sc-Sr alloys (2 wt.% Sc, and 2 wt.% Sr) were manufactured by casting method using high purity Mg (Goodfellow, Pittsburgh, PA, USA), Sc (Luciteria science, Olympia, WA, USA), and Sr (Luciteria science, Olympia, WA, USA). The Mg, Sc, and Sr were melted at 750 °C in a graphite crucible. After 10 min, the melted alloy was cast in a cylindrical mold. The melt was protected by an inert high-purity argon (Ar) atmosphere. Next, the Mg-Sc-Sr rods were heat-treated at 400 °C for 2 h. They were then cut into 12 mm diameter and 5 mm height disks for microstructure characterizations, osteogenesis, and angiogenesis evaluation as described below. Finally, discs were wet ground with successive grades of silicon carbide (SiC) abrasive paper (400, 600, 1200 grit) and ultrasonically cleaned (50 Hz and 150 W) in ethanol for 15 min and air-dried for further analysis.

2.2. Microstructure Characterization

The phase constitutions of the Mg and Mg-Sc-Sr disks were detected by X-ray diffraction (XRD) using an X-ray diffractometer (PANalytical Empyrean, Netherlands) with Cu-Kα radiation (λ = 0.154 nm, 40 kV) at a step size of 0.02°/s over the angular range of 10–90° (2θ value). The international center of diffraction data (ICDD) PDF-4 database was used for phase identification. The elemental phase compositions of the Mg and Mg-Sc-Sr disks were detected by X-ray Fluorescence (XRF) using an X-ray fluorescence spectrometer (PANalytical Epsilion, Netherlands) with a 50 KV silver anode.

The surface chemistry and chemical compositions were examined using X-ray photoelectron spectroscopy (XPS, Physical electronics PHI 5802). A non-monochromatic aluminum Kα X-ray source (51,486.6 eV, 250 W) with an acceleration voltage of 20 kV and amperage of 15 mA was used for all measurements. Measurements were performed under vacuum at <10^{-8} mbar. Core-level XPS spectra for C1s, Mg2p, and O1s were recorded. The binding energy values were referenced to a C1s peak (hydrocarbon C–C, C–H) at 284.6 eV, ascribed to adventitious carbon. A smart (Thermo Avantage v5.9902, Thermo Fisher) fitted background was applied to all collected spectra.

To observe sample microstructure and grain character, a picric acid-based etchant composed of 50 mL ethanol (100% v/v, Decon Labs, Montgomery, PA, USA), 5 mL acetic acid (Fisher, Pittsburgh, PA, USA), and 3 mL picric acid (LabChem, Zelienople, PA, USA) was used for etching the discs.

Finally, the microstructure was observed using an optical microscope (Amscope FMA-50, Irvine, CA, USA), and the line intercept method (ASTM, 2013) was used to calculate the average grain sizes for each disc [31].

2.3. Angiogenesis

Human umbilical vein/vascular endothelial cells (HUVECs) were obtained from American Type Culture Collection (ATCC, PCS-100-010, Manassas, WA, USA). HUVECs were cultured in a complete endothelial cell growth medium containing Vascular Cell Basal Medium (ATCC, PCS-100-030, Manassas, WA, USA) supplemented with Endothelial Cell Growth Kit-BBE (ATCC, PCS-100-040, Manassas, WA, USA) and incubated in a humidified atmosphere under the condition of Atm (atmospheric) O_2 and 5% CO_2 at 37 °C and incubated until cells reached to 80–90% confluency. To prepare endothelial cells extract, Mg and Mg-Sc-Sr disks were immersed in a complete endothelial cell growth medium (disc weight/medium volume ratio was 0.2 g/mL) [11]. The tube formation assay was performed as described before [9,11].

Briefly, a 96-well plate was coated with 50 µL of ice-cold Matrigel in each well (BD Bioscience, San Jose, CA, USA) on the ice. Then, the well plate was kept in the humidified atmosphere under the condition of Atm O_2 and 5% CO_2 for 30 min at 37 °C to allow gelation of the Matrigel. Next, the HUVECs (3×10^4/well) were seeded onto Matrigel-coated wells and cultured in a complete endothelial cell growth medium. Cells were incubated in a humidified atmosphere under Atm O_2 and 5% CO_2 for 4 h, 24 h, and 48 h at 37 °C after adding 200 μL of Mg and Mg-Sc-Sr extract. HUVECs were cultured with a complete endothelial cell growth medium without an extraction medium as a control.

Cells were then imaged at each time point using an optical microscope. The ability of HUVECs to form tubes was quantified by measuring the sum of tube length in each image using ImageJ software and reported as total tube length. Also, the total number of completed circular meshes in each image was counted and reported as the number of meshes. Five images per experimental group were analyzed.

To study the wound healing ability of Mg-Sc-Sr alloy, HUVECs (5×10^5/well) were seeded in six-well plates and cultured with a complete endothelial cell growth medium in a humidified atmosphere under the condition of Atm O_2 and 5% CO_2 at 37 °C for 24 h. A straight line was scratched on the confluent monolayer the next day by scraping with sterile 1000 μL pipette tips. Cells were then rinsed three times with PBS, and the prepared Mg and Mg-Sc-Sr alloy extracts were added to each well (1000 μL/well). As controls, HUVECs were cultured with a complete endothelial cell growth medium. Finally, cells were imaged after 4 h, 24 h, and 48 h using an optical microscope. To quantify the wound healing ability and cell migration activities of HUVECs, the number of migrated cells and scratch width were measured for five images per experimental group. The scratch width was measured using ImageJ software (version 1.53k).

To measure the VEGF expression from HUVECs, cells (3×10^4/well) were seeded in 96-well plates and cultured with a complete endothelial cell growth medium in a humidified atmosphere under the condition of Atm O_2 and 5% CO_2 at 37 °C for 24 h. Then, 100 μL Mg or Mg-Sc-Sr alloy extracts were added to each well. The HUVECs cultured with a complete endothelial cell growth medium were considered as control. The concentration of VEGF in the supernatant of HUVECs for each group was measured using a Human VEGF ELISA Kit (Boster, Pleasanton, CA, USA) after 4 h, 10 h, and 24 h of incubation, according to the manufacturer's instruction. Briefly, 100 μL supernatant of each experimental group was added to each well of the ELISA kit well plate. Then, the well plate was covered with the plate sealer and incubated for 90 min at 37 °C. Next, 100 μL of the Biotinylated Anti-Human VEGFA antibody was added to each well, and the well plate was incubated for 60 min at 37 °C. Afterward, the wells were washed with wash buffer, 100 μL of the Avidin-Biotin-Peroxidase Complex was added to each well, and the well plate was incubated at 37 °C for 30 min. Finally, 90 μL of Color Developing Reagent was added to each well for 15 min at 37 °C, and the reaction stopped by adding 100 μL of Stop Solution. The optical density of each well was determined by using a microplate reader (Bio-Rad 680, Hercules, CA, USA) at 450 nm.

2.4. Osteogenesis

Human bone marrow-derived mesenchymal stromal cells (hBM-MSCs, ATCC, PCS-500-012, Gaithersburg, MD, USA) were cultured in a complete cell culture medium composing Low glucose Dulbecco's Modified Eagle's medium (DMEM; glucose concentration: 1 g/L, Invitrogen, Carlsbad, CA, USA) supplemented with 50 U/mL penicillin-50 μg/mL streptomycin (Lonza, Basel, Switzerland) and 10% fetal bovine serum (FBS; Invitrogen, Carlsbad, CA, USA) at 37 °C under the condition of Atm O_2 and 5% CO_2 in a humidified atmosphere. The cell culture medium was refreshed every three days until cells reached 80–90% confluency. Then, to prepare Mg and Mg-Sc-Sr differentiation extracts, disks were first sterilized with ethanol, washed three times with sterile PBS, and sterilized with ultraviolet light for 2 h. Next, the discs were immersed in a differentiation medium (complete cell culture medium supplemented with 10 mM β-glycerophosphate, 50 mM ascorbic acid, and 100 nM dexamethasone (Sigma-Aldrich, Burlington, MA, USA)) [32] and incubated for three days in a humidified atmosphere at 37 °C under the condition of Atm O_2 and 5% CO_2 to obtain extract [33]. The discs' weight/medium volume ratio was 0.2 g/mL [33]. The acquired extracts of each disc were centrifuged, the solution was collected and stored separately at 4 °C, and the collected differentiation extracts were then used for Von Kossa staining and ALP staining.

To perform ALP staining and Von Kossa staining, hBM-MSCs (5 × 10^4 cells/well) and HUVECs (5 × 10^4 cells/well) were mixed and seeded in 12-well plates and cultured in a 1:1 mix of the complete cell growth medium and complete endothelial cell growth medium for 1 day at 37 °C in a humidified atmosphere under the normal condition of Atm O_2 and 5% CO_2. Then, the culture medium was replaced and hBM-MSCs-HUVECs were cultured (7 days for ALP staining and 21 days for Von Kossa staining) with a 1:1 mix of differentiation extract and endothelial cell extract for Mg and Mg-Sc-Sr groups. The medium was refreshed every 3 days. The mix of HUVECs and hBM-MSCs cultured with a 1:1 mix of differentiation medium and endothelial cell growth medium with no sample extract is considered a control.

The ALP staining was performed using Alkaline Phosphatase Staining Kit (Abcam, Boston, MA, USA). On day 7, the hBM-MSCs and HUVECs were washed with wash buffer. Then, 150 μL of ALP Staining Reagent solution was added to each well and incubated at 37 °C for 20 min. Finally, the cells were imaged using BZ-X810 Keyence.

Von Kossa staining was performed using a Von Kossa Stain Kit (Stablab, McKinney, TX, USA). On day 21, the medium was removed, hBM-MSCs and HUVECs were rinsed with DI water, and a 5% silver nitrate solution was added to each well. Then, cells were irradiated with ultraviolet light for 15 min. Once completed, cells were rinsed in DI water, and the 5% sodium thiosulfate solution was added for 1 min. Finally, cells were incubated in the nuclear fast red stain for 5 min, washed, and imaged (BZ-X810, Keyence, Itasca, IL, USA). To quantify the Von Kossa staining images, the number of black precipitates was counted as the number of nodules in five images per group using ImageJ software (version 1.53k) [34].

2.5. Statistical Analysis

Metallography, osteogenesis, and angiogenesis were performed in $n = 5$ samples per group. The XRD and XPF were performed in $n = 1$, and XPS was done in $n = 3$ samples per group. Results were reported as mean ± standard deviation. One-way ANOVA followed by a post-hoc Tukey test (Astatsa.com, Online and Free Web Statistical Calculators) and two-way ANOVA using GraphPad Prism software were used to obtain statistical significance between groups. The p-value of <0.05 denotes a statistically significant result.

3. Results

3.1. Microstructure Characterization

The chemical compositions of Mg and Mg-Sc-Sr alloys are presented in Table 1. The phase analysis of Mg and Mg-Sc-Sr discs was evaluated using X-ray diffraction (XRD). Figure 1 shows the XRD pattern of Mg that includes an α-Mg phase with diffraction peaks at 32.17°, 34.4°, 36.59°, 47.78°, 57.30°, 63.03°, 68.57°, 69.93°, 72.44°, 77.75°, and 81.09° (JCPDS No. 35-0821). Also, as shown in Figure 1, the Mg-Sc-Sr pattern indicated the α-Mg phase and $Mg_{17}Sr_2$ intermetallic phase at 67.25°, representing the formation of $Mg_{17}Sr_2$ in Mg-Sc-Sr discs (JCPDS No. 18-1275) [35]. Due to the high dissolution of Sc in Mg (24.61 wt.%), Sc element is well-dispersed in Mg, and no separate peaks were detected. As Figure 1 shows, adding more than 0.11 wt.% of the Sr (e.g., 2 wt.%) as an alloying element in Mg resulted in the alloying of some fraction of Sr content into the Mg, with the remainder contributing to the formation of an intermetallic phase ($Mg_{17}Sr_2$) [31].

Table 1. The chemical compositions of Mg and Mg-Sc-Sr alloys.

Element Concentration (%)	Mg	Sc	Sr	P	Cl	Ca	Si	Zn	Cr	Ti	Mn	Fe
Mg	98.4340	-	-	0.7580	0.4790	0.2270	0.0294	0.0048	0.0158	-	0.0322	0.0188
Mg-Sc-Sr	94.120	1.369	2.003	1.223	0.657	0.488	-	-	0.0074	0.0058	0.0824	0.0355

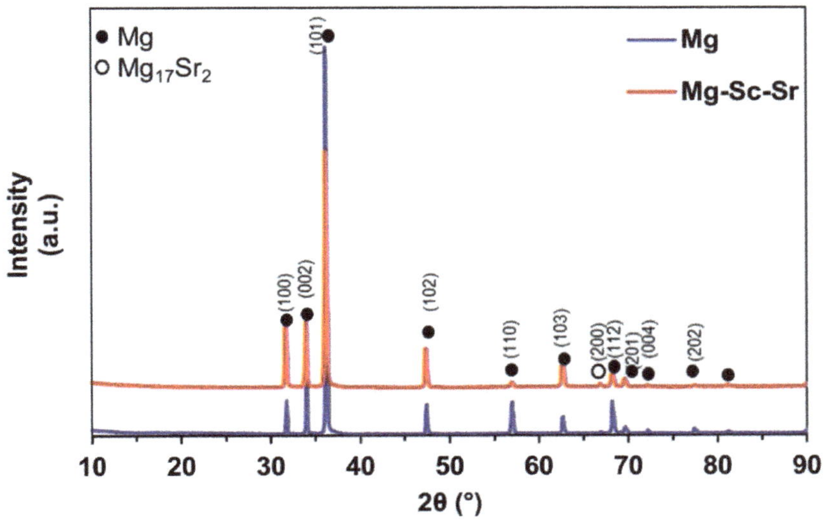

Figure 1. XRD patterns of the Mg and Mg-Sc-Sr alloy showing α-Mg phase for Mg sample and α-Mg and $Mg_{17}Sr_2$ intermetallic phase for Mg-Sc-Sr alloy.

XPS analysis was performed to evaluate the chemical composition of the surface layers on Mg and Mg-Sc-Sr discs. As shown in the de-convoluted XPS spectra (Figure 2A,B), peaks fitted in the binding energy range specific to Mg (1s) were ascribed to metallic (Mg^0) and oxide (Mg^{2+}) chemical states for both Mg sample (Figure 2A) and Mg-Sc-Sr sample (Figure 2B). The presence of both states is typical of a material surface wherein Mg is a major phase (or parent material), given the material's negative reduction potential. In the case of the Mg-Sc-Sr alloy sample, the ratio of fitted peak areas (measurement intensity area under-the-curve values) for Mg^0:Mg^{2+} was lower (0.8) compared to the pure Mg (2.5). Along with this, a shift in binding energy was observed to higher values, for both Mg1s-associated peaks, in the alloyed sample (+Δ0.4 and 0.6 eV, for Mg^0 and Mg^{2+}, respectively). This shift thereby suggests greater stability of related chemical states at the surface of the Mg-Sc-Sr alloy sample.

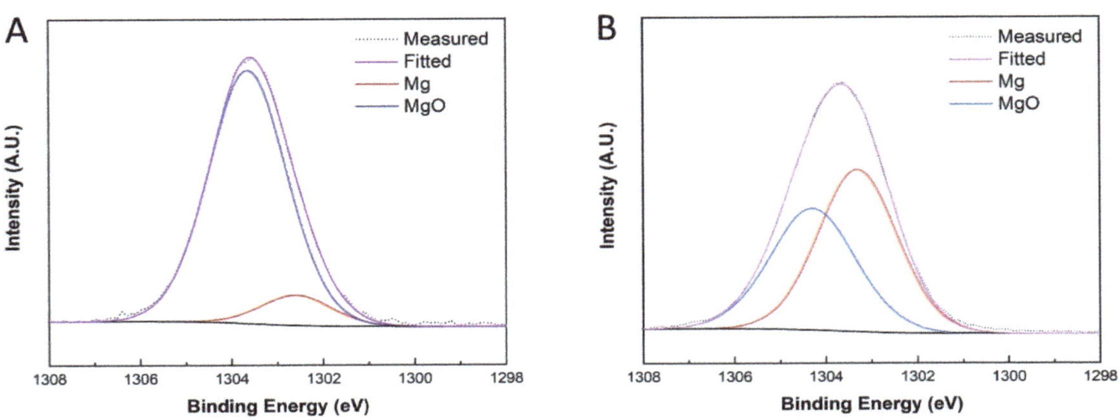

Figure 2. XPS spectra: (**A**) Mg 1s for Mg sample, (**B**) Mg 1s for Mg-Sc-Sr sample. We observed a lower oxide proportion relative to the Mg-Sc-Sr sample's metallic state.

Figure 3A–C showed that the average grain size for Mg-Sc-Sr (124.86 ± 22.61 µm) was significantly lower compared to Mg (539.16 ± 164.58 µm) due to the grain refinement effect of Sc [36–38] and Sr [35,39,40]. Intermetallic particles in the Mg-Sc-Sr discs could also cause a reduction in grain size through the grain-growth restriction mechanism [31,41,42].

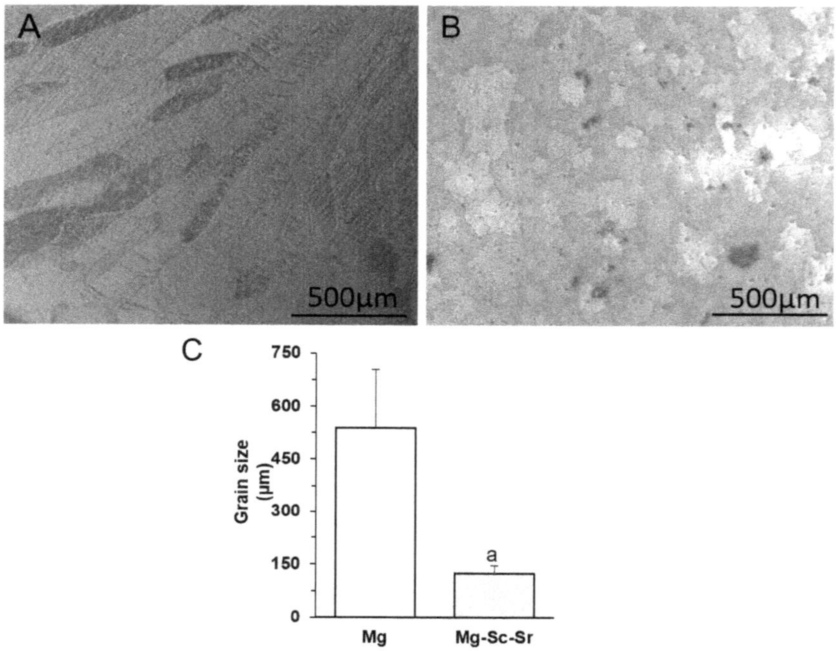

Figure 3. Metallography images: (**A**) Mg, (**B**) Mg-Sc-Sr, (**C**) average grain sizes of the Mg and Mg-Sc-Sr discs. In comparing the Mg-Sc-Sr grain size to the Mg grain size, a significant reduction in grain size for Mg-Sc-Sr was observed. Significant differences: a: $p < 0.05$ Mg-Sc-Sr vs. Mg.

3.2. Angiogenesis

To detect the expression of VEGF secreted from HUVECs cultured alone as control or HUVECs cultured with Mg, and Mg-Sc-Sr extracts, a VEGF ELISA was performed after 4 h, 10 h, and 24 h of incubation (Figure 4). After 4 h of incubation, the VEGF concentration of HUVECs cultured with Mg-Sc-Sr extract (58.83 ± 8.89 pg/mL) and Mg (52.66 ± 15.22 pg/mL) showed insignificant changes ($p > 0.05$) compared to VEGF concentration in the control group (46.93 ± 11.96 pg/mL). After 10 h of incubating HUVECs with the control, Mg, and Mg-Sc-Sr extract, the VEGF concentration for Mg (279.53 ± 14.66 pg/mL) and Mg-Sc-Sr (464.55 ± 32.31 pg/mL) obtained significantly higher than the control (187.90 ± 18.36 pg/mL, $p < 0.05$). The measured VEGF concentration of the HUVECs cultured with Mg-Sc-Sr was higher than the VEGF concentration of HUVECs cultured with Mg ($p < 0.05$). A significantly higher concentration of VEGF in the supernatant of HUVECs cultured with Mg-Sc-Sr extract (514.33 ± 44.90 pg/mL) compared to Mg (350.90 ± 50.25 pg/mL) and control (270.33 ± 35.61 pg/mL) after 24 h of incubation indicated that Mg-Sc-Sr extract potentiates VEGF secretion. These results also showed that the VEGF concentration in all groups increased over time, and the Mg-Sc-Sr alloy has a higher pro-angiogenic potential relative to pure Mg.

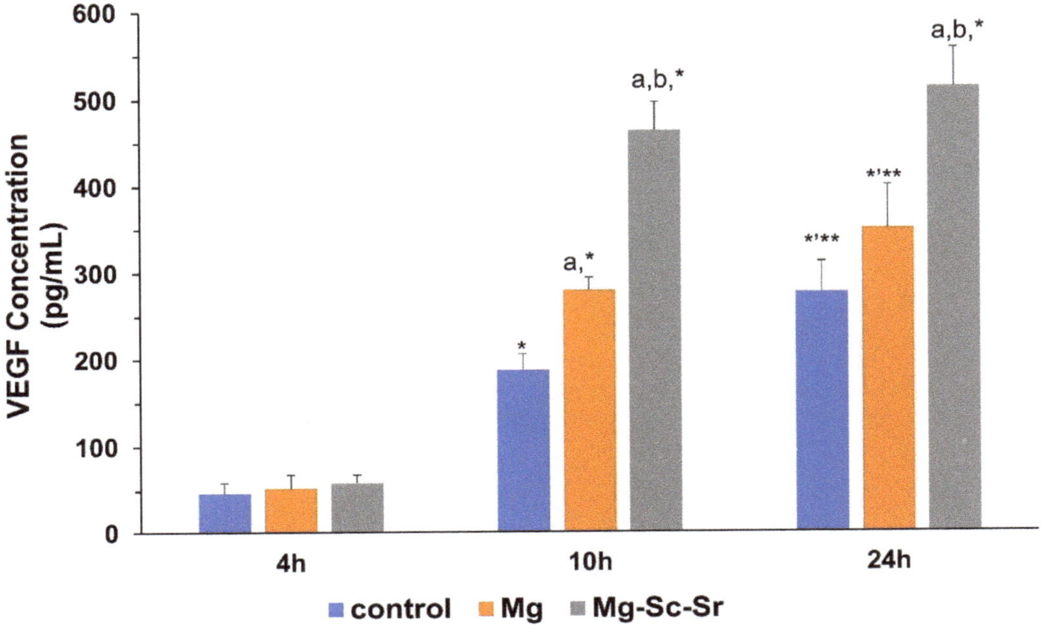

Figure 4. VEGF expression from HUVECs after 4 h, 10 h, and 24 h culturing with the control, Mg, and Mg-Sc-Sr extracts in vitro showing a significant increase in VEGF concentration for Mg-Sc-Sr-treated HUVECs compared to control. Significant differences based on two-way A: a: $p < 0.05$ Mg and Mg-Sc-Sr vs. control, b: $p < 0.05$ Mg-Sc-Sr vs. Mg. *: $p < 0.05$ 24 h and 10 h vs. 4 h, **: $p < 0.05$ 24 h vs. 10 h for each experimental group.

In Mg extract-treated samples, the VEGF concentration at 10 h and 24 h of incubation were significantly higher than the VEGF concentration at 4 h, and the highest concentration of VEGF was obtained after 24 h. Similarly, for HUVECs cultured with Mg-Sc-Sr extract, the VEGF concentration at 10 h and 24 h of incubation was significantly higher than the VEGF concentration at 4 h ($p < 0.05$). However, no significant changes ($p > 0.05$) were observed between 10 h and 24 h.

As shown in Figure 5A–C,J, the quantified total tube length for HUVECs cultured in the Mg extract (4708.8 ± 422.3) and the Mg-Sc-Sr extract (5020.2 ± 460.9) after 4 h of incubation was significantly higher compared to the control group (1758.0 ± 576.0) ($p < 0.05$). Figure 5D–F,J show tube formation after 24 h incubation of HUVECs with control, Mg, and Mg-Sc-Sr extract. The total tube length for Mg-Sc-Sr (10,413.6 ± 1014.6 μm) obtained significantly higher than the control (4723.2 ± 328.5 μm) and Mg (5759.6 ± 429.6 μm) ($p < 0.05$). A similar trend was obtained in HUVECs incubated with the control, Mg, and Mg-Sc-Sr extract after 48 h of incubation (Figure 5G–J). After 48 h, the measured total tube length of HUVECs cultured with Mg-Sc-Sr (10,352.8 ± 1140.0 μm) was significantly higher than the total tube length of HUVECs cultured with Mg (5988.2 ± 669.7 μm) and control (4698.4 ± 507.6 μm) that indicates the potential of Mg-Sc-Sr extract to enhance tube formation. In the control, the Mg and Mg-Sc-Sr groups, the total tube length after 24 h and 48 h of incubation was significantly higher than the total tube length at 4 h ($p < 0.05$). However, at 24 h and 48 h of incubation, the differences in total tube length between the Mg extracts and control were insignificant ($p > 0.05$).

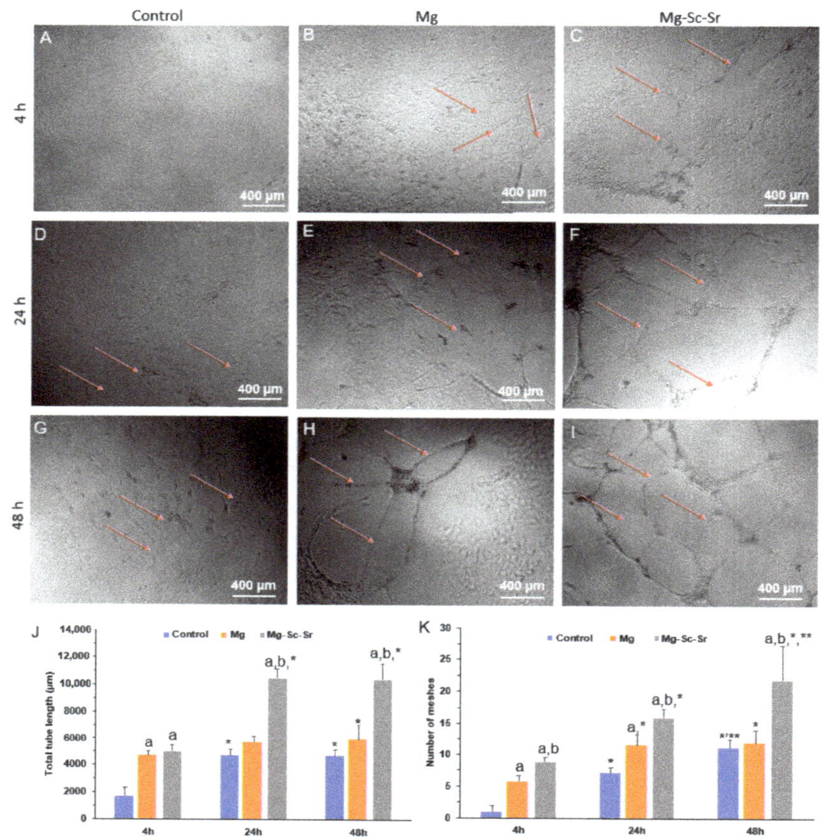

Figure 5. Tube formation assay for HUVECs cultured with the control, Mg, and Mg-Sc-Sr extracts in vitro. Tube formation images of HUVECs after culturing with the control group, Mg extract, and Mg-Sc-Sr extract after (**A–C**) 4 h, (**D–F**) 24 h, (**G–I**) 48 h, (**J**) quantification of total tube length formed by HUVECs, (**K**) quantification of the number of formed meshes by HUVECs, for control, Mg, and Mg-Sc-Sr at 4 h, 24 h, and 48 h. Significant differences based on two-way Anova: a: $p < 0.05$ Mg and Mg-Sc-Sr vs. control, b: $p < 0.05$ Mg-Sc-Sr vs. Mg, *: $p < 0.05$ 48 h and 24 h vs. 4 h, **: $p < 0.05$ 48 h vs. 24 h for each experimental group. Red arrows show newly formed tubes.

The number of HUVECs meshes at 4 h of HUVECs culture with Mg extract (5.8 ± 0.8), and Mg-Sc-Sr extract (8.8 ± 0.8) was significantly higher compared to the control group (1.0 ± 1.0) (Figure 5K). After 24 h, the number of meshes for HUVECs cultured with Mg-Sc-Sr extract (16.0 ± 1.2) was significantly higher than the control group (7.2 ± 0.8) and Mg (11.6 ± 1.7). In addition, the number of meshes for HUVECs treated with Mg-Sc-Sr extract was significantly higher than HUVECs cultured with Mg extract. After 48 h of incubation, the number of meshes quantified for HUVECs cultured in Mg-Sc-Sr extract (21.8 ± 5.4) was significantly higher than the control group (11.2 ± 1.3) and Mg (12.0 ± 2.0). No significant difference was obtained for the number of meshes in Mg and control groups at 48 h. In the Mg-Sc-Sr group, the number of meshes at 48 h was significantly higher than the number of meshes at 24 h and 4 h ($p < 0.05$).

To evaluate the migration potential and wound-healing ability of HUVECs cultured with the control, Mg, and Mg-Sc-Sr extracts, a wound-healing (scratch) assay was performed. As indicated in Figure 6A–E, the scratch width in the samples cultured with Mg-Sc-Sr extract

(605.6 ± 52.5 μm) and Mg (666.4 ± 88.2) was significantly lower than the control (826.4 ± 36.4 μm) after 4 h of incubation. In addition, the scratch width obtained for the Mg-Sc-Sr-treated group was significantly lower than that of the Mg-treated group. The 'wounds' were almost fully recovered in the Mg-Sc-Sr extract-treated HUVEC at 24 h. The calculated scratch width in HUVECs cultured with Mg-Sc-Sr and Mg (332.8 ± 68.2 μm) was higher than in the control group (448.6 ± 73.2 μm) groups, indicating a faster wound closure. In addition, the wound size in the Mg-Sc-Sr group was significantly reduced compared to the Mg group. No scratches were found in the control group, Mg, and Mg-Sc-Sr after 48 h of incubation (Figure 6D). The scratch size for the Mg-Sc-Sr-treated group was significantly reduced after 24 h of incubation compared to the Mg-Sc-Sr-treated group for 4 h.

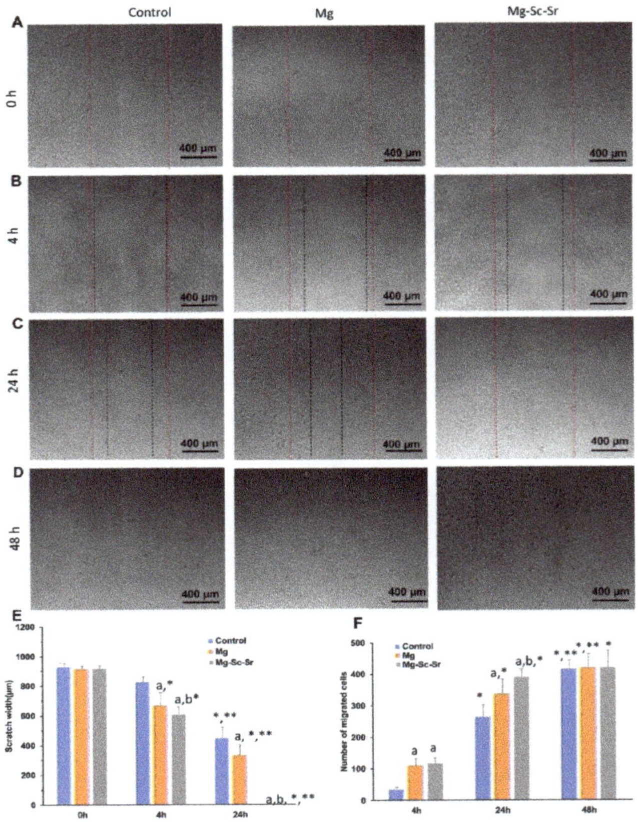

Figure 6. Scratch assay of HUVECs cultured with the control, Mg, and Mg-Sc-Sr extractions in vitro. (**A**) Images of initial wound scratch. Representative migration images of HUVECs after culturing with the control group, Mg extract, and Mg-Sc-Sr extract after (**B**) 4 h, (**C**) 24 h, (**D**) 48 h, (**E**) quantification of scratch width, (**F**) Quantification of the number of migrated cells. The width between red dots shows the primary scratch and the width between black dots shows the scratch width after incubation with or without extracts. Significant differences based on two-way ANOVA: a: $p < 0.05$ Mg and Mg-Sc-Sr vs. control, b: $p < 0.05$ Mg-Sc-Sr vs. Mg. For scratch width: *: $p < 0.05$ 4 h and 24 h vs. 0 h, **: $p < 0.05$ 24 h vs. 4 h for each experimental group. For the number of migrated cells: *: $p < 0.05$ 24 h and 48 h vs. 4 h, **: $p < 0.05$ 48 h vs. 24 h for each experimental group.

As presented in Figure 6F, the numbers of migrated HUVECs for samples cultured with Mg-Sc-Sr extract (115.8 ± 18.4) and Mg extract (110.8 ± 21.4) were significantly higher than the control values (36.0 ± 7.2) after 4 h of incubation. No significant difference was obtained between Mg and Mg-Sc-Sr. After 24 h, the number of migrated HUVECs for cells cultured with Mg-Sc-Sr (390.0 ± 25.9) and Mg (337.2 ± 47.2) obtained significantly higher than the control (265.1 ± 36.4). In addition, the number of migrated cells for the Mg-Sc-Sr-treated group was significantly higher compared to the Mg-treated group. After 48 h, no significant difference obtained between experimental groups ($p > 0.05$), and the wound was fully recovered. The number of migrated HUVECs in the Mg-Sc-Sr group was significantly higher after 24 h and 48 h of incubation compared to 4 h, but the number of migrated cells between 24 h and 48 h was not significant. Data from HUVECs treated with Mg-Sc-Sr extracts show improvement in cell migration ability.

3.3. Osteogenesis

ALP and Von Kossa staining were used to identify osteogenic properties (Figure 7A–G). Figure 7A–C showed that the ALP staining on day 7 indicated higher ALP expression in hBM-MSC-HUVECs incubated with Mg-Sc-Sr extract than in cells incubated with Mg extract or the control group. ALP expression in hBM-MSCs-HUVECs was also higher in cells incubated with Mg extract compared to ALP stain intensity in the control group. The ALP staining result in cells incubated with Mg-Sc-Sr extract showed that this group had the highest ALP activity.

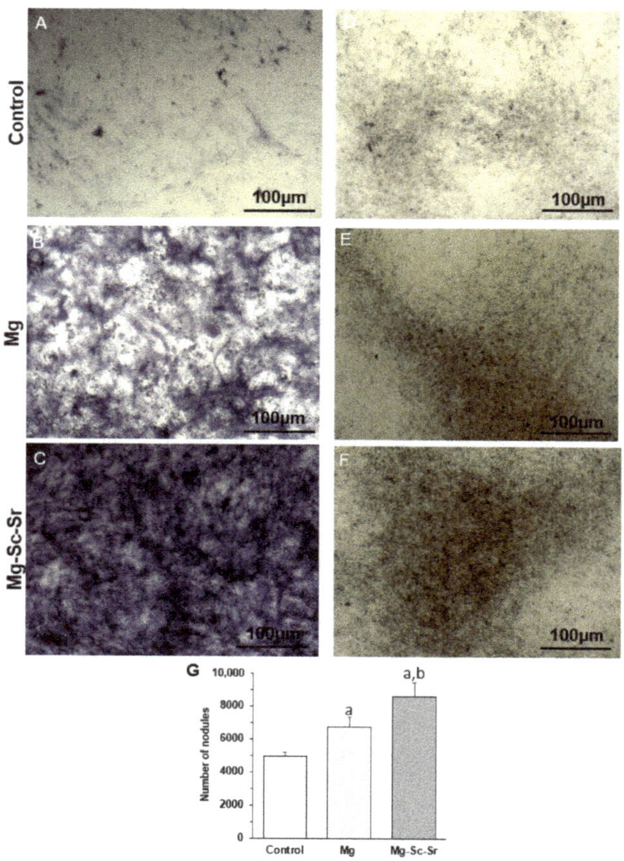

Figure 7. In vitro assay of osteogenic differentiation of hBM-MSCs and HUVECs cultured with alloy extracts: ALP staining of hBM-MSCs and HUVECs cultured with (**A**) control (differentiation medium

without disc extract), (**B**) Mg disc extract, (**C**) Mg-Sc-Sr disc extract for seven days. A dark blue stain indicates more ALP activity by sample. Von Kossa staining of hBM-MSCs and HUVECs cultured with (**D**) Control (differentiation medium without disc extract), (**E**) Mg extract, (**F**) Mg-Sc-Sr disc extract for 21 days. More black precipitation indicates higher Ca deposition in the sample. (**G**) Quantitative analysis of Von Kossa staining, significant differences: a: $p < 0.05$ Mg and Mg-Sc-Sr vs. control, b: $p < 0.05$ Mg-Sc-Sr vs. Mg.

To examine the calcium deposition, hBM-MSC and HUVECs were cultured with the control, Mg, and Mg-Sc-Sr in differentiation extracts for 21 days. The ability of the disc extracts to promote bone formation could be directly correlated to the intensity of von Kossa staining. Figure 7D–G indicated higher Von Kossa staining (calcified nodules stained black) in cells cultured with Mg-Sc-Sr disc extracts compared to the cells cultured alone or with Mg disc extracts that included lower black stained nodes. As shown in Figure 7G, the quantified number of calcified nodules that stained black confirmed a significant increase ($p < 0.05$) in the mineral deposition by cells cultured with Mg-Sc-Sr extract (8562.21 $\pm$ 853.54) and Mg extract (6756.60 $\pm$ 567.50), compared to control (4938.02 $\pm$ 231.51). In addition, the number of nodules in cells cultured with Mg-Sc-Sr extract was significantly higher than in cells cultured with Mg ($p < 0.05$). As shown in the results, Mg-Sc-Sr extract promoted calcium deposition and improved osteogenic properties.

4. Discussion

We developed an Mg-Sc-Sr alloy that simultaneously releases Mg^{2+} and Sr^{2+} to enhance new bone formation and new vessel formation. This will further facilitate the application of Mg alloys in osteogenic fixation. In this study, we evaluated the effects of adding Sc and Sr elements to Mg matrices on microstructure, angiogenic character, and osteogenic properties. Herein, 2 wt.% Sr was used to enhance the angiogenesis and osteogenesis of the Mg alloy. Previous studies reported the addition of 0–2 wt.% Sr reduces the material corrosion rate and increases the corrosion uniformity in Sr-containing alloys [31,35,43,44]. Therefore, 2 wt.% Sr was added to maintain corrosion uniformity and maximize angiogenic and osteogenic properties. Many alloying elements have been used to limit the high corrosion of Mg [12]. However, avoiding elements such as Be, Nb, Cr, Ni, and Co is crucial to avoid long-term systemic inflammatory reactions [11]. Therefore, 2 wt.% Sc was added to increase the corrosion resistance without noticeable toxicity [45]. The effects of Sr and Sc addition on microstructure are discussed in the next paragraph.

The XRD pattern of Mg indicated the dominant α-Mg phase peaks. In the Mg-Sc-Sr alloy, the $Mg_{17}Sr_2$ intermetallic phase and α-Mg phase were observed. Based on the limited solubility of Sr (0.11%) in Mg, the $Mg_{17}Sr_2$ intermetallic phase forms in Sr-containing alloys [2]. The $Mg_{17}Sr_2$ formation refines the Mg grains via a grain-growth restriction mechanism during the solidification stages in the casting process [31,41,42]. As mentioned earlier, the elemental Mg and Sr can improve osteogenesis and angiogenesis. The combination of them ($Mg_{17}Sr_2$ intermetallic phase) may also be the reason for the increased osteogenesis and angiogenesis. The lower degradation rate of $Mg_{17}Sr_2$ could cause a longer systematic release of Mg and Sr from the Mg-Sc-Sr alloy. In addition, the structural grain refinement in alloys can change cellular differentiation, cell adhesion, and gene expression [46]. In our study, we performed indirect cell culture using alloy extracts. In future studies, direct cell culture experiments can be performed to evaluate the grain refinement effect of Mg-Sc-Sr alloy on the osteogenic differentiation of cells, tube formation, and scratch closure.

Moreover, the chemical composition of the surface layers for Mg and Mg-Sc-Sr discs showed the existence of Mg in the elemental metallic form (Mg) and oxidized state (MgO) in XPS measurements. The MgO layer can prolong the implant's life span and work as a protective barrier on the implant's surface to increase corrosion resistance [47,48]. This MgO layer on the surface of Mg-Sc-Sr could be a potential reason for reducing the corrosion rate by adding a protective layer on the surface of discs [49–51]. However, other studies found that increasing the MgO content can have an adverse effect on corrosion resistance [47,48] because the hydrophilic behavior of MgO increases the wetting ability of the surface by water and exposes the Mg-Sc-Sr implant to water for an extended period, resulting in the reduction of corrosion resistance [52].

Our results showed a lower amount of MgO on the surface of Mg-Sc-Sr than Mg. This low amount of MgO on the surface of Mg-Sc-Sr could be a potential reason for improving the corrosion resistance by adding a protective and uniform layer on the surface of the discs [49–51]. In addition, during solidification, MgO can promote heterogeneous nucleation and increase grain refinement, which improves corrosion resistance. The formation of MgO can upregulate angiogenesis and osteogenesis-related gene expressions, such as VEGF, BMP2, and RUNX2, that enhance the differentiation of stem cells to osteoblasts [52]. In the following discussion, the effect of Sr on the osteogenic properties of Mg is discussed.

Mg can enhance extracellular matrix mineralization, ALP activity, and osteogenesis-related gene expression, indicating improved osteogenesis character [9,53–57]. The ALP activity assessment and Von Kossa staining were performed to verify the osteogenic property of developed Mg-Sc-Sr. Significantly higher ALP staining of Mg-Sc-Sr compared to Mg and the control group after seven days of incubating HUVECs and hBM-MSCs confirmed the effect of Sr on Mg to significantly enhance the ALP activity and osteogenesis, which is consistent with the previous studies on Sr-containing alloys [58–61]. Our Von Kossa staining results also showed a higher amount of deposited calcium nodes in cells incubated with Mg-Sc-Sr extract, indicating a high potential for new bone formation. The possible molecular mechanisms underlying osteogenesis promotion by Mg are activation of mitogen-activated protein kinases/extracellular signal-regulated kinases (MAPK/ERK) signaling pathway and/or Wnt signaling pathways leading to an increase in osteogenic differentiation of stem cells to osteoblasts [62,63]. Additionally, the Sr-containing alloys showed upregulation of osteogenic-related gene expression, including Osterix (OSX), Runt-related transcription factor 2 (Runx2), Osteopontin (OPN), Osteocalcin (OCN), and Collagen I (COL I) [64–66]. The Sr addition can enhance the release of prostaglandin E-2 through cyclooxygenase 2 (COX2) expression and thereby increase osteoblastic differentiation of hBM-MSCs [67]. In addition, additional molecular pathways, such as the fibroblast growth factor/fibroblast growth factor receptor (FGF/FGFR) signaling pathway, have been indicated to be involved in strontium-mediated osteogenesis [68]. To accelerate the osteogenesis of a bone implant, improving the angiogenesis in addition to osteogenesis is vital [9,19]. The effect of Sr on angiogenesis potentiation by Mg alloys is discussed in the following discussion.

Angiogenic properties are necessary to repair large bone defects because new vessel formation can improve the transport of required nutrients and oxygen for osteogenesis at the fracture site and facilitate the transfer of cellular waste [25,69]. Higher angiogenic properties, such as upregulated related protein expression, endothelial tube formation, cell migration, and neovascularization in bone defect models, are reported for Sr-incorporated alloys [27,28]. Here, we measured VEGF expression as an essential regulator of angiogenesis that increases the proliferation and migration of HUVECs and plays an indirect role in osteogenesis by promoting angiogenesis [70–72]. The upregulation of VEGF in Mg-Sc-Sr extract-treated HUVECs confirmed our alloy's impact on early angiogenesis. Then, scratch tests were performed to prove the critical effect of HUVECs migration in enhancing angiogenic character. The potential of HUVECs migration was assessed as the number of migrated HUVECs and the scratch width at each time point. As results showed, the number of migrated HUVECs was significantly higher for cells cultured with Mg-Sc-Sr extract than cells cultured alone or with Mg. The increased cell number was additionally reflected in the rapid decrease in scratch width and the extent of wound closure. The scratch was fully closed for Mg-Sc-Sr-treated HUVECs after 24 h of incubation. In contrast, scratches treated with Mg extract were not completely closed until 48 h of incubation.

Additionally, tube formation quantification was performed as an in vitro descriptor of tubulogenic ability. The significant increase in the total number of tubes and mesh numbers showed the tubulogenic ability of Mg-Sc-Sr extract. Our results in the angiogenic evaluations were consistent with previous results. Although the angiogenesis mechanism of Sr is not clear, most of the studies showed that Sr enhances the expression of angiogenic factors such as vascular endothelial growth factor (VEGF), fibroblast growth factor (bFGF), and hypoxia-induced factor α (HIF-α) that enhance angiogenesis [27,28].

Thus, our study showed that Mg-Sc-Sr is a promising Mg alloy for bone repair applications. The Mg-Sc-Sr alloy can be used for bone defect restoration, such as pins, screws, and tissue engineering scaffolds, by showing improved osteogenic and angiogenic properties in vitro. The Mg-Sc-Sr alloy could enhance bone formation and early vascularization in vitro. In the future, FESEM assessment, wetting properties, scratch properties, and degradation rate measurement in vitro and in vivo experiments will be required on Mg-Sc-Sr to obtain its potential as an implant in the human body. In addition, assessing the effect of bioactive coatings, such as calcium silicate [73] or adding MgO nanoparticles [74], on the surface of the Mg-Sc-Sr alloy could be beneficial.

5. Conclusions

The Mg and Mg-Sc-Sr alloy were comprehensively investigated to reveal their microstructural refinement, angiogenesis, and osteogenesis. The XRD pattern showed α-Mg phase for Mg and α-Mg and $Mg_{17}Sr_2$ intermetallic phases for Mg-Sc-Sr alloy. The metallography results showed the grain size for Mg-Sc-Sr significantly decreased (76.84%) compared to Mg. A significant osteogenic differentiation of hBM-MSCs co-cultured with HUVECs cultured with Mg-Sc-Sr extract was obtained by ALP staining and the number of nodules in Von Kossa staining (73% increase) compared to the control. Moreover, the HUVECs cultured with Mg-Sc-Sr extracts demonstrated enhanced angiogenesis measured by the number of migrated cells (1.5-fold increase) than in the control after 24 h of incubation. Also, the scratch width was reduced significantly for the HUVECs cultured with Mg-Sc-Sr extracts compared to the control, and the scratch completely closed after 24 h of incubation. We concluded that the total tube length (1.2 increase) and the number of formed meshes (0.9 increase) were enhanced for the HUVECs cultured with the Mg-Sc-Sr extracts than those in the control group after 48 h of incubation.

Author Contributions: Conceptualization: N.A., S.S. and M.R.; methodology: N.A., S.S. and M.R., formal analysis: N.A., C.J.N., S.S. and M.R.; resources: M.R.; data curation: N.A. and C.J.N.; writing and draft preparation: N.A. and C.J.N.; writing—review and editing: N.A., S.S. and M.R.; visualization, N.A. and C.J.N.; supervision: M.R. and S.S.; project administration: N.A., C.J.N. and M.R.; Funding Acquisition: M.R. All authors have read and agreed to the published version of the manuscript.

Funding: We thank UCF for providing the UCF Start-Up Funding and UCF Seed Funding: ER2. Shared Research Facility Usage Award. Additionally, this material is based upon work supported by the National Science Foundation under Grant No. (2142610). We thank the Burnett School of Biomedical Sciences at UCF for providing core facilities and Advanced Material Processing and Analysis at UCF, Kirk Scammon. We also appreciate Thomas Kean for providing additional vials of hBM-MSCs and HUVECs, Cristina Fernandez-Valle for providing the optical microscope, and Melanie Coathup for providing the Keyence microscope. We also thank NSF MRI ECCS (grant no. 1726636) for funding XPS measurements. C.N. acknowledges funding from the UCF P3 post-doctoral program.

Institutional Review Board Statement: Not applicable.

Informed Consent Statement: Not applicable.

Data Availability Statement: The corresponding author will make the data available upon reasonable request.

Conflicts of Interest: The authors declare no conflict of interest.

References

1. Yang, Y.; He, C.; Dianyu, E.; Yang, W.; Qi, F.; Xie, D.; Shen, L.; Peng, S.; Shuai, C.J.M. Design, Mg bone implant: Features, developments and perspectives. *Mater. Des.* **2020**, *185*, 108259. [CrossRef]
2. Munir, K.; Lin, J.; Wen, C.; Wright, P.F.; Li, Y.J.A.B. Mechanical, corrosion, and biocompatibility properties of Mg-Zr-Sr-Sc alloys for biodegradable implant applications. *Acta Biomater.* **2020**, *102*, 493–507. [CrossRef] [PubMed]

3. Wei, F.; Neal, C.J.; Sakthivel, T.S.; Kean, T.; Seal, S.; Coathup, M.J. Multi-functional cerium oxide nanoparticles regulate inflammation and enhance osteogenesis. *Mater. Sci. Eng. C* **2021**, *124*, 112041. [CrossRef]
4. Kiernan, C.; Knuth, C.; Farrell, E. *Endochondral Ossification: Recapitulating Bone Development for Bone Defect Repair, Developmental Biology and Musculoskeletal Tissue Engineering*; Elsevier: Amsterdam, The Netherlands, 2018; pp. 125–148.
5. Zhang, N.; Wang, W.; Zhang, X.; Nune, K.C.; Zhao, Y.; Liu, N.; Misra, R.; Yang, K.; Tan, L.; Yan, J. The effect of different coatings on bone response and degradation behavior of porous magnesium-strontium devices in segmental defect regeneration. *Bioact. Mater.* **2020**, *6*, 1765–1776. [CrossRef] [PubMed]
6. Qin, H.; Zhao, Y.; An, Z.; Cheng, M.; Wang, Q.; Cheng, T.; Wang, Q.; Wang, J.; Jiang, Y.; Zhang, X. Enhanced antibacterial properties, biocompatibility, and corrosion resistance of degradable Mg-Nd-Zn-Zr alloy. *Biomaterials* **2015**, *53*, 211–220. [CrossRef]
7. Xu, T.; Yang, Y.; Peng, X.; Song, J.; Pan, F. Overview of advancement and development trend on magnesium alloy. *J. Magnes. Alloy.* **2019**, *7*, 536–544. [CrossRef]
8. Li, W.; Qiao, W.; Liu, X.; Bian, D.; Shen, D.; Zheng, Y.; Wu, J.; Kwan, K.Y.; Wong, T.M.; Cheung, K.M. Biomimicking bone–implant interface facilitates the bioadaption of a new degradable magnesium alloy to the bone tissue microenvironment. *Adv. Sci.* **2021**, *8*, 2102035. [CrossRef]
9. Gao, P.; Fan, B.; Yu, X.; Liu, W.; Wu, J.; Shi, L.; Yang, D.; Tan, L.; Wan, P.; Hao, Y. Biofunctional magnesium coated Ti6Al4V scaffold enhances osteogenesis and angiogenesis in vitro and in vivo for orthopedic application. *Bioact. Mater.* **2020**, *5*, 680–693. [CrossRef]
10. Chen, X.B.; Nisbet, D.R.; Li, R.W.; Smith, P.; Abbott, T.B.; Easton, M.A.; Zhang, D.-H.; Birbilis, N. Controlling initial biodegradation of magnesium by a biocompatible strontium phosphate conversion coating. *Acta Biomater.* **2014**, *10*, 1463–1474. [CrossRef]
11. Sun, K.; Fu, R.; Liu, X.; Xu, L.; Wang, G.; Chen, S.; Zhai, Q.; Pauly, S. Osteogenesis and angiogenesis of a bulk metallic glass for biomedical implants. *Bioact. Mater.* **2021**, *8*, 253–266. [CrossRef]
12. Li, L.; Zhang, M.; Li, Y.; Zhao, J.; Qin, L.; Lai, Y. Corrosion and biocompatibility improvement of magnesium-based alloys as bone implant materials: A review. *Regen. Biomater.* **2017**, *4*, 129–137. [CrossRef]
13. al Alawi, A.M.; Majoni, S.W.; Falhammar, H. Magnesium and human health: Perspectives and research directions. *Int. J. Endocrinol.* **2018**, *2018*, 9041694. [CrossRef]
14. Guan, R.-G.; Cipriano, A.F.; Zhao, Z.-y.; Lock, J.; Tie, D.; Zhao, T.; Cui, T.; Liu, H. Development and evaluation of a magnesium-zinc-strontium alloy for biomedical applications—Alloy processing, microstructure, mechanical properties, and biodegradation. *Mater. Sci. Eng. C* **2013**, *33*, 3661–3669. [CrossRef]
15. Lin, Z.; Wu, S.; Liu, X.; Qian, S.; Chu, P.K.; Zheng, Y.; Cheung, K.M.; Zhao, Y.; Yeung, K.W. A surface-engineered multifunctional TiO_2 based nano-layer simultaneously elevates the corrosion resistance, osteoconductivity and antimicrobial property of a magnesium alloy. *Acta Biomater.* **2019**, *99*, 495–513. [CrossRef]
16. Zocchi, M.; Locatelli, L.; Zuccotti, G.V.; Mazur, A.; Béchet, D.; Maier, J.A.; Castiglioni, S. Magnesium homeostasis in myogenic differentiation—A focus on the regulation of TRPM7, MagT1 and SLC41A1 transporters. *Int. J. Mol. Sci.* **2022**, *23*, 1658. [CrossRef]
17. Adams, R.H.; Alitalo, K. Molecular regulation of angiogenesis and lymphangiogenesis. *Nat. Rev. Mol. Cell Biol.* **2007**, *8*, 464–478. [CrossRef]
18. Liu, W.; Guo, S.; Tang, Z.; Wei, X.; Gao, P.; Wang, N.; Li, X.; Guo, Z. Magnesium promotes bone formation and angiogenesis by enhancing MC3T3-E1 secretion of PDGF-BB. *Biochem. Biophys. Res. Commun.* **2020**, *528*, 664–670. [CrossRef]
19. Wei, X.; Zhou, W.; Tang, Z.; Wu, H.; Liu, Y.; Dong, H.; Wang, N.; Huang, H.; Bao, S.; Shi, L. Magnesium surface-activated 3D printed porous PEEK scaffolds for in vivo osseointegration by promoting angiogenesis and osteogenesis. *Bioact. Mater.* **2023**, *20*, 16–28. [CrossRef]
20. Maier, J.A.; Bernardini, D.; Rayssiguier, Y.; Mazur, A. High concentrations of magnesium modulate vascular endothelial cell behaviour in vitro. *Biochim. Biophys. Acta BBA Mol. Basis Dis.* **2004**, *1689*, 6–12. [CrossRef]
21. Liu, Y.; Li, H.; Xu, J.; TerBush, J.; Li, W.; Setty, M.; Guan, S.; Nguyen, T.D.; Qin, L.; Zheng, Y. Biodegradable metal-derived magnesium and sodium enhances bone regeneration by angiogenesis aided osteogenesis and regulated biological apatite formation. *Chem. Eng. J.* **2020**, *410*, 127616. [CrossRef]
22. Antoniac, I.; Miculescu, M.; Mănescu, V.; Stere, A.; Quan, P.H.; Păltânea, G.; Robu, A.; Earar, K. Magnesium-based alloys used in orthopedic surgery. *Materials* **2022**, *15*, 1148. [CrossRef] [PubMed]
23. Tian, P.; Liu, X. Surface modification of biodegradable magnesium and its alloys for biomedical applications. *Regen. Biomater.* **2015**, *2*, 135–151. [CrossRef] [PubMed]
24. Li, D.; Yuan, Q.; Yu, K.; Xiao, T.; Liu, L.; Dai, Y.; Xiong, L.; Zhang, B.; Li, A. Mg–Zn–Mn alloy extract induces the angiogenesis of human umbilical vein endothelial cells via FGF/FGFR signaling pathway. *Biochem. Biophys. Res. Commun.* **2019**, *514*, 618–624. [CrossRef] [PubMed]
25. Mao, L.; Xia, L.; Chang, J.; Liu, J.; Jiang, L.; Wu, C.; Fang, B. The synergistic effects of Sr and Si bioactive ions on osteogenesis, osteoclastogenesis and angiogenesis for osteoporotic bone regeneration. *Acta Biomater.* **2017**, *61*, 217–232. [CrossRef] [PubMed]
26. Li, T.; He, Y.; Zhou, J.; Tang, S.; Yang, Y.; Wang, X.J.M.L. Microstructure and mechanical property of biodegradable Mg–1.5 Zn–0.6 Zr alloy with varying contents of scandium. *Mater. Lett.* **2018**, *229*, 60–63. [CrossRef]
27. Cheng, S.; Ke, J.; Yao, M.; Shao, H.; Zhou, J.; Wang, M.; Ji, X.; Zhong, G.; Peng, F.; Ma, L. Improved osteointegration and angiogenesis of strontium-incorporated 3D-printed tantalum scaffold via bioinspired polydopamine coating. *J. Mater. Sci. Technol.* **2021**, *69*, 106–118. [CrossRef]

28. Zhao, F.; Lei, B.; Li, X.; Mo, Y.; Wang, R.; Chen, D.; Chen, X. Promoting in vivo early angiogenesis with sub-micrometer strontium-contained bioactive microspheres through modulating macrophage phenotypes. *Biomaterials* **2018**, *178*, 36–47. [CrossRef]
29. Lin, Y.-H.; Lee, A.K.-X.; Ho, C.-C.; Fang, M.-J.; Kuo, T.-Y.; Shie, M.-Y. The effects of a 3D-printed magnesium-/strontium-doped calcium silicate scaffold on regulation of bone regeneration via dual-stimulation of the AKT and WNT signaling pathways. *Mater. Sci. Eng. C* **2022**, *133*, 112660. [CrossRef]
30. Li, T.; He, H.; Yang, Z.; Wang, J.; Zhang, Y.; He, G.; Huang, J.; Song, D.; Ni, J.; Zhou, X. Strontium-doped gelatin scaffolds promote M2 macrophage switch and angiogenesis through modulating the polarization of neutrophils. *Biomater. Sci.* **2021**, *9*, 2931–2946. [CrossRef]
31. Wang, Y.; Tie, D.; Guan, R.; Wang, N.; Shang, Y.; Cui, T.; Li, J. Microstructures, mechanical properties, and degradation behaviors of heat-treated Mg-Sr alloys as potential biodegradable implant materials. *J. Mech. Behav. Biomed. Mater.* **2018**, *77*, 47–57. [CrossRef]
32. Lin, Z.; Zhao, Y.; Chu, P.K.; Wang, L.; Pan, H.; Zheng, Y.; Wu, S.; Liu, X.; Cheung, K.M.; Wong, T. A functionalized TiO_2/Mg_2TiO_4 nano-layer on biodegradable magnesium implant enables superior bone-implant integration and bacterial disinfection. *Biomaterials* **2019**, *219*, 119372. [CrossRef]
33. Song, G.; Song, S. A possible biodegradable magnesium implant material. *Adv. Eng. Mater.* **2007**, *9*, 298–302. [CrossRef]
34. Borciani, G.; Montalbano, G.; Baldini, N.; Vitale-Brovarone, C.; Ciapetti, G. Protocol of co-culture of human osteoblasts and osteoclasts to test biomaterials for bone tissue engineering. *Methods Protoc.* **2022**, *5*, 8. [CrossRef]
35. Gu, X.; Xie, X.; Li, N.; Zheng, Y.; Qin, L. In vitro and in vivo studies on a Mg–Sr binary alloy system developed as a new kind of biodegradable metal. *Acta Biomater.* **2012**, *8*, 2360–2374. [CrossRef]
36. Li, T.; Zhang, H.; He, Y.; Wen, N.; Wang, X. Microstructure, mechanical properties and in vitro degradation behavior of a novel biodegradable Mg–1.5 Zn–0.6 Zr–0.2 Sc alloy. *J. Mater. Sci. Technol.* **2015**, *31*, 744–750. [CrossRef]
37. Zhang, X.; Wang, H.; Yan, B.; Zou, C.; Wei, Z. The effect of grain refinement and precipitation strengthening induced by Sc or Er alloying on the mechanical properties of cast Al-Li-Cu-Mg alloys at elevated temperatures. *Mater. Sci. Eng. A* **2021**, *822*, 141641. [CrossRef]
38. Suresh, M.; Sharma, A.; More, A.; Nayan, N.; Suwas, S. Effect of Scandium addition on evolution of microstructure, texture and mechanical properties of thermo-mechanically processed Al-Li alloy AA2195. *J. Alloys Compd.* **2018**, *740*, 364–374. [CrossRef]
39. Wang, J.; Ma, Y.; Guo, S.; Jiang, W.; Liu, Q. Effect of Sr on the microstructure and biodegradable behavior of Mg–Zn–Ca-Mn alloys for implant application. *Mater. Des.* **2018**, *153*, 308–316. [CrossRef]
40. Sadeghi, A.; Hasanpur, E.; Bahmani, A.; Shin, K.S. Corrosion behaviour of AZ31 magnesium alloy containing various levels of strontium. *Corros. Sci.* **2018**, *141*, 117–126. [CrossRef]
41. Cheng, R.; Pan, F.; Jiang, S.; Li, C.; Jiang, B.; Jiang, X. Effect of Sr addition on the grain refinement of AZ31 magnesium alloys. *Prog. Nat. Sci. Mater. Int.* **2013**, *23*, 7–12. [CrossRef]
42. StJohn, D.; Easton, M.; Qian, M.; Taylor, J. Grain refinement of magnesium alloys: A review of recent research, theoretical developments, and their application. *Met. Mater. Trans. A* **2013**, *44*, 2935–2949. [CrossRef]
43. Li, W.; Shen, Y.; Shen, J.; Shen, D.; Liu, X.; Zheng, Y.; Yeung, K.W.; Guan, S.; Kulyasova, O.B.; Valiev, R. In vitro and in vivo studies on pure Mg, Mg–1Ca and Mg–2Sr alloys processed by equal channel angular pressing. *Nano Mater. Sci.* **2020**, *2*, 96–108. [CrossRef]
44. Fan, Y.; Wu, G.H.; Zhai, C.Q. Effect of strontium on mechanical properties and corrosion resistance of AZ91D. In *Materials Science Forum*; Trans Tech Publications Ltd.: Bäch, Switzerland, 2007; pp. 567–570.
45. Liu, J.; Lin, Y.; Bian, D.; Wang, M.; Lin, Z.; Chu, X.; Li, W.; Liu, Y.; Shen, Z.; Liu, Y. In vitro and in vivo studies of Mg-30Sc alloys with different phase structure for potential usage within bone. *Acta Biomater.* **2019**, *98*, 50–66. [CrossRef] [PubMed]
46. Medvedev, A.; Neumann, A.; Ng, H.; Lapovok, R.; Kasper, C.; Lowe, T.; Anumalasetty, V.; Estrin, Y. Combined effect of grain refinement and surface modification of pure titanium on the attachment of mesenchymal stem cells and osteoblast-like SaOS-2 cells. *Mater. Sci. Eng. C* **2017**, *71*, 483–497. [CrossRef] [PubMed]
47. Fayomi, O.; Joseph, O.; Mubiayi, M.; Durodola, B.; Gabriel, O. Microstructural evolution and characterization of super-induced MgO composite on zinc-rich coatings. *Egypt. J. Basic Appl. Sci.* **2016**, *3*, 1–9. [CrossRef]
48. Khalajabadi, S.Z.; Kadir, M.R.A.; Izman, S.; Marvibaigi, M. The effect of MgO on the biodegradation, physical properties and biocompatibility of a Mg/HA/MgO nanocomposite manufactured by powder metallurgy method. *J. Alloys Compd.* **2016**, *655*, 266–280. [CrossRef]
49. Yao, H.-L.; Xia, J.; Yi, D.-L.; Yang, C.; Zhang, M.-X.; Bai, X.-B.; Chen, Q.-Y.; Wang, H.-T.; Li, S.-B. Microstructure and corrosion properties of biodegradable Mg/MgO composite coating on Mg alloy prepared by high velocity suspension flame spraying. *J. Therm. Spray Technol.* **2021**, *30*, 1544–1556. [CrossRef]
50. Stippich, F.; Vera, E.; Wolf, G.; Berg, G.; Friedrich, C. Enhanced corrosion protection of magnesium oxide coatings on magnesium deposited by ion beam-assisted evaporation. *Surf. Coat. Technol.* **1998**, *103*, 29–35. [CrossRef]
51. Banerjee, P.; Hasda, R.; Murmu, M.; Hirani, H. MgO as corrosion inhibitor. In *Inorganic Anticorrosive Materials*; Elsevier: Amsterdam, The Netherlands, 2022; pp. 183–210.
52. Ke, D.; Tarafder, S.; Vahabzadeh, S.; Bose, S. Effects of MgO, ZnO, SrO, and SiO_2 in tricalcium phosphate scaffolds on in vitro gene expression and in vivo osteogenesis. *Mater. Sci. Eng. C* **2019**, *96*, 10–19. [CrossRef]

53. Zhang, X.; Chen, Q.; Mao, X. Magnesium enhances osteogenesis of BMSCs by tuning osteoimmunomodulation. *BioMed Res. Int* **2019**, *2019*, 7908205. [CrossRef]
54. Lu, W.-C.; Pringa, E.; Chou, L. Effect of magnesium on the osteogenesis of normal human osteoblasts. *Magnes. Res.* **2017**, *30*, 42–52 [CrossRef]
55. Li, D.; Zhang, D.; Yuan, Q.; Liu, L.; Li, H.; Xiong, L.; Guo, X.; Yan, Y.; Yu, K.; Dai, Y. In vitro and in vivo assessment of the effect of biodegradable magnesium alloys on osteogenesis. *Acta Biomater.* **2022**, *141*, 454–465. [CrossRef]
56. Li, Z.; Gu, X.; Lou, S.; Zheng, Y. The development of binary Mg–Ca alloys for use as biodegradable materials within bone *Biomaterials* **2008**, *29*, 1329–1344. [CrossRef]
57. Liu, C.; Fu, X.; Pan, H.; Wan, P.; Wang, L.; Tan, L.; Wang, K.; Zhao, Y.; Yang, K.; Chu, P.K. Biodegradable Mg-Cu alloys with enhanced osteogenesis, angiogenesis, and long-lasting antibacterial effects. *Sci. Rep.* **2016**, *6*, 27374. [CrossRef]
58. Aimaiti, A.; Maimaitiyiming, A.; Boyong, X.; Aji, K.; Li, C.; Cui, L. Low-dose strontium stimulates osteogenesis but high-dose doses cause apoptosis in human adipose-derived stem cells via regulation of the ERK1/2 signaling pathway. *Stem Cell Res. Ther.* **2017**, *8*, 282. [CrossRef]
59. Park, J.W.; Kim, Y.J.; Jang, J.H.; Song, H. Positive modulation of osteogenesis-and osteoclastogenesis-related gene expression with strontium-containing microstructured Ti implants in rabbit cancellous bone. *J. Biomed. Mater. Res. Part A* **2013**, *101*, 298–306 [CrossRef]
60. Li, J.; Yang, L.; Guo, X.; Cui, W.; Yang, S.; Wang, J.; Qu, Y.; Shao, Z.; Xu, S. Osteogenesis effects of strontium-substituted hydroxyapatite coatings on true bone ceramic surfaces in vitro and in vivo. *Biomed. Mater.* **2017**, *13*, 015018. [CrossRef]
61. Zhang, W.; Cao, H.; Zhang, X.; Li, G.; Chang, Q.; Zhao, J.; Qiao, Y.; Ding, X.; Yang, G.; Liu, X. A strontium-incorporated nanoporous titanium implant surface for rapid osseointegration. *Nanoscale* **2016**, *8*, 5291–5301. [CrossRef]
62. Zhou, H.; Liang, B.; Jiang, H.; Deng, Z.; Yu, K. Magnesium-based biomaterials as emerging agents for bone repair and regeneration From mechanism to application. *J. Magnes. Alloy.* **2021**, *9*, 779–804. [CrossRef]
63. Wang, Y.; Geng, Z.; Huang, Y.; Jia, Z.; Cui, Z.; Li, Z.; Wu, S.; Liang, Y.; Zhu, S.; Yang, X. Unraveling the osteogenesis of magnesium by the activity of osteoblasts in vitro. *J. Mater. Chem. B* **2018**, *6*, 6615–6621. [CrossRef]
64. Li, M.; Yang, X.; Wang, W.; Zhang, Y.; Wan, P.; Yang, K.; Han, Y. Evaluation of the osteo-inductive potential of hollow three-dimensional magnesium-strontium substitutes for the bone grafting application. *Mater. Sci. Eng. C* **2017**, *73*, 347–356. [CrossRef] [PubMed]
65. Li, M.; He, P.; Wu, Y.; Zhang, Y.; Xia, H.; Zheng, Y.; Han, Y. Stimulatory effects of the degradation products from Mg-Ca-Sr alloy on the osteogenesis through regulating ERK signaling pathway. *Sci. Rep.* **2016**, *6*, 32323. [CrossRef] [PubMed]
66. Han, Y.; Zhou, J.; Lu, S.; Zhang, L. Enhanced osteoblast functions of narrow interligand spaced Sr-HA nano-fibers/rods grown on microporous titania coatings. *RSC Adv.* **2013**, *3*, 11169–11184. [CrossRef]
67. Wang, Q.; Tang, P.; Ge, X.; Li, P.; Lv, C.; Wang, M.; Wang, K.; Fang, L.; Lu, X. Experimental and simulation studies of strontium/zinc-codoped hydroxyapatite porous scaffolds with excellent osteoinductivity and antibacterial activity. *Appl. Surf. Sci.* **2018**, *462*, 118–126. [CrossRef]
68. Boda, S.K.; Thrivikraman, G.; Panigrahy, B.; Sarma, D.D.; Basu, B. Competing roles of substrate composition, microstructure, and sustained strontium release in directing osteogenic differentiation of hMSCs. *ACS Appl. Mater. Interfaces* **2017**, *9*, 19389–19408 [CrossRef]
69. Yu, Y.; Jin, G.; Xue, Y.; Wang, D.; Liu, X.; Sun, J. Multifunctions of dual Zn/Mg ion co-implanted titanium on osteogenesis, angiogenesis and bacteria inhibition for dental implants. *Acta Biomater.* **2017**, *49*, 590–603. [CrossRef]
70. Yan, R.; Li, J.; Wu, Q.; Zhang, X.; Hu, L.; Deng, Y.; Jiang, R.; Wen, J.; Jiang, X. Trace element-augmented titanium implant with targeted angiogenesis and enhanced osseointegration in osteoporotic rats. *Front. Chem.* **2022**, *10*, 839062. [CrossRef]
71. Ramasamy, S.K.; Kusumbe, A.P.; Wang, L.; Adams, R.H. Endothelial Notch activity promotes angiogenesis and osteogenesis in bone. *Nature* **2014**, *507*, 376–380. [CrossRef]
72. Diomede, F.; Marconi, G.D.; Fonticoli, L.; Pizzicanella, J.; Merciaro, I.; Bramanti, P.; Mazzon, E.; Trubiani, O. Functional relationship between osteogenesis and angiogenesis in tissue regeneration. *Int. J. Mol. Sci.* **2020**, *21*, 3242. [CrossRef]
73. Qin, W.; Kolooshani, A.; Kolahdooz, A.; Saber-Samandari, S.; Khazaei, S.; Khandan, A.; Ren, F.; Toghraie, D. Coating the magnesium implants with reinforced nanocomposite nanoparticles for use in orthopedic applications. *Colloids Surf. A Physicochem. Eng. Asp.* **2021**, *621*, 126581. [CrossRef]
74. Sahmani, S.; Saber-Samandari, S.; Khandan, A.; Aghdam, M.M. Influence of MgO nanoparticles on the mechanical properties of coated hydroxyapatite nanocomposite scaffolds produced via space holder technique: Fabrication, characterization and simulation. *J. Mech. Behav. Biomed. Mater.* **2019**, *95*, 76–88. [CrossRef]

Journal of
Functional Biomaterials

MDPI

Article

Photobiomodulation and Inorganic Bovine Bone in Guided Bone Regeneration: Histomorphometric Analysis in Rats

Nicole Rosa de Freitas [1], Luísa Belluco Guerrini [1], Luis Augusto Esper [2], Michyele Cristhiane Sbrana [2], Caroline Chepernate Vieira dos Santos [3] and Ana Lúcia Pompéia Fraga de Almeida [2,4,*]

1 Postgraduate Program, Bauru School of Dentistry, University of São Paulo, Bauru 17012-901, Brazil; nicolefreitas@usp.br (N.R.d.F.)
2 Periodontics Sector, Hospital for Rehabilitation of Craniofacial Anomalies, University of São Paulo, Bauru 17012-900, Brazil
3 Postgraduate Program, Hospital for Rehabilitation of Craniofacial Anomalies, University of São Paulo, Bauru 17012-901, Brazil
4 Department of Prosthodontics and Periodontics, Bauru School of Dentistry, University of São Paulo, Bauru 17012-901, Brazil
* Correspondence: analmeida@usp.br; Tel.: +55-14-32358300

Abstract: The objective of this study was to evaluate the efficacy of photobiomodulation in the bone regeneration of critical-sized defects (CSD) filled with inorganic bovine bone associated or not with collagen membranes. The study has been conducted on 40 critical defects in the calvaria of male rats, divided into four experimental groups ($n = 10$): (1) DBBM (deproteinized bovine bone mineral); (2) GBR (DBBM+collagen membrane); (3) DBBM+P (DBBM+photobiomodulation); and (4) GBR+P (GBR+photobiomodulation). At 30 days postoperative, the animals were euthanized, and after the tissue had been processed, histological, histometric, and statistical analyses were performed. The analyses have taken into account newly formed bone area (NBA), linear bone extension (LBE), and residual particle area (RPA) as variables. The Kruskal-Wallis test has been performed, followed by the Dwass-Steel-Critchlow-Fligner test for comparison between groups ($p < 0.05$). When the DBBM+P group was compared to the DBBM group, it was possible to observe significant statistical differences in all the variables analyzed ($p < 0.05$). The application of photobiomodulation in guided bone regeneration (GBR+P) has shown a decrease in the median value for the RPA variable (26.8) when compared to the GBR group (32.4), with a significant statistical difference; however, for NBA and LBE, the therapy has not provided significant results.

Keywords: bone transplantation; lasers; regeneration

1. Introduction

Although autogenous bone is considered the gold standard in reconstructive surgeries due to its osteogenic, osteoinductive, and osteoconductive properties [1], its lack of antigenicity, natural structure, and the presence of type I collagen [2] present some disadvantages, such as high morbidity and uncontrolled resorption rates, which can negatively impact postoperative results [3,4].

Thus, biomaterials are widely used as an alternative to autogenous bone in post-extraction dental alveoli, angular bone defects in teeth, maxillary sinus lift, alveolar ridge augmentation, and other procedures. Ideally, they should be easily handled during the surgical procedure, and they should also have antigenic characteristics, biocompatibility, sterilizability, and space maintainability [4].

Biomaterials may come from various origins. Among them, xenogen, especially Bio-Oss®, is the material with the most scientific evidence [3,5]. These materials are derived from bovine bone, whose organic portion is eliminated during laboratory processing in order to avoid rejection after implantation in the receptor bed. After this procedure, the material loses its osteoinductive capacity and only retains its osteoconductive features [3]. However, the original structureof the bone is maintained [4], with similar macro and micropores to those of human spongy bone, to provide an excellent bone structure for the formation of new bone [3]. There are disagreements in the literature regarding the potential for resorption of these substitutes. Thus, given the absence of osteogenic properties and the potential for uncontrolled resorption of the biomaterials currently used, in addition to the delay in repair, photobiomodulation has been widely researched [6–15].

The use of biological membranes in guided bone regeneration (GBR) provides stabilization of the grafting material in the dependencies of the bone defect as well as excluding epithelial migration towards its interior, maintaining space for osteogenic cells of the guide bone to repopulate the area of the host defect and accelerate bone repair [16–18].

By associating photobiomodulating therapy with inorganic bovine bone grafts, the effects related to collagen synthesis and hydroxyapatite result in an improvement in bone repair [9,13,19]. Photobiomodulation (PBM) may increase the expression of bone matrix proteins, accelerating all phases of bone formation, including the inflammatory phase, with a reduction in its infiltrate, improvement in periosteum development, and a significant increase in the formation of the trabecular matrix. In addition to it, precocious deposition of osteocalcin and osteopontin on the edges of the defect in rats irradiated by laser provides a faster and more organized bone repair process [20].

Although many studies have used photobiomodulation, there is no consensus as to the protocol to be used. The aim of this study was to evaluate the effect of this therapy, using a dose of 6 J in an intraoperative application, on the bone regeneration of critical sized defects (CSD) filled with inorganic bovine bone associated or not with collagen membranes.

2. Materials and Methods

2.1. Experimental Model

The sample consisted of 40 male rats (*Rattus norvegicusalbinus*, Wistar) from the Central Animal Laboratory of the Bauru School of Dentistry—Bauru, São Paulo, Brazil, weighing between 250 and 300 g and randomly divided into four experimental groups (n = 10). Critical size defects (CSD) of 5 mm in diameter have been performed surgically in the calvaria using a trephine bur (JJGC Indústria e Comércio de Materiais Dentários S.A., Curitiba, PR, Brazil), at low speed, under thorough cooling with sterile saline. Extreme care has been taken to protect the dura mater and brain during the surgical procedure [7]. After that, the defects were filled in according to treatment groups. For the experimental procedures, the rats have been anesthetized by an intramuscular injection of ketamine hydrochloride (Vetbrands, Paulínia, Brazil) (0.4 mL/kg) and xylazine (Vetbrands, Paulínia, Brazil) (0.02 mL/kg). The surgical procedure has been performed by only one operator, as follows: (1) antisepsis and trichotomy of the calvaria of each animal; (2) semilunar incision on the calvaria with a full-thickness flap raised; (3) confection of L-shaped marks 2 mm anteriorly and 2 mm posteriorly to the surgical defect margins with conical carbide bur FG-700 (Microdont Micro Usinagem de Precisão Ltda., São Paulo, Brazil); and (4) filling the L-shaped marks with amalgam (the marks are possible to be observed in Figure 1). These marks have been used to identify the center of the original surgical defect during laboratory processing as well as to locate the original bone margins during histometric analysis [7,8,14,15].

After creating the defects, each animal received its respective treatment according to the randomization performed prior to the surgical procedure: (1) DBBM Group (deproteinized bovine bone mineral, Bio-Oss®, 0.25–1 mm; Geistlich Pharma AG, Wolhusen, Switzerland); (2) GBR (deproteinized bovine bone mineral + collagen membrane, BioGide®, Geistlich Pharma AG, Wolhusen, Switzerland); (3) DBBM+P Group (deproteinized bovine bone mineral + photobiomodulation); (4) GBR+P Group (deproteinized bovine bone mineral + photobiomodulation + collagen membrane).

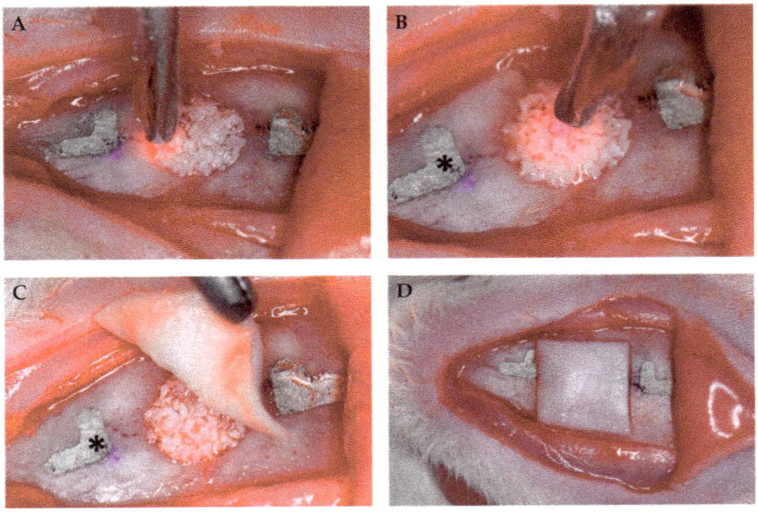

Figure 1. (**A**) Application of photobiomodulation therapy at the edges of the surgical defect and (**B**) in the central point on the bone graft; (**C**) placement of the collagen membrane after photobiomodulation; (**D**) positioned collagen membrane. * L-shaped marks filled with amalgam.

The flap was repositioned and sutured with silk suture 4-0 (Ethicon, Johnson and Johnson do Brasil Indústria e Comércio de Produtos para Saúde Ltda., São José dos Campos, Brazil), and each animal received a dose of antibiotic (intramuscular injection of 24,000 units of penicillin G-benzathine, Fort Dodge® Saúde Animal Ltda., Campinas, São Paulo, Brazil) [7] and anti-inflammatory (intramuscular dose of 0.3 mg/kg of Ketoprofen, Laboratório Teuto Brasileiro S.A., Anápolis, Goiás, Brazil).

During the period of the study, the rats were kept in an environment with a 12-h light cycle and a temperature between 22 and 24 °C and fed with selected solid food and water ad libitum.

2.2. Photobiomodulation

In the DBBM+P group, the defect had been previously filled by the bone graft, and photobiomodulation was performed. In the GBR+P group, after filling the defect and performing photobiomodulation, the collagen membrane was positioned and the flap was sutured. (Figure 1).

Photobiomodulation has been performed in a single application during the transoperative period with a gallium-aluminum-arsenide laser (GaAlAs) (TheraLase DMC®, São Carlos, São Paulo, Brazil), in continuous emission mode, with a beam area of 0.028 cm^2, a 730 nm wave length, 100 mW power, 210 J/cm^2 energy density, 6 J per point, and an irradiation time of 60 s [15], at four points at the edges of the surgical defect created (12 h, 3 h, 6 h, and 9 h), in addition to a central point on the bone graft [7] (Figure 2).

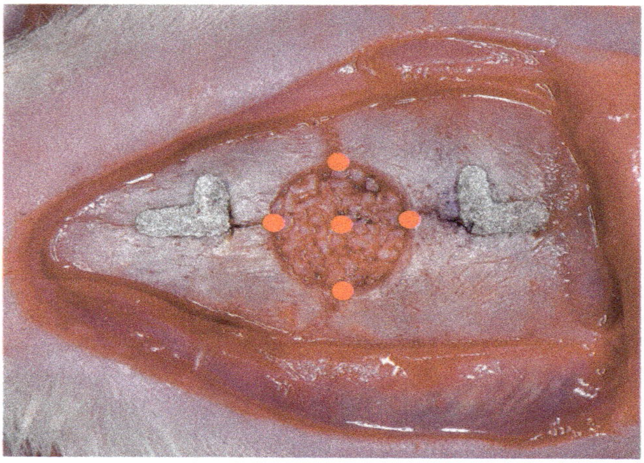

Figure 2. Scheme of application of photobiomodulatory therapy at the edges of the surgical defect in addition to a central point on the bone graft.

2.3. Tissue Processing

At 30 days postoperative, the animals were euthanized with an excessive dose of anesthetic (5 mg/mL of ketamine hydrochloride and xylazine). The original defect area and the surrounding tissues were removed en bloc, the specimens were processed, and they were longitudinally divided exactly along the center of the CSD, using the L-shaped marks filled with amalgam as a reference. Serial longitudinal cuts were made at 6 μm thick. The specimens were stained using the technique of hematoxylin and eosin (H.E.) for analysis in light microscopy. The tissue processing has been described in detail in previous studies carried out by the group [7,8,14,15].

2.4. Histomorphometric Analysis

Images from histological sections representing the central area of the original surgical defect have been captured by a SPOT RT3-2540 Color Slider 2.0 Mp camera (SPOT Imaging-Solutions, Diagnostic Instruments, Inc., Sterling Heights, MI, USA) coupled to an Olympus BX50 microscope (Olympus Corporation, Tokyo, Japan) with a 4× amplification and saved in a computer. With the magnification that has been used, it was not possible to capture the entire calvaria in the same image, so a digital image was created by combining three smaller images based on reference structures (blood vessels, bone trabeculae, and particles of grafted material) in each histological section.

Histometric analysis has been performed using a computer image evaluation system, ImageLab 2000 software (Diracon Bio Informática Ltda., Vargem Grande do Sul, São Paulo, Brazil), following the criteria adopted by de Almeida et al. [7] and Cunha et al. [8].

For each image, the area corresponding to the region of the calvaria bone where the defect had been originally created was delimited and measured in mm^2, representing 100% of the analyzed area (total area—TA). Taking into account the total length of the specimen of 9 mm and 2 mm as measured from each end to determine the limits of the original surgical defect (5 mm), the areas of newly formed bone area (NBA), residual particle area (RPA) of the implanted materials [7,8,14,15], and linear bone extension (LBE), measured in mm^2 and calculated as percentages, have been delimited within the TA (Figure 3).

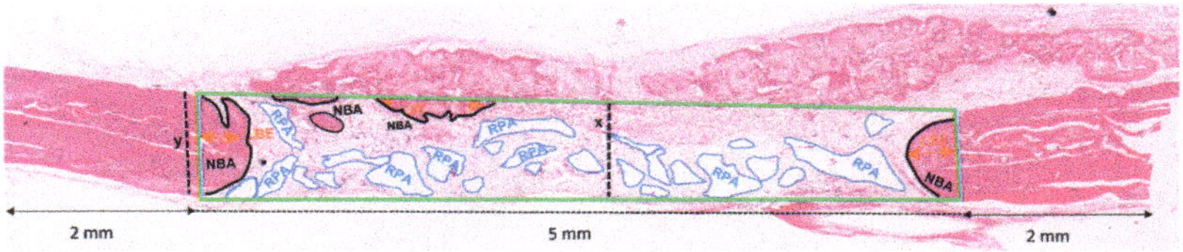

Figure 3. Photomicrograph of a histological section (4x—hematoxylin and eosin) to outline the measurements performed. TA, delimited by the green line, corresponds to the height of the calvaria where the defect has been surgically created. The TA height (x) corresponds to the thickness of the original calvaria bone (y). The NBA was delimited by the black line, the RPA by the blue line, and the LBE by the red arrow.

2.5. Statistical Analysis

For each animal, the NBA, LBE, and ARP values have been represented by the arithmetic mean of the four most central sections of the calvaria. The NBA variable has not passed the normality test, so the Kruskal-Wallis test has been performed, followed by the Dwass-Steel-Critchlow-Fligner test for comparison between groups. The results were considered statistically significant when $p < 0.05$.

3. Results

During laboratory processing, one GBR specimen has been lost.

3.1. Qualitative Histological Analysis

The original thickness of the calvaria has been maintained in all groups, but with variable amounts of new bone formation (Figure 4).

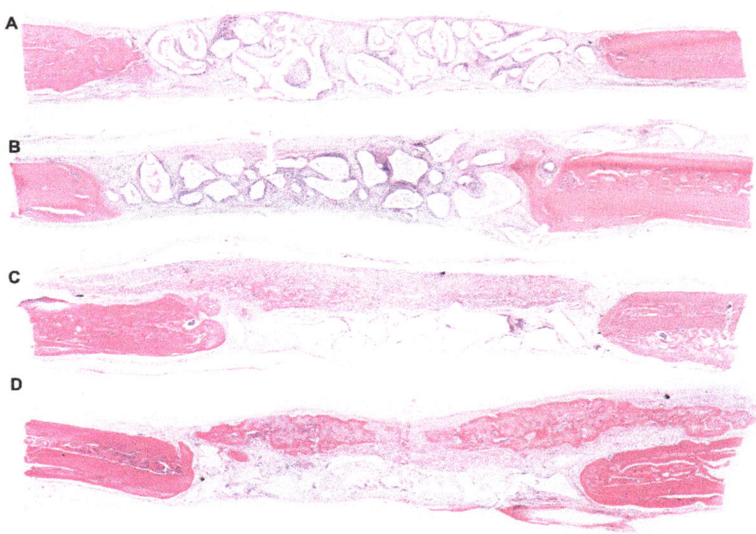

Figure 4. Photomicrographs of panoramic views of surgical defects (4×—hematoxylin and eosin). **(A)**—DBBM Group; **(B)**—DBBM+P Group; **(C)**—GBR Group; **(D)**—GBR+P Group.

The DBBM group presented minor bone formation at the end of the defect that did not extend to the center in any specimen. In some specimens, the presence of osteoclasts was observed in close contact with residual particles, and the connective tissue was well organized (Figure 5).

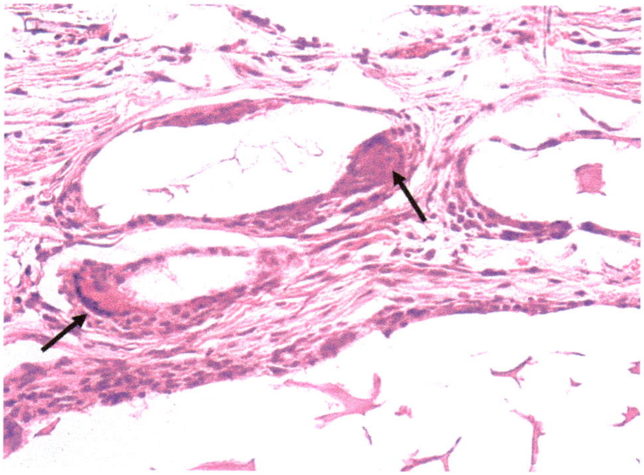

Figure 5. Photomicrograph of the DBBM group (H.E. 40×). Osteoclast on the bovine bone periphery (arrow).

In the GBR group, bone formation has been observed with variable extension between specimens and membrane ossification towards the center of the defect. Additionally, the presence of residual particles and well-organized connective tissue has been observed (Figure 6).

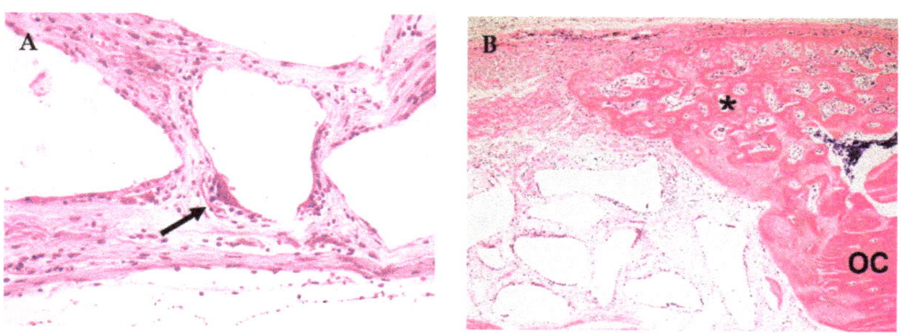

Figure 6. Photomicrograph of the GBR group. (**A**) Osteoclast on the bovine bone particle periphery (arrow) (H.E.—40×). (**B**) Ossification of the collagen membrane towards the center of the defect (asterisk *); OC—original calvaria (H.E.—10×).

In the DBBM+P group, several specimens have presented new bone formation that extended towards the center of the defect and resulted in in-line closure in three specimens without complete closure in the total area. The presence of residual particles inside the defect allowed maintenance of the thickness of the original calvaria and the beginning of bone formation inside the particles in some specimens, as well as the presence of osteoclasts around them. These remaining particles have been encircled by thick bundles of collagen fibers and the osteoid matrix (Figure 7).

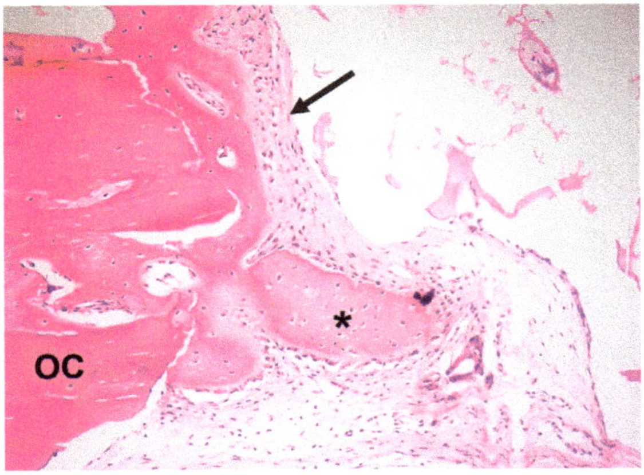

Figure 7. Photomicrographs of the DBBM+P group. Presence of well-organized connective tissue; formation of new bone towards the defect center (asterisk *); and presence of osteoid matrix (arrow); OC—original calvaria (H.E.—10×).

The GBR+P group has displayed membrane ossification above its limits. Furthermore, new bone that extended to the center of the defect has been detected in several specimens without total closure of the defect. The biomaterial particles have been maintained close to the defect, with the presence of surrounding bone formation in some specimens. The connective tissue has been well vascularized, with parallel fibers and several fibroblasts (Figure 8).

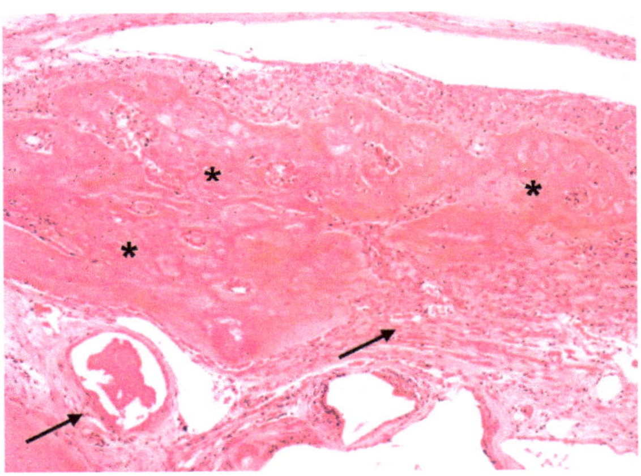

Figure 8. Photomicrograph of the GBR+P group. Ossified collagen membrane (asterisks *); presence of well-organized connective tissue (arrows) (H.E.—10×).

3.2. Histometric and Statistical Analysis

The analysis using the median, first, and second quartiles is described in Table 1 and in comparison between groups in Table 2.

Table 1. Means, standard deviations, medians, Q25, and Q75 values of newly formed bone area (NBA), linear bone extension (LBE), and residual particle area (RPA) in %.

	Group	NBA	LBE	RPA
N	DBBM	10	10	10
	DBBM+P	10	10	10
	GBR	9	9	9
	GBR+P	10	10	10
Mean	DBBM	11.4	27.4	38.7
	DBBM+P	48.6	76.2	16.7
	GBR	22.0	62.9	32.3
	GBR+P	19.0	48.9	24.2
Satandard deviation	DBBM	7.8	15.5	6.9
	DBBM+P	28.2	29.0	15.2
	GBR	10.7	21.6	6.2
	GBR+P	13.7	21.7	6.0
Median	DBBM	9.4	23.6	37.7
	DBBM+P	42.2	79.4	22.4
	GBR	19.9	71.9	32.4
	GBR+P	13.6	57.1	26.8
Q 25	DBBM	6.5	18.7	36.2
	DBBM+P	35.9	51.9	0
	GBR	15.8	48.6	29.9
	GBR+P	9.6	34.1	21.2
Q75	DBBM	11.4	31.1	42.2
	DBBM+P	74.6	96.8	27.5
	GBR	24.6	77.3	37.4
	GBR+P	28.2	62.4	28.3

Table 2. Comparison between groups (Dwass-Steel-Critchlow-Fligner test, $p < 0.05$).

	NBA	LBE	RPA
DBBM × DBBM+P	0.017 *	0.005 *	0.008 *
DBBM × GBR	0.068	0.013 *	0.238
DBBM × GBR+P	0.525	0.148	0.003 *
DBBM+P × GBR	0.238	0.761	0.054
DBBM+P × GBR+P	0.106	0.232	0.666
GBR × GBR+P	0.713	0.663	0.017 *

The presence of * indicates a statistically significant difference.

When the DBBM+P group was compared to the bovine bone used alone (DBBM group), it was possible to observe that the former had higher medians for new bone area (42.2) and linear bone length (79.4). This comparison showed a lower median value for the residual particle area variable (22.4), with a statistically significant difference ($p < 0.05$) in all analyzed variables, while the second had the lowest medians, both for NBA and LBE (9.4 and 23.6, respectively), and the highest median for particle residual area (37.7).

By associating the membrane with the grafting material (GBR), the results show improvement in all variables, with a median of 19.9 for NBA and LBE of 71.9, with a statistically significant difference ($p = 0.013$), and a slight decrease in the median of RPA (32.4), but without a significant difference. The application of photobiomodulatory therapy to these materials (GBR+P group) did not improve the NBA and LBE variables but showed a decrease in the median value of the RPA variable (26.8) when compared to the GBR and the DBBM groups ($p = 0.017$ and $p = 0.003$, respectively) with a significant statistical difference.

4. Discussion

The effects of photobiomodulation on critical size defects in rat calvaria treated with bovine bone graft and collagen membrane have been evaluated. Bone formation has been observed in the collagen membrane region that exceeds the limits of the defect originally created. Furthermore, the therapy has shown positive results on demineralized bovine bone and provided greater bone formation in new bone areas while accelerating the resorption of residual particles.

The 30-day analysis has been carried out due to the fact that photobiomodulation acts in the initial phases of regeneration. At this period, there are a large number of differentiating cells, while in late stages, these cells decrease in number and therefore may reduce the effectiveness of the PBM in stimulating bone formation [21].

There are discrepancies about the ideal photobiomodulation protocol. Several investigators have evaluated the effectiveness of various low-intensity lasers, including GaAlAs, at different wavelengths, doses, output power, and treatment protocols [13,19,21–24]. The protocol used in the present study was based on previous publications, which used a dose of 6 J in a single application and achieved positive results in CSD regeneration [7,8,14,15].

We suggest that an application in the transoperative period, with an ideal amount of energy applied to the edges of the surgical defect and on the bone graft, is sufficient to observe the effects of the photobiomodulatory therapy. The stage at which PBM is used may influence the amount of newly formed bone, suggesting that the mechanism accelerates the process of bone formation in the first stages of healing in which there is greater cellular experience [23], mesenchymal cell differentiation [25], and higher alkaline phosphatase activity (ALP), providing an increase in the number of differentiated cells that express markers of osteoblastic differentiation generated in bone formation [6].

The biostimulatory effect of the laser occurs when, upon reaching the cell, the light is absorbed by specific chromophores in the mitochondria. This provides electron excitation, increases adenosine triphosphate (ATP) production, controls the release of reactive oxygen species, and regulates the production of transcription factors, which are molecules that regulate the activity of other cells, including proliferation, differentiation, and the secretion of growth factors [26]. Although several prior studies have used repeated applications [9,10,12,21,27,28], there are already reports of positive results with a single application [7,8,13–15,29].

In order to translate our experiments into clinical practice, a single application during the surgical procedure is believed to facilitate the use of the therapy. Repeated sessions may be possible in research settings and/or universities; however, in daily clinical practice, it may not be that easy [30], as they might increase the time and cost of dental treatment.

The application of therapy to the edges of the defect has been performed in order to stimulate the bone cells at the margins of the defect surgically created. The application of therapy to the grafting material is not clearly shown in the literature [19]. Thus, in an attempt to optimize the results and considering the possibility of therapy improvement of the osteoconductive power of the material, photobiomodulation has been applied to the central point of the defect, over the graft, before positioning the membrane in order to avoid blocking the passage of the light.

In the present study, a positive effect of photobiomodulation on DBBM has been observed. The combination of therapy with bovine bone graft (DBBM+P group) provided better repair, with significant statistical differences among all variables observed (NBA, LBE, and RPA). When the xenogen material is used alone, effects related to collagen and hydroxyapatite synthesis occur more slowly [8,13,22], but when the photobiomodulatory therapy is used in conjunction with the biomaterial, excellent results can be obtained, allowing for greater production of collagen from fibroblasts and osteoid matrix [21]. In addition, PBM can provide organization of collagen fibers, corroborating our histological findings in the DBBM+P and GBR+P groups, which showed better fiber organization when compared to the other groups.

The PBM can increase the osteoconductive potential of the biomaterial and maintain bone volume until the end of the analysis [9]. In addition, it may provide greater expression of the OPG and RANKL biomarkers, indicating that the presence of the biomaterial could induce controlled ossification in surgical defects [13].

Acceleration of the resorption of residual particles enabling the formation of new bone (DBBM+P group) has been observed as a positive effect of PBM, which corroborates the findings of Cunha et al. [8] and Sbrana et al. [15]. This can be explained by photobiomodulation's actions in stimulating osteoblastic and osteoclastic activity without changing the bone structure [31]. However, our results differ from those of Bosco et al. [22], who did not identify a relationship between the use of low-intensity lasers and residual particle resorption. These may occur due to differences in materials and photobiomodulation protocols between the studies. However, the role of photobiomodulatory therapy on residual particles is still unclear [21].

In all specimens of the study, it was possible to observe the maintenance of calvaria thickness and the presence of biomaterial particles, which were sometimes surrounded by osteoid matrix and other times by osteoclasts. These findings corroborate those of other studies that reported that bovine bone graft granules undergo slow resorption. Notably, instead of being reabsorbed and entering the physiological process of bone resorption, they remained surrounded by newly formed bone [32,33]. This provided a structure for bone formation that kept tissues in position and prevented the resorption of the defect. There are authors who consider Bio-Oss® (BO) non-resorbable and harmful to the repair; once non-resorbable granules are present, they may negatively interfere with the repair of defects and compromise the osseointegration of implants [34,35]. Nonetheless, when extraction sockets were clinically treated with BO, there was bone formation in sufficient quality and quantity for implant installation in the correct position with maintenance of socket thickness [36]. Oliveira et al. [9] found approximately 60% more bone in the control group, but the microtomography of the groups that used bovine bone showed mineralized bone. The presence of osteoconductive material in the defect may even delay the repair and the formation of new bone; however, it may improve bone quality, and it is very important to maintain the bone morphology.

Bosco et al. [22] have stated that regardless of the presence of the biomaterial, photobiomodulation is capable of improving bone formation. In this study, a photobiomodulation group has not been included because the intention was to observe the effect of therapy on the biomaterial. Previous studies conducted by our group have also shown reliable results from the therapy used alone, although there has been no maintenance of the original thickness of the calvaria, which directly affects the results when they are transposed into clinical practice because maintaining the thickness is extremely important for rehabilitation and therefore the success of the implant.

The collagen membrane is used in GBR in order to provide space for the osteogenic cells of the host bone to repopulate the region. It prevents connective tissue cells from migrating to regions near the defect and compromising bone regeneration [16,17]. Its use also provides greater stability for the grafting material in the region near the defect [37].

The association of PBM therapy in GBR could further optimize the results by increasing the concentration of calcium hydroxyapatite, the main substrate for bone structure, which provides greater bone maturation [19,23]. However, in the present study, there was no significant difference related to this association. The histological analysis revealed a greater amount of new bone formation below, above, and inside the collagen membrane in the GBR+P group. In the GBR group, membrane ossification was observed but in smaller proportions, which corroborates the findings of Freitas et al. [14]. Membrane ossification can be explained by the findings of Elgali et al. [37], who observed the presence of different cell phenotypes in the membrane region at different periods of the study, with a high prevalence of osteogenic cells and macrophages/monocytes that migrated to the inside of the membrane. This indicates that instead of the membrane acting as a barrier, it has properties that promote cell migration, cell involvement, and differentiation, regardless of the presence or absence of bone substitutes. Bone neoformation beyond the limits of defects has not been assessed in the present study, which may have influenced the results.

The closure of the defect, expressed by the variable LBE, showed greater bone formation in the DBBM+P group (79.4). This result can be explained by the fact that PBM stimulates vascularization and acts early in the inflammatory phase, favoring bone neoformation [22,29]. This is extremely important in situations where the regenerative challenge is more complex, as in the case of smoking patients or even those who require grafting materials that can delay bone formation [29]. Notably, regardless of the application of a low-intensity laser and the use of a collagen membrane, good bone formation results have been observed. Furthermore, when analyzing histological images, the GBR+P group displayed greater membrane mineralization and bone formation beyond the thickness of the calvaria compared to the GBR group. The NBA and LBE observed beyond the delimitation of the TA were not taken into account in this study due to the methodology employed. Thus, significant statistical differences have not been observed when these groups were compared (GBR vs. GBR+P). However, there have been reports showing that photobiomodulating therapy in conjunction with GBR increases the concentration of calcium hydroxyapatite, the main substrate for bone structure, and provides greater bone maturation [18,19].

In the present study, only the histomorphometric analysis has been performed, which is a limitation of the same. Analyses that take into account bone formation beyond the limits of the defect, as well as immunohistochemistry and microtomography techniques, were necessary to identify the bone markers that were present in bone repair as well as bone density. This is especially important for verifying the potential of therapy to improve bone quality in the groups that received photobiomodulation.

5. Conclusions

The results of the present study indicate that photobiomodulation in guided bone regeneration of critical-size defects does not provide greater new bone formation for closuring the defect. However, bone formation has been observed in the region of the collagen membrane that exceeds the limits of the originally created defect. In addition, the therapy has shown positive results on demineralized bovine bone and provided greater bone formation in the area, as well as, to a concomitant extent, the resorption of residual particles.

Author Contributions: N.R.d.F. has written the manuscript and performed the analysis and interpretation of the histological and statistical data. L.B.G. and C.C.V.d.S. have contributed to the acquisition and interpretation of histological data. L.A.E. and A.L.P.F.d.A. have been involved in the conception and design of the work, as well as the supervision and administration of the project. L.A.E. has performed the surgical procedures and the euthanasia, and M.C.S. has performed the laboratory preparation of the parts. All authors have read and agreed to the published version of the manuscript.

Funding: This research was supported by São Paulo Research Foundation—FAPESP (#2010/10538-8, #2010/13170-3, #2011/00779-2) and CAPES (-001).

Institutional Review Board Statement: This study has been approved by the Ethics Committee on Teaching and Research in Animals of the Bauru School of Dentistry, University of São Paulo (CEEPA—015/2016).

Informed Consent Statement: Not applicable.

Data Availability Statement: The data presented in this study are openly available in FigShare at doi 10.6084/m9.figshare.22858028.

Conflicts of Interest: The authors declare no conflict of interest.

References

1. Brunsvold, M.A.; Mellonig, J.T. Bone grafts and periodontal regeneration. *Periodontol. 2000* **1993**, *1*, 80–91. [CrossRef] [PubMed]
2. Hoexter, D.L. Bone regeneration graft materials. *J. Oral. Implantol.* **2002**, *28*, 290–294. [CrossRef] [PubMed]

3. Hallman, M.; Thor, A. Bone substitutes and growth factors as an alternative/complement to autogenous bone for grafting in implant dentistry. *Periodontol. 2000* **2008**, *47*, 172–192. [CrossRef] [PubMed]
4. Baldini, N.; De Sanctis, M.; Ferrari, M. Deproteinized bovine bone in periodontal and implant surgery. *Dent. Mater.* **2011**, *27*, 61–70. [CrossRef]
5. Shamsoddin, E.; Houshmand, B.; Golabgiran, M. Biomaterial selection for bone augmentation in implant dentistry: A systematic review. *J. Adv. Pharm. Technol. Res.* **2019**, *10*, 46–50. [CrossRef]
6. Ozawa, Y.; Shimizu, N.; Kariya, G.; Abiko, Y. Low-energy laser irradiation stimulates bone nodule formation at early stages of cell cultures in rat calvarial cells. *Bone* **1998**, *22*, 347–354. [CrossRef]
7. de Almeida, A.L.P.F.; Medeiros, I.L.; Cunha, M.J.; Sbrana, M.C.; de Oliveira, P.G.; Esper, L.A. The effect of low-level laser on bone healing in critical size defects treated with or without autogenous bone graft. An experimental study in rat calvaria. *Clin. Oral Implant. Res.* **2014**, *25*, 1131–1136. [CrossRef]
8. Cunha, M.J.; Esper, L.A.; Sbrana, M.C.; de Oliveira, P.G.; do Valle, A.L.; de Almeida, A.L.P.F. Effect of low-level laser on bone defects treated with bovine or autogenous bone grafts: In Vivo study in rat calvaria. *BioMed Res. Int.* **2014**, *2014*, 104230. [CrossRef]
9. de Oliveira, G.J.P.L.; Aroni, M.A.T.; Medeiros, M.C.; Marcantonio, E., Jr.; Marcantonio, R.A.C. Effect of low-level laser therapy on the healing of sites grafted with coagulum, deproteinized bovine bone, and biphasic ceramic made of hydroxyapatite and β-tricalcium phosphate. In vivo study in rats. *Lasers Surg. Med.* **2018**, *50*, 651–660. [CrossRef]
10. Jonasson, T.H.; Zancan, R.; de Oliveira Azevedo, L.; Fonseca, A.C.; Silva, M.C.D.; Giovanini, A.F.; Zielak, J.C.; Araujo, M.R. Effects of low-level laser therapy and platelet concentrate on bone repair: Histological, histomorphometric, immunohistochemical, and radiographic study. *J. Craniomaxillofac. Surg.* **2017**, *45*, 1846–1853. [CrossRef]
11. Rosso, M.P.d.O.; Buchaim, D.V.; Pomini, K.T.; Della Coletta, B.B.; Reis, C.H.B.; Pilon, J.P.G.; Júnior, G.D.; Buchaim, R.L. Photobiomodulation Therapy (PBMT) Applied in Bone Reconstructive Surgery Using Bovine Bone Grafts: A Systematic Review. *Materials* **2019**, *12*, 4051. [CrossRef] [PubMed]
12. Rufato, F.C.T.; de Sousa, L.G.; Scalize, P.H.; Gimenes, R.; Regalo, I.H.; Rosa, A.L.; Beloti, M.M.; Regalo, S.C.H.; Siéssere, S. Texturized P(VDF-TrFE)/BT membrane enhances bone neoformation in calvaria defects regardless of the association with photobiomodulation therapy in ovariectomized rats. *Clin. Oral. Investig.* **2022**, *26*, 1053–1065. [CrossRef] [PubMed]
13. Torquato, L.C.; Suárez, E.A.C.; Bernardo, D.V.; Pinto, I.L.R.; Mantovani, L.O.; Silva, T.I.L.; Jardini, M.A.N.; Santamaria, M.P.; De Marco, A.C. Bone repair assessment of critical size defects in rats treated with mineralized bovine bone (Bio-Oss®) and photobiomodulation therapy: A histomorphometric and immunohistochemical study. *Lasers Med. Sci.* **2021**, *36*, 1515–1525. [CrossRef]
14. Freitas, N.R.; Guerrini, L.B.; Esper, L.A.; Sbrana, M.C.; Dalben, G.S.; Soares, S.; de Almeida, A.L.P.F. Evaluation of photobiomodulation therapy associated with guided bone regeneration in critical size defects. In Vivo study. *J. Appl. Oral. Sci.* **2018**, *26*, e20170244. [CrossRef] [PubMed]
15. Sbrana, M.C.; Esper, L.A.; de Freitas, N.R.; Guerrini, L.B.; Hernandes, P.H.J.; de Almeida, A.L.P.F. Evaluation of Different Photobiomodulation Therapy Protocols as Adjuncts in the Healing of Bone Defects Grafted with Inorganic Bovine Bone. *Int. J. Oral. Maxillofac. Implant.* **2022**, *37*, 1244–1249. [CrossRef]
16. Buser, D.; Brägger, U.; Lang, N.P.; Nyman, S. Regeneration and enlargement of jaw bone using guided tissue regeneration. *Clin. Oral. Implant. Res.* **1990**, *1*, 22–32. [CrossRef]
17. Kitayama, S.; Wong, L.O.; Ma, L.; Hao, J.; Kasugai, S.; Lang, N.P.; Mattheos, N. Regeneration of rabbit calvarial defects using biphasic calcium phosphate and a strontium hydroxyapatite-containing collagen membrane. *Clin. Oral. Implant. Res.* **2016**, *27*, 206–214. [CrossRef] [PubMed]
18. Dimitriou, R.; Mataliotakis, G.I.; Calori, G.M.; Giannoudis, P.V. The role of barrier membranes for guided bone regeneration and restoration of large bone defects: Current experimental and clinical evidence. *BMC Med.* **2012**, *10*, 81. [CrossRef]
19. Gerbi, M.E.; Pinheiro, A.L.; Marzola, C.; Limeira Júnior, F.A.; Ramalho, L.M.; Ponzi, E.A.; Soares, A.O.; Carvalho, L.C.; Lima, H.V.; Gonçalves, T.O. Assessment of bone repair associated with the use of organic bovine bone and membrane irradiated at 830 nm. *Photomed. Laser Surg.* **2005**, *23*, 382–388. [CrossRef]
20. Sella, V.R.; do Bomfim, F.R.; Machado, P.C.; da Silva Morsoleto, M.J.; Chohfi, M.; Plapler, H. Effect of low-level laser therapy on bone repair: A randomized controlled experimental study. *Lasers Med. Sci.* **2015**, *30*, 1061–1068. [CrossRef]
21. de Oliveira Gonçalves, J.B.; Buchaim, D.V.; de Souza Bueno, C.R.; Pomini, K.T.; Barraviera, B.; Júnior, R.S.F.; Andreo, J.C.; de Castro Rodrigues, A.; Cestari, T.M.; Buchaim, R.L. Effects of low-level laser therapy on autogenous bone graft stabilized with a new heterologous fibrin sealant. *J. Photochem. Photobiol. B* **2016**, *162*, 663–668. [CrossRef]
22. Bosco, A.F.; Faleiros, P.L.; Carmona, L.R.; Garcia, V.G.; Theodoro, L.H.; de Araujo, N.J.; Nagata, M.J.H.; de Almeida, J.M. Effects of low-level laser therapy on bone healing of critical-size defects treated with bovine bone graft. *J. Photochem. Photobiol. B* **2016**, *163*, 303–310. [CrossRef]
23. Pinheiro, A.L.; Gerbi, M.E. Photoengineering of bone repair processes. *Photomed. Laser Surg.* **2006**, *24*, 169–178. [CrossRef]
24. da Silva, J.P.; da Silva, M.A.; Almeida, A.P.; Lombardi Junior, I.; Matos, A.P. Laser therapy in the tissue repair process: A literature review. *Photomed. Laser Surg.* **2010**, *28*, 17–21. [CrossRef]

25. Scalize, P.H.; Sousa, L.G.; Regalo, S.C.H.; Semprini, M.; Pitol, D.L.; Da Silva, G.A.; Coelho, J.A.; Coppi, A.A.; Laad, A.A.B.L.; Prado, K.F.B.; et al. Low-level laser therapy improves bone formation: Stereology findings for osteoporosis in rat model. *Lasers Med. Sci.* **2015**, *30*, 1599–1607. [CrossRef]
26. Khalil, N.M.; Noureldin, M.G. Comparison of Single Versus Multiple Low-Level Laser Applications on Bone Formation in Extraction Socket Healing in Rabbits (Histologic and Histomorphometric Study). *J. Oral. Maxillofac. Surg.* **2019**, *77*, 1760–1768. [CrossRef]
27. Weber, J.B.; Pinheiro, A.L.; de Oliveira, M.G.; Oliveira, F.A.; Ramalho, L.M. Laser therapy improves healing of bone defects submitted to autologous bone graft. *Photomed. Laser Surg.* **2006**, *24*, 38–44. [CrossRef]
28. Meme', L.; Gallusi, G.; Coli, G.; Strappa, E.; Bambini, F.; Sampalmieri, F. Photobiomodulation to Reduce Orthodontic Treatment Time in Adults: A Historical Prospective Study. *Appl. Sci.* **2022**, *12*, 11532. [CrossRef]
29. Nunes, C.M.M.; Ferreira, C.L.; Bernardo, D.V.; Oblack, G.B.; Longo, M.; Santamaria, M.P.; Jardini, M.A.N. The influence of LLLT applied on applied on calvarial defect in rats under effect of cigarette smoke. *J. Appl. Oral. Sci.* **2019**, *27*, e20180621. [CrossRef]
30. Tunér, J. Photobiomodulation: How Many Sessions and How Often? *Photomed. Laser Surg.* **2018**, *36*, 59–60. [CrossRef]
31. Nicola, R.A.; Jorgetti, V.; Rigau, J.; Pacheco, M.T.; dos Reis, L.M.; Zângaro, R.A. Effect of low-power GaAlAs laser (660 nm) on bone structure and cell activity: An experimental animal study. *Lasers Med. Sci.* **2003**, *18*, 89–94. [CrossRef] [PubMed]
32. Tetè, S.; Zara, S.; Vinci, R.; Zizzari, V.; Di Tore, U.; Mastrangelo, F.; Cataldi, A.; Gherlone, E. Vascular endothelial growth factor and e-nitric oxide synthase-mediated regenerative response occurring upon autologous and heterologous bone grafts. *Int. J. Immunopathol. Pharmacol.* **2009**, *22*, 1105–1116. [CrossRef] [PubMed]
33. Zizzari, V.L.; Zara, S.; Tetè, G.; Vinci, R.; Gherlone, E.; Cataldi, A. Biological and clinical aspects of integration of diferente bone substitutes in oral surgery: A literature review. *Oral. Surg. Oral. Med. Oral. Pathol. Oral. Radiol.* **2016**, *122*, 392–402. [CrossRef] [PubMed]
34. Tamimi, F.M.; Torres, J.; Tresguerres, I.; Clemente, C.; Lopez-Cabarcos, E.; Blanco, L.J. Bone augmentation in rabbit calvariae: Comparative study between bio-oss and a novel betatcp/dcpd granulate. *J. Clin. Periodontol.* **2006**, *33*, 922–928. [CrossRef]
35. Ezirganli, S.; Polat, S.; Baris, E.; Tatar, I.; Celik, H.H. Comparative investigation of the effects of different materials used with a titanium barrier on new bone formation. *Clin. Oral. Implant. Res.* **2013**, *24*, 312–319. [CrossRef]
36. Carmagnola, D.; Adriaens, P.; Berglundh, T. Healing of human extraction sockets filled with Bio-Oss. *Clin. Oral. Implant. Res.* **2003**, *14*, 137–143. [CrossRef]
37. Elgali, I.; Turri, A.; Xia, W.; Norlindh, B.; Johansson, A.; Dahlin, C.; Thomsen, P.; Omar, O. Guided bone regeneration using resorbable membrane and different bone substitutes: Early histological and molecular events. *Acta Biomater.* **2016**, *29*, 409–423. [CrossRef]

Journal of
*Functional
Biomaterials*

MDPI

Article

Manganese-Implanted Titanium Modulates the Crosstalk between Bone Marrow Mesenchymal Stem Cells and Macrophages to Improve Osteogenesis

Kuicai Ye [1,2], Xianming Zhang [1], Li Shangguan [1,3], Xingdan Liu [1,2], Xiaoshuang Nie [1,2] and Yuqin Qiao [1,*]

1 State Key Laboratory of High-Performance Ceramics and Superfine Microstructure, Shanghai Institute of Ceramics, Chinese Academy of Sciences, Shanghai 200050, China; kcyeh@mail.ustc.edu.cn (K.Y.)
2 Center of Materials Science and Optoelectronics Engineering, University of Chinese Academy of Sciences, Beijing 100049, China
3 School of Materials Science, Shanghai University, Shanghai 200444, China
* Correspondence: qiaoyq@mail.sic.ac.cn

Abstract: Manganese (Mn) is an essential micronutrient in various physiological processes, but its functions in bone metabolism remain undefined. This is partly due to the interplay between immune and bone cells because Mn plays a central role in the immune system. In this study, we utilized the plasma immersion ion implantation and deposition (PIII&D) technique to introduce Mn onto the titanium surface. The results demonstrated that Mn-implanted surfaces stimulated the shift of macrophages toward the M1 phenotype and had minimal effects on the osteogenic differentiation of mouse bone marrow mesenchymal stem cells (mBMSCs) under mono-culture conditions. However, they promoted the M2 polarization of macrophages and improved the osteogenic activities of mBMSCs under co-culture conditions, indicating the importance of the crosstalk between mBMSCs and macrophages mediated by Mn in osteogenic activities. This study provides a positive incentive for the application of Mn in the field of osteoimmunology.

Keywords: manganese; plasma immersion ion implantation and deposition; crosstalk; mBMSCs; macrophages

1. Introduction

The process of osseointegration, in which chemical bonding between biomaterials and bone tissue occurs, is a highly complex and dynamic mechanism [1]. Inflammatory reactions play a crucial role in the recovery and reconstruction process of damaged bone tissue through a series of intricate cellular and molecular interactions [2,3]. Recently, modulating the immune responses has been recognized as a pivotal strategy to regulate or control the process of bone remodeling and offers a prospect and challenges for developments in bone biomedical devices [4].

Macrophages, one of the first cells to arrive at the site of injury, are present throughout the whole repair and regeneration process [5]. They can not only defend against pathogens but also, more importantly, secrete an array of cytokines and chemokines in response to certain stimuli and recruit other cells to promote inflammation [6]. Generally, macrophages are divided into at least two subgroups, the classically activated M1 phenotype and alternatively the M2 phenotype [7]. M1 macrophages, with a typical surface marker C-C chemokine receptor type 7 (CCR7, or CD197), usually feature a pro-inflammatory phenotype and tend to secrete pro-inflammatory cytokines including interleukin 6 (IL-6) and tumor necrosis factor-α (TNF-α), as well as chemokines such as C-C motif chemokine ligand 3 (CCL3), colony-stimulating factor 2 (CSF2), and C-C motif chemokine ligand 5 (CCL-5). And M2 macrophages, with the typical surface marker of mannose receptor 206 (CD206), are associated with anti-inflammatory responses that can accelerate the tissue healing process through anti-inflammatory cytokines such as interleukin 4 (IL-4), interleukin 10 (IL-10), and arginase 1 (Arg-1) [7,8]. This molecular mechanism suggests that biomaterials which can stimulate the macrophage polarization from M1 to M2 may enhance macrophage-mediated osteogenesis [9]. Thus, a lot of efforts have been spent on optimizing the physicochemical properties of biomaterials to induce M2 transition of macrophages in previous studies [10,11]. It is widely recognized that excessive inflammation can result in implant failure and complications such as peri-implantitis [12]. However, with the increasing understanding of macrophages, emerging evidence suggests that the osseointegration is intricately mediated by a complex interplay of various factors in the immune microenvironment, including not only M2 macrophages but also M1 macrophages [13–15]. And insufficient inflammation may also lead to slow osseointegration or poor bone-to-implant contact. Moreover, inconsistencies between in vitro and in vivo studies are not uncommon, indicating that the mechanisms which modulate the materials' capacity to regulate osteogenic activities in the immunological microenvironment are not well understood [16]. Therefore, it is necessary to investigate the role of the interactions between immune- and bone-related cells mediated by biomaterials in osteogenesis.

Some nutrient elements have been found to play key roles in bone regeneration and remodeling and also elicit significant immune responses [17–20]. For example, copper-incorporated biomaterials (bioactive glass, ceramics, and titanium alloys) have been found to improve osteogenic activities and angiogenic properties [21,22]. In a recent study, Huang et al. showed that a Cu-containing micro/nano-topographical bio-ceramic surface promoted macrophages to the M1 pro-inflammatory phenotype by activating Cu-transport signaling, which can enhance the phagocytic ability on bacteria [23]. Among these elements, Mn also serves as a cofactor in various enzymatic reactions and is crucial for the proper functioning of many enzymes, such as manganese superoxide dismutase (Mn-SOD) [24]. And comparatively, Mn plays a more significant role in the modulation of immune responses rather than in bone metabolism [18,25]. Increasing evidence has mounted to suggest that Mn is an immunostimulatory agent which augments the inflammatory response through the cGAS-STING signaling pathway [26]. Although some Mn-incorporated biomaterials have been reported to stimulate the osteogenic differentiation of pre-osteoblasts or mesenchymal stem cells, there is still little evidence on how Mn-incorporated biomaterials affect osteogenic activities in the immune microenvironment [27–29]. Therefore, further investigations are focused on the manganese-mediated regulation of the immune system in the context of osteogenic effects.

In this study, we employed the plasma immersion ion implantation and deposition (PIII&D) technique to introduce Mn onto the surface of the titanium. This method is superior because it can incorporate elements into various medical devices with irregular shapes and minimize the effects of other confounding factors on the biological properties [30]. The Raw264.7 macrophage cell line was utilized to investigate the impact of materials on cellular immune responses in our study due to its ease of culture and strong phenotypic stability [31]. Raw264.7 cells can elicit a robust and well-known inflammatory response upon challenges with increasing dosages of stimulants, such as lipopolysaccharide (LPS), which is often employed to screen biomaterials and to predict their potential in regulating immune responses [31]. This may ultimately activate a series of events, including the production of inflammatory and anti-inflammatory cytokines, which can be used as biomarkers to screen for possible anti-inflammatory and immunomodulatory compounds or biomaterials [32]. Both mono- and co-culture experiments of Raw264.7 cells and mBMSCs were conducted to deepen the understanding of the crosstalk between macrophages and mBMSCs mediated by Mn-implanted surfaces.

2. Materials and Methods

2.1. Preparation of Mn-Implanted Titanium

In this study, two distinct sizes of titanium plates (Cp Ti, TA1, purity > 99.85%) were employed: 20 mm × 10 mm × 1 mm plates were utilized for zeta potential measurement, and 10 mm × 10 mm × 1 mm plates were used in the remaining experiments. The titanium plates were initially polished with silicon carbide abrasive papers and then ultrasonically washed with acetone and alcohol. Prior to plasma immersion ion implantation and deposition (PIII&D), titanium plates were cleaned using radio frequency argon ions for 10 min at a bias of 800 V. The Mn cathode (purity > 99.9%) was obtained from Qi Jin New Material Co., Ltd. (Quanzhou, China). Then, Mn ions were implanted onto the pretreated titanium surface at 15 kV for 60 min (denoted as Mn60) and 120 min (denoted as Mn120). During PIII&D, the sample stage was cooled by circulating water to keep the sample temperature at 25 °C. The specific parameters are outlined in Table 1.

Table 1. The primary parameters used in the plasma ion immersion implantation.

Parameters	Mn60	Mn120
Target voltage pulse duration (μs)	500	500
Cathodic arc voltage pulse duration (μs)	800	800
Pulsing frequency (Hz)	5	5
Voltage (kV)	−15	−15
Time (min)	60	120
Pressure (Pa)	5.0×10^{-3}	5.0×10^{-3}

2.2. Characterization

2.2.1. Surface Structure and Physicochemical Characterization

Field emission scanning electron microscopy (FE-SEM; Magellan 400, Hillsboro, OR, USA) was used to analyze the surface morphologies of materials at a 5 kV acceleration voltage. The elemental valences, contents, and distribution of Mn-implanted surfaces were analyzed using X-ray photoelectron spectroscopy equipped with a monochromatic Al Kα source (XPS; K-Alpha, Thermo Scientific, Waltham, MA, USA). The nano-hardness and elastic modulus of Mn-implanted surfaces were determined using a nanoindenter (G200, Agilent, Santa Clara, CA, USA), and the average value of each sample was obtained from five measurements. Each experiment was independently performed twice.

2.2.2. Surface Wettability

The surface wettability was measured with a surface contact angle meter (SL200B, Solon, Shanghai, China). The water droplets were captured and analyzed after 2 μL of ultrapure water was vertically dropped onto surfaces. The experiment was performed twice independently with three replicates.

2.2.3. Dynamic Potential Polarization Test

An electrochemical workstation (CHI760, Shanghai Chenhua, Shanghai, China) was used for electrochemical analysis, and the electrolyte was a 0.9% NaCl solution. The measurement was performed by utilizing a conventional and electrochemical cell with three-electrodes. The reference, counter, and working electrodes were a saturated calomel electrode, graphite, and a test sample, respectively. The scanning rate was set as 0.01 V/s. The experiment was performed twice independently.

2.2.4. Surface Zeta Potential

The surface zeta potential of the plates was measured using a SurPASS 3 electric analyzer (AntonPaar, Graz, Austria). Two 20 mm × 10 mm × 1 mm samples were put on the test cell. The space between two parallel samples was set to 100 ± 5 μm to guarantee that the electrolyte (0.001 mol/L KCl solution) went through the gap. The pH of the measurement was set from 10.0 to 5.0 and adjusted by automatic machine titration of 0.05 M HCl. The final value was taken as the average of four measurements for each measurement point. The calculation formula was determined according to Helmholtz–Smoluchowski equation [33]:

$$\zeta = \frac{dU}{dP}\frac{\eta}{\varepsilon\varepsilon_0}K$$

in which ζ is the zeta potential, dU/dP stands for the slope of the streaming potential versus differential pressure, and η, ε, ε_0, and K denote the electrolyte viscosity, vacuum permittivity, dielectric constant of the electrolyte, and electrical conductivity, respectively.

2.2.5. Concentration of Released Mn ions

Mn-implanted samples were immersed in 5 mL PBS (phosphate-buffered saline, pH = 7.4) and kept at 37 °C. At the end of 1, 4, 7, and 14 days, PBS solution was collected and replaced with fresh 5 mL PBS. The Mn ion concentration was determined using inductively coupled plasma mass spectrometry (ICP-MS, X Series 2, Thermo Scientific, Waltham, MA, USA). Each group of samples was tested in duplicate.

2.3. Biological Evaluations

2.3.1. Cell Culture

The effects of different surfaces on the behavior of macrophages were evaluated using mouse mononuclear macrophage leukemia cells (Raw264.7, obtained from Cell Bank of the Chinese Academy of Sciences, Shanghai, China), which were cultured in DMEM medium (high glucose, Gibco, Waltham, MA, USA) supplemented with 15% fetal bovine serum (FBS, Gibco, Waltham, MA, USA) and 1% penicillin/streptomycin (100 U/mL, Gibco, Waltham, MA, USA) at 37 °C in a humidified environment of 5% CO_2. Cells were passaged at a 1:3 ratio every three days and maintained up to 10 passages.

The osteogenic activity of various samples was evaluated using mouse bone marrow mesenchymal stem cells (mBMSCs, obtained from Shanghai Zhong Qiao Xin Zhou Biotechnology Co., Ltd., Shanghai, China). The mBMSCs were maintained in a basal medium composed of Mesenchymal Stem Cell Medium (MSCM, Sciencell, Carlsbad, CA, USA), 25 mL of FBS (Sciencell, Carlsbad, CA, USA), 5 mL of mesenchymal stem cell growth supplement (Sciencell, Carlsbad, CA, USA), and 5 mL of penicillin/streptomycin solution (Sciencell, Carlsbad, CA, USA). For osteogenic differentiation of mBMSCs, the osteogenic differentiation media were composed of DMEM medium (high glucose, Gibco, Waltham, MA, USA), 15% FBS (Gibco, Waltham, MA, USA), 1% penicillin/streptomycin serum (100 U/mL, Gibco, Waltham, MA, USA), 100 nM dexamethasone (Sigma-Aldrich, Burlington, MA, USA), 50 μg/mL L-ascorbic acid (Sigma-Aldrich, Burlington, MA, USA), and 10 mM β-glycerophosphate (Sigma-Aldrich, Burlington, MA, USA). The cells were passaged at a ratio of 1:3 every three days in a humidified atmosphere of 5% CO_2 at 37 °C and were used in this study within 5 passages.

In this work, co-culture experiments were performed to investigate the interaction effects between Raw264.7 cells and mBMSCs. The experimental methods were in reference to the literature report [34]. The experimental schematic is shown in Figure S1. Firstly, Raw264.7 cells and mBMSCs were incubated on different samples (four plates per group) for 1 day and 7 days, respectively. Then, 2 samples from each group, which had been seeded with Raw264.7 cells and mBMSCs, were put in the same well of a 6-well plate to co-culture for an additional 3 days. An osteogenic differentiation medium was used in the co-culture condition. At the end of the incubation time, samples seeded with Raw264.7 cells and mBMSCs were separately transferred to different new plates for further biological evaluations.

2.3.2. Cell Proliferation and Viability

AlamarBlue™ (Thermo Fisher Scientific, Waltham, MA, USA) was used to detect cell viability. At the end of different incubation time points, the medium was replaced by 500 μL of fresh culture medium containing 10% AlamarBlue™ and incubated at 37 °C for 2 h. Then, medium aliquots (100 μL per well) were transferred to a black 96-well plate to measure the fluorescence intensity (E_x = 560 nm, E_m = 590 nm) using a multifunctional microplate reader (Cytation 5, BioTek, Winooski, VT, USA).

Specifically, Raw264.7 cells were inoculated into a 24-well plate at a density of 1×10^5 cells per well, and cell viability was measured at 4 h and 1 and 4 days. The mBMSCs were inoculated into a 24-well plate at a density of 2×10^4 cells per well, and the measured time points were at 1, 4, and 7 days. Four parallel samples were set in each group. The experiment was performed twice independently for each experiment.

2.3.3. Cell Morphology

Raw264.7 cells were seeded at a density of 1×10^5 cells per well on samples (four replicates) in a 24-well plate. The cells were cultured for 4 days. Then macrophages were washed with PBS and fixed with 2.5% glutaraldehyde at 4 °C overnight. Following this, the cells were dehydrated in a series of ethanol solutions (30, 50, 75, 90, 95%, and anhydrous ethanol, *v*/*v*) for 10 min each and dried in the hexamethyl disilylamine/ethanol solution series (1: 2, 1: 1, 2: 1 and 3:0, *v*/*v*). All samples were treated with gold spraying (120 s). Then, the cell morphologies were observed using SEM (accelerating voltage: 5 kV, S-3400 N Type I, Hitachi, Tokyo, Japan).

mBMSCs were seeded at a density of 2×10^4 per well in 24-well plates and incubated for 4, 12, and 24 h. After being fixed with 4% paraformaldehyde overnight at 4 °C in the dark, the cells were permeabilized with 0.1% Triton X-100 (Amresco, Solon, OH, USA) for 2 min, blocked with 1% Bovine Serum Albumin (BSA, Sigma, Saint-Louis, MO, USA) for 30 min, and subsequently stained with rhodamine phalloidin (Sigma, Saint-Louis, MO, USA) at room temperature for 1 h and 4′,6-diamidino-2-phenylindole (DAPI, Sigma-Aldrich, Burlington, MA, USA) for an additional 3 min. Finally, the F-actin of mBMSCs was observed using fluorescence microscopy (IX71, Olympus, Tokyo, Japan).

2.3.4. Immunofluorescence Staining of Macrophages

Mannose receptor 206 (CD206, M2 marker, in red) and inducible nitric oxide synthase (iNOS, M1 marker, in green) were selected to evaluate the polarization state of macrophages. Firstly, macrophages were cultured on different sample surfaces at a density of 1×10^5 per well for 4 h and 1 and 4 days. At the end of incubation, cells were washed with PBS three times, fixed with 4% paraformaldehyde (PFA, Sigma, Saint-Louis, MO, USA) at 4 °C for 10 min, permeabilized by 0.1% Triton X-100 (Amresco, Solon, OH, USA) for 2 min, and blocked with 3% BSA (Sigma, Saint-Louis, MO, USA) in PBS for 1 h. Then, the cells were incubated with rabbit anti-mouse iNOS (1: 50, Cell Signaling Technology, Beverly, MA, USA) and goat anti-mouse CD206 (1:40, R&D Systems, Minneapolis, MN, USA) at 4 °C overnight. Subsequently, the cells were incubated with secondary antibodies donkey anti-rabbit IgG H&L Alexa Fluor 488 (1:200, Invitrogen, Thermo Fisher Scientific, Waltham, MA, USA) and donkey anti-goat IgG H&L Alexa Fluor 594 (1: 200, Abcam, Cambridge, UK) for 2 h at room temperature. The nuclei of the cells were stained with DAPI for 3 min in the dark. The immunofluorescence staining was visualized using a fluorescence microscope (IX71, Olympus, Tokyo, Japan). The experiment was performed twice independently.

2.3.5. Flow Cytometry Analysis of Macrophages

C-C chemokine receptor type 7 (CCR7) and CD206 were used to characterize M1 and M2 phenotypes of macrophages, their expression levels were detected using flow cytometry, and F4/80 was selected as the marker of Raw264.7 cells. Macrophages were seeded at a density of 1×10^5 per well on samples (four plates per replicate and three replicates per group) in a 24-well plate for 4 days. Then, the cells were detached by trypsin-EDTA (0.05%), washed with PBS, and resuspended in PBS. Single-stained cells served as compensation controls, while unstained cells were used to set the negative gate. The resuspended cells were blocked with mouse Fc block (BD Pharmingen, San Diego, CA, USA) for 10 min at room temperature. Then, the blocked cells were sequentially stained with PE/Cy7-labeled anti-mouse F4/80 antibody (BD Pharmingen, San Diego, CA, USA),

APC-labeled anti-mouse CD206 antibody (Biolegend, San Diego, CA, USA), and PE-labeled anti-mouse CD197 (CCR7) antibody (Biolegend, San Diego, CA, USA) for 30 min at 4 °C in the dark. Then, they were washed and resuspended with PBS and detected on the flow cytometer (Celesta, BD Biosciences, San Jose, CA, USA).

2.3.6. Real-Time Quantitative Polymerase Chain Reaction (RT-qPCR) Analysis

RT-qPCR was utilized to measure the immune- and osteogenic-related gene expression. Four samples per group were placed in a 24-well plate and then seeded with Raw264.7 cells at a density of 1×10^5 cells per well and mBMSCs at a density of 2×10^4 cells per well. Raw264.7 cells were cultured for 4 days and mBMSCs were cultured for 10 days. At the end of incubation, samples were moved to a fresh six-well plate, washed with PBS twice, and adhered cells were blown off with 1 mL Trizol™ (Invitrogen, Thermo Fisher Scientific, Waltham, MA, USA). Complementary DNA (cDNA) was synthesized from total RNA using a transcriptor first-strand cDNA synthesis kit (Roche, Basel, Switzerland) according to the manufacturer's protocol. RT-qPCR was performed on the LightCycler® 480 system II (Roche, Basel, Switzerland) utilizing SYBR Green I Master (Roche, Basel, Switzerland). The level of each target gene was calculated using the $2^{-\Delta\Delta Ct}$ method, and GAPDH was used as a reference for normalization. The primers used in RT-qPCR are given in Table S1 and were obtained from BioTNT (Shanghai, China). Each group of samples was analyzed in triplicate.

2.3.7. Alkaline Phosphatase (ALP) Activity of mBMSCs

mBMSCs were seeded on different samples (four replicates per group) at a density of 2×10^4 cells per well in 24-well plates. The cells were cultured in osteogenic differentiation medium for 10 days. Cells were lysed for 40 min on ice using a lysis buffer containing 1% protease inhibitor (Sigma-Aldrich, Burlington, USA), 1% IGEPAL CA-630 (Beyotime-Biotech Co., Shanghai, China), 10 mM Tris-HCl (Sinopharm, Shanghai, China, pH = 7.5), and 1 mM $MgCl_2$ (Sinopharm, Shanghai, China) [35]. Then, cell lysates were centrifuged at 8000 rpm for 10 min at 4 °C, and the supernatant was collected. After that, p-nitrophenyl phosphate (*p*NPP) (Sigma-Aldrich, Burlington, MA, USA) was added and incubated for 30 min at 37 °C. Then, 1 M NaOH (Sinopharm, Shanghai, China) solution was added to terminate the reaction. The ALP activity was measured using a multifunctional microplate reader (Cytation 5, BioTek, Winooski, VT, USA) at an absorbance of 405 nm for detecting the production of *p*-nitrophenol [36]. Total intracellular protein concentration was quantified using a bicinchoninic acid (BCA) kit (Thermo Fisher Scientific, Waltham, MA, USA) and used to normalize the ALP activity.

For ALP staining, a 5-bromo-4-chloro-3-indolylphosphate/nitrobluetetra-zoliumchloride (BCIP/NBT) alkaline phosphatase color development kit (Beyotime-Biotech Co., Shanghai, China) was used. After incubation for 10 days, cells were treated with BCIP/NBT working solutions for 2 h in the dark and washed twice with ultrapure water. A fluorescent microscope (IX71, Olympus, Tokyo, Japan) was used to observe the stained cells.

2.3.8. Collagen Secretion of mBMSCs

mBMSCs were seeded on different samples (four replicates per group) at a density of 2×10^4 cells per well in a 24-well plate and were cultured in osteogenic differentiation medium for 10 days. After finishing the culture, cells were fixed with 4% PFA (Sigma, Saint-Louis, MO, USA) for 10 min and stained with 0.1% Sirius Red solution (dissolved in saturated picric acid) for 18 h [34]. Afterward, the cells were washed with 0.1 M acetic acid to remove the excess dye. Then, the images were captured using a fluorescent microscope (IX71, Olympus, Tokyo, Japan) under a brightness field. The quantitative results were obtained by dissolving the stain in a mixed solution (0.5 mL per well, 0.2 M NaOH: methanol = 1:1, *v*/*v*), and the absorbance of the stain was measured at 540 nm using a multifunctional microplate reader (Cytation 5, BioTek, Winooski, VT, USA). The experiment was performed twice independently.

2.3.9. Extracellular Matrix Mineralization (ECM) of mBMSCs

mBMSCs were seeded at a density of 2×10^4 cells per well on samples (four replicates) in 24-well plates and cultured in an osteogenic differentiation medium for 10 days. The culture medium was changed every 2 days. After that, the cells were fixed with 75% ethanol solution for 1 h and subsequently stained with 40 mM Alizarin Red (pH = 4.2) for 10 min. The excess dye was removed with distilled water before taking images with a fluorescent microscope (IX71, Olympus, Tokyo, Japan) under a brightness field. An eluent (0.5 mL per well, 10% cetylpyridinium chloride (CPC) diluted in 10 mM sodium phosphate, pH = 7) was applied to dissolve the stain, and the absorbance of the stain at 620 nm was evaluated to quantify the ECM mineralization using a multifunctional microplate reader (Cytation 5, BioTek, Winooski, VT, USA). The experiment was performed twice independently.

2.4. Statistical Analysis

Statistical analysis was conducted utilizing GraphPad Prism 6.0 (GraphPad Software, La Jolla, CA, USA). Quantitative data were presented as the mean ± standard deviation (SD). The significance of differences (p-value) was analyzed through the use of one-way analysis of variance (ANOVA), two-way ANOVA, and Tukey's multiple comparison tests. A p-value less than 0.05 was considered statistically significant and denoted by the symbol "*", a p-value less than 0.01 was denoted by "**", and a p-value less than 0.001 was denoted by "***". The "ns" means not statistically significant.

3. Results

3.1. Surface Characteristics

The SEM images of Mn-implanted surfaces are shown in Figure 1. After the polishing process, the Ti surface was structurally flat. No obvious fluctuation in topography was observed. After Mn implantation, there were no visible structural changes on the surfaces of Mn60 and Mn120.

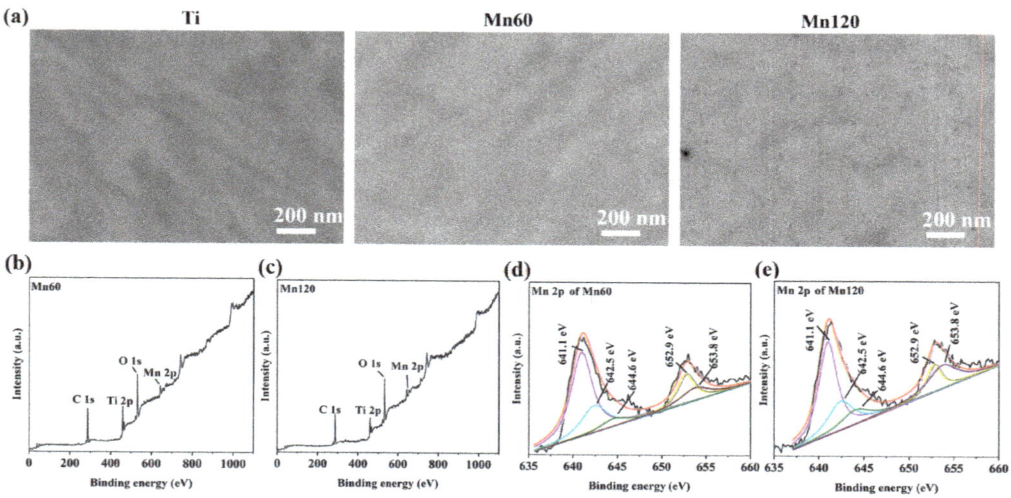

Figure 1. (**a**) SEM images of the surface morphologies of various samples; (**b**) XPS full spectrum of Mn60; (**c**) XPS full spectrum of Mn120; (**d**) XPS high-resolution spectrum of Mn 2p obtained from Mn60; (**e**) XPS high-resolution spectrum of Mn 2p obtained from Mn120.

XPS was conducted to determine the elemental composition of the sample surfaces. The relative amounts of Mn were 3.20% and 4.09% for Mn60 and Mn120, respectively (Table 2). Due to the interference of adsorbed C on the surfaces, the amount of implanted

Mn in the two samples cannot be compared directly according to their relative elemental percentages. The Mn 2p peaks could be found from Mn-implanted surfaces (Figure 1b,c). As shown in the Mn 2p high-resolution spectra (Figure 1d,e), the peaks at 642.5 eV and 653.8 eV correspond to the $2p_{3/2}$ and $2p_{1/2}$ peaks of MnO_2, respectively [37], the peaks at 641.1 eV and 652.9 eV correspond to the $2p_{3/2}$ and $2p_{1/2}$ peaks of MnO_x/Mn, respectively, and the peak at 644.6 eV is considered to be the broad satellite peak of MnO_x [38,39]. The results demonstrated that Mn was present on the surface in the form of manganese oxides.

Table 2. The elemental contents of Mn-implanted samples.

Sample Name	C 1s (at%)	O 1s (at%)	Ti 2p (at%)	Mn 2p (at%)
Mn60	56.97	32.12	7.71	3.20
Mn120	51.99	37.01	6.91	4.09

When performing PIII&D, Mn implantation and deposition took place simultaneously on the surfaces of Mn60 and Mn120 by modulating the pulse width of the cathode to be bigger than that of the target (Table 1), which may affect the mechanical properties of titanium. In this study, changes in nano-hardness and elastic modulus within the 110 nm region were measured (Figure 2a,b). As shown in Figure 2a, compared to pure titanium, the surface nano-hardness of the Mn120 exhibited an obvious increase compared to titanium within the 110 nm region, while the nano-hardness of Mn60 only improved within the range of 15 nm to 30 nm. Figure 2b illustrates the variation in the specific elastic modulus on the modified surfaces. Different from the tendency of Mn implantation on nano-hardness, the surface elastic modulus of Mn60 and Mn120 exhibited a slight variation (Figure 2b).

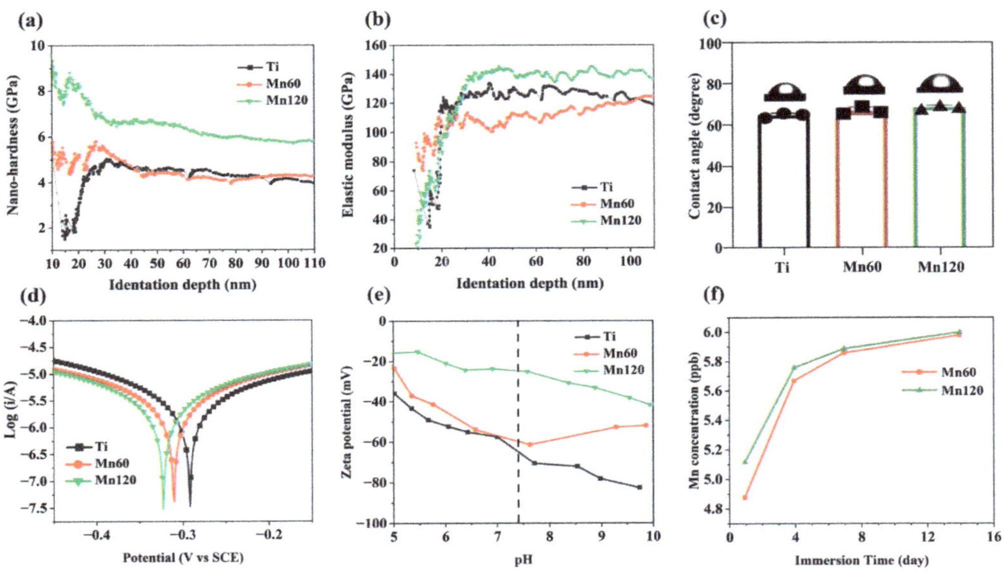

Figure 2. Physicochemical characterization of Mn-implanted surfaces: (**a**) nano-hardness curves of various samples; (**b**) elastic modulus curves of various samples; (**c**) water contact angles measured in various samples; (**d**) polarization curves of various samples; (**e**) surface zeta potential of various samples; (**f**) the concentrations of Mn ions in PBS solution.

The relative surface wettability of the coatings was determined according to the water contact angle measurement. As shown in Figure 2c, Mn implantation did not alter the contact angles of Mn60 and Mn120 compared to Ti. Figure 2d and Table 3 exhibit Tafel curves and relevant data for different samples. The corrosion potentials of the Mn60 and Mn120 displayed a slight negative shift (0.017 eV for Mn60 and 0.035 eV for Mn120).

Table 3. Corrosion potentials and currents of various samples.

Sample Name	Ti	Mn60	Mn120
I_{corr} (A·cm^{-2})	3.387×10^{-8}	4.235×10^{-8}	2.997×10^{-8}
E_{corr} (V) vs. SCE	−0.292	−0.309	−0.327

The role of surface charge in regulating cell responses has been extensively reported. As shown in Figure 2e, the zeta potentials of Mn-implanted surfaces presented a decreasing trend in a pH range from 5.0 to 7.4. At a pH of 7.4, Mn60 (−57.9 mV) and Mn120 (−24.3 mV) presented a more positive value than Ti (−61.2 mV). Previous studies have reported that a high concentration of manganese ions was detrimental to cells and can also influence the behavior of immune cells. The accumulative Mn ion release is shown in Figure 2f. During a 14-day period, the accumulative profiles of Mn60 and Mn120 showed similar Mn ion release characteristics in PBS, which include a burst release followed by a low-level continuous release. Although Mn120 displayed a higher release amount, both Mn-implanted surfaces released Mn ions at ppb level (0–10 ppb). This release profile can reflect that the amount of Mn implanted in Mn120 was higher than that implanted in Mn60.

3.2. Effects of Mn Implantation on Cellular Immune Response

The effects of Mn-implanted surfaces on macrophage polarization and plasticity were investigated using Raw264.7 cells. The cell viability results showed that Mn implantation did not exert any negative effects on cell proliferation (Figure 3a). When cultured for 4 days, round macrophages with long filopodia grew in grape-like clusters on all surfaces (Figure 3b), suggesting that Mn implantation had no obvious effects on the adhesion and spreading of macrophages.

The polarization state and activation process of macrophages were evaluated at both gene and protein level using immunofluorescent staining, flow cytometry, and RT-PCR methods. As shown in Figure 3c, macrophages cultured on the Mn120 sample for 4 days expressed higher levels of iNOS (green) and similar levels of CD206 (red) compared to those on Ti and Mn60. The proportion of the M1 and M2 phenotypes of macrophages was quantitatively measured using flow cytometry (Figure 3c). The results demonstrated that the average proportion of CD206$^+$ / CCR7$^-$ (M2) macrophages of Ti was 60.4 ± 10.8%. The proportion of CD206$^+$ / CCR7$^-$ (M2) macrophages of Mn60 and Mn120 was 53.4 ± 1.2% and 50.5 ± 1.0%, respectively. Although there was no significant difference among the three groups, there was still a decreasing trend of M2 macrophages on Mn-implanted surfaces compared to those on Ti. To further investigate the influence of Mn implantation on macrophages, the expressions of macrophage marker genes were quantified using RT-PCR after being cultured for 4 days (Figure 3d). Furthermore, growth factors, cytokines, and chemokines released in the microenvironment may force macrophages to undergo substantial phenotypic and functional changes. Cytokines such as C-C motif chemokine ligand 3 (CCL3), colony-stimulating factor 2 (CSF2), and C-C motif chemokine ligand 5 (CCL5) have been reported to induce a transition in the phenotype of macrophages into a pro-inflammatory type [40–43]. The RT-qPCR results showed that the mRNA expressions of pro-inflammatory-related gene tumor necrosis factor-α (TNF-α) and cytokines including CCL3, CSF2, and CCL5 were significantly upregulated (Figure 3e) on different surfaces with the trend Mn120 > Mn60 > Ti, while those of anti-inflammatory-related genes, including interleukin 10 (IL-10), CD206, arginase 1 (Arg-1), and interleukin 4 (IL-4), were downregulated.

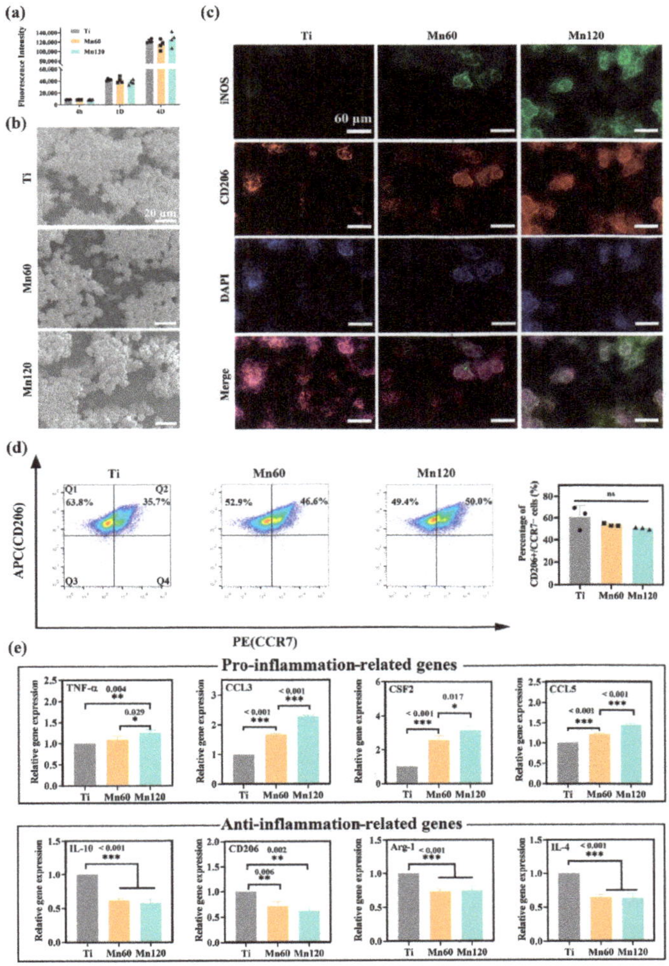

Figure 3. Mono-cultured Raw264.7 cells on various samples after incubation for 4 days: (**a**) cell viability; (**b**) SEM morphology; (**c**) immunofluorescent staining images; (**d**) flow cytometry analyses; (**e**) relative mRNA expression levels of the related genes.

3.3. *Little Effects of Mn Implantation on Osteogenic Differentiation*

It is well known that cell adhesion provides cells with the ability to sense the mechanical properties of the substratum, especially surface elastic modulus [44]. Figure 4a shows the F-actin filaments of mBMSCs on different surfaces. After seeding for 4 h, F-actin spread and most of the cells had a round shape, while a small portion of the cells exhibited star or spindle shapes. After seeding for 12 h and 24 h, a majority of cells were stretched out lengthwise, and linear F-actin filaments with a ruffle formation were observed on both Ti and Mn-implanted surfaces. There was no obvious difference in F-actin and the shape of mBMSCs among surfaces, indicating that the variation in elastic modulus between Ti and Mn120 did not exhibit an obvious impact on the process of cell attachment. When cultured on these surfaces for 4 or 7 days, there was no obvious change in the proliferation of mBMSCs, and the difference was not statistically significant ($p > 0.05$) (Figure 4b).

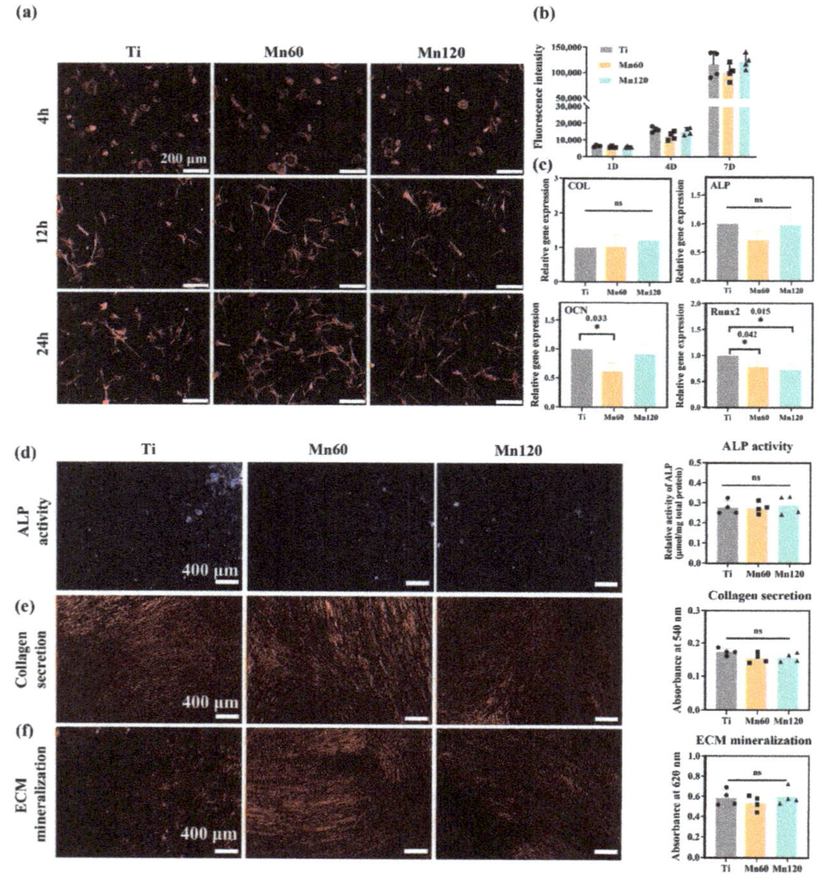

Figure 4. Osteogenic activity of mono-cultured mBMSCs on various samples after incubation for 10 days: (**a**) images of F-actin filaments; (**b**) cell viability assay; (**c**) relative mRNA expression levels of the related genes; (**d**) ALP staining images and quantitative results; (**e**) collagen secretion and the quantitative results; (**f**) extracellular matrix mineralization and the quantitative results.

To investigate the influence of Mn implantation on osteogenic differentiation, RT-qPCR was used to quantify the mRNA expressions of osteogenic-related genes, including early markers, such as alkaline phosphatase (ALP), type I collagen (COL) and runt-related transcription factor 2 (Runx2), as well as later marker osteocalcin (OCN) (Figure 4c). Runx2 is a key bone-specific transcription factor which plays an essential role in the commitment of pluripotent mesenchymal cells to the osteoblastic lineage [45,46]. Comparatively, the mRNA expressions of OCN and Runx2 were obviously downregulated on Mn60 compared to on Ti ($p < 0.05$), and Mn120 with a higher amount of Mn did not further enhance the decreasing trend.

The osteogenic differentiation was further examined at the protein level. As shown in Figure 4d–f, the qualitative and quantitative results demonstrated that Mn implantation exhibited neglected effects on ALP activity, collagen secretion, and ECM mineralization ($p > 0.05$). These observations demonstrated that the osteogenic activities of mBMSCs were barely unaffected by Mn ions at lower concentrations (0–10 ppb).

3.4. Positive Immunomodulatory Effects of Mn Implantation on Osteogenic Differentiation

As Mn plays an important role in immunological functions, and the immune response has a far-reaching impact on osteogenic activities, it is possible that Mn can exert an effect on osteogenic activities through its modulation of macrophages. After mBMSCs were co-cultured with Raw264.7 cells for 4 days, the RT-qPCR data revealed that Mn ions upregulated the mRNA expressions of osteogenic-related genes, such as OCN, Runx2, ALP, and COL (Figure 5a). Furthermore, ALP activities, collagen secretion, and ECM mineralization were also significantly promoted by Mn120 under co-culture conditions (Figure 5b–d).

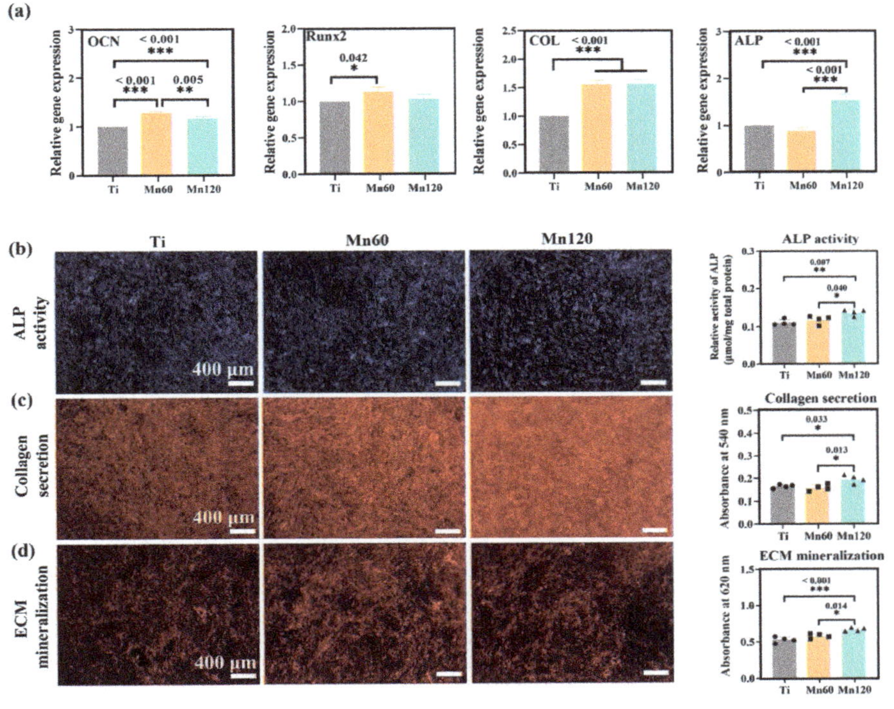

Figure 5. Osteogenic activities of mBMSCs on various samples after co-culturing for 10 days: (**a**) relative mRNA expression levels of osteogenic-related genes; (**b**) ALP staining and quantitative results; (**c**) collagen secretion staining and quantitative results; (**d**) extracellular matrix mineralization staining and the quantitative results.

The effects of Mn-induced osteogenic activities on macrophages were also investigated under co-culture conditions for 4 days. As shown in Figure 6a, Raw264.7 cells had a round morphology on both Ti and Mn-implanted surfaces, which were similar to the morphology of macrophages under mono-culture conditions. The phenotype of Raw264.7 cells was identified using immunofluorescent staining and RT-qPCR experiments (Figure 6b,c). Immunofluorescent staining results showed that the macrophages on all surfaces expressed similar intensities of M1 marker iNOS (green) and M2 marker CD206 (red). The mRNA expressions of inflammatory markers and cytokines were evaluated using RT-qPCR. As shown in Figure 6c, the Mn-implanted surface Mn120 led to a remarkably increased mRNA expression of anti-inflammatory markers, including interleukin 10 (IL-10), CD206, arginase 1 (Arg-1), and interleukin 4 (IL-4). At the same time, both Mn60 and Mn120 significantly downregulated the expressions of TNF-α, CCL3, CSF2, and CCL5.

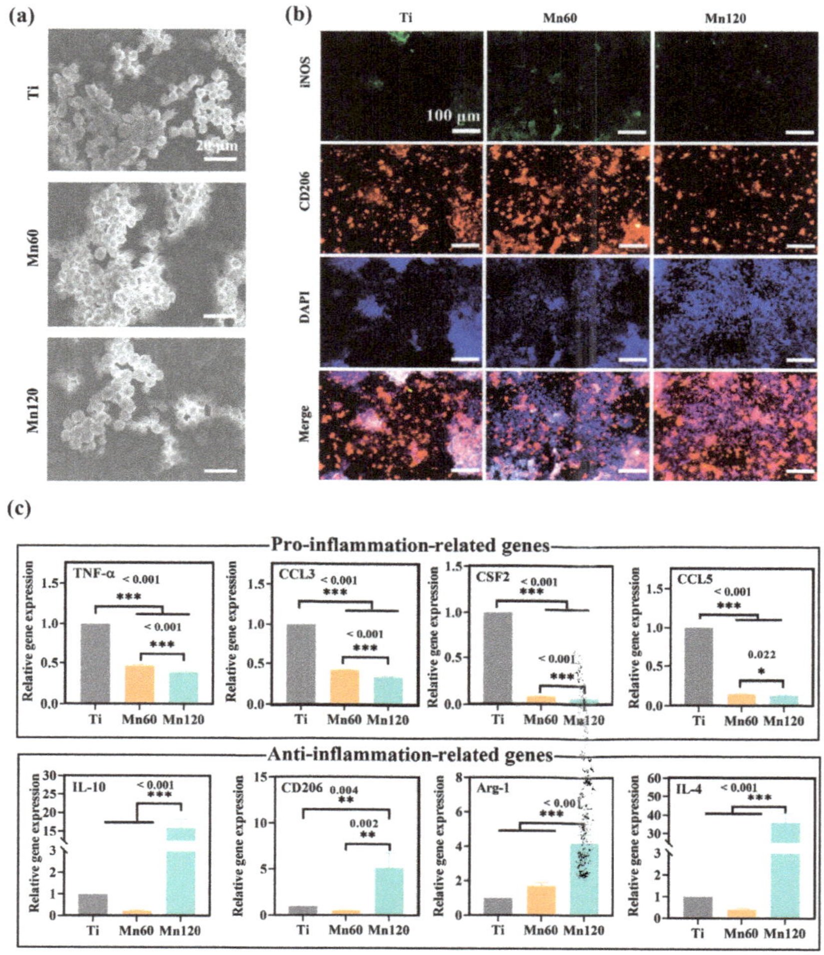

Figure 6. Raw264.7 cells on various samples under co-culture condition for 4 days: (a) SEM morphology; (b) immunofluorescent staining images; (c) relative mRNA expression levels of the related genes.

4. Discussion

The modulatory effects of Mn-doped biomaterials have attracted much attention due to the roles of Mn in nutritional immunity [18,47]. In vitro studies afford the opportunity to simulate complex interactions among cells in relatively controlled environments. This work sought to investigate the cellular immunomodulation impact of Mn-implanted surfaces and its influence on the osteogenic differentiation of mBMSCs in vitro. Previous studies have reported that stimulation with Mn can significantly upregulate the expression of genes such as IFN-β in Raw264.7 cell lines, indicating manganese's potential to activate the immune response [48]. However, it has been reported that the cytokine of Raw264.7 cells to stimulants was different from that of human leukocytes to a certain degree; thus, caution should still be taken when extending the results of this experiment to the immune responses related to human cells.

The roles of surface properties such as surface topography, wettability, and chemistry in osteogenic activities and macrophage polarization are well recognized [49,50]. Therefore, the characterization of the surface properties of Mn-implanted surfaces is essential to elucidate osteogenic activities and immunological responses. Although the amount of Mn implanted on the two surfaces could not be directly determined using XPS, the ICP-MS results showed that the concentration of Mn released from Mn120 was higher than that from Mn60, indicating that the amount of Mn implanted in Mn120 was larger than that in Mn60. It has been reported that the Ti-Mn system exhibits a pure Ti phase when the Mn content is below 50% [51]. After implantation, Mn may exist in the form of a solid solution in a titanium matrix. This explains why no visible changes in surface topography were observed [52]. Referring to the model established by Jones et al., the etching process and the deposition process of the modified layer could exist simultaneously during the PIII&D, and the amount of implanted ions on the surface layer is significantly less than that in the deep deposited layer within the modified range [53]. This could explain why the surface of Mn60, despite the presence of a certain deposition layer, showed more etching characteristics and a decreased elastic modulus compared to those of Ti and Mn120. More specifically, the disruption of the surface structure caused by the initial ion implantation prior to the formation of a sufficiently thick deposited layer may be considered as another influencing factor [54,55]. In addition, ion implantation may have produced residual stress on the surfaces, which resulted in higher surface nano-hardness. With the extended implantation time, the deposited layer of Mn120 may be thicker, leading to its obviously enhanced nano-hardness compared to that of Ti and Mn60 [55]. As demonstrated in previous studies, cells can sense the mechanical cues, such as elastic modulus, dimensionality, and pattern, and then control the function of cells [56,57]. It can be seen that Mn implantation had little effect on the adhesion and morphologies of Raw264.7 cells and mBMSCs (Figures 3b and 4a); thus, the differences in surface elastic modulus and nano-hardness were considered to exert limited effects on cells. Figure 2e shows that Mn-implanted samples, especially Mn120, displayed a tendency progressing towards less negative zeta potential values (−24.3 mV) than Ti (−61.2 mV) at pH 7.4, which may lead to increased adsorption of negatively charged proteins in the culture medium and body fluids [58].

Previous studies have established that a low concentration of manganese is optimal for the modulation of immune cells, while excessive amounts can result in toxicity [59]. In this study, PIII&D has been employed to introduce a small amount of Mn to a Ti surface, which released Mn ions at a concentration of 0–10 ppb.

Mn is an essential micronutrient required for diverse biological activities and has been found to exhibit potential osteogenic effects on bone mineralization [60,61]. However, this study found that Mn-implanted surfaces had little effect on the expressions of osteogenic-related markers at both mRNA level (ALP, COL, OCN, and Runx2) and protein levels (ALP activities, collagen secretion, and ECM mineralization) under mono-culture conditions. Therefore, it is proposed that this osteogenic effect may be indirectly mediated by the regulation of manganese when co-cultured with macrophages. As shown in Figure 5, the osteogenic-related markers were found to be significantly upregulated at both the gene and protein levels under the co-culture conditions. Given the complexity of the osteogenic process and in conjunction with other experimental data, it can be inferred that the overall results obtained are conducive to promoting cellular osteogenic activities. Therefore, further investigation is warranted to fully comprehend the mechanisms underlying the changes produced by macrophages under co-culture conditions, particularly the role of Mn in this process.

Macrophages play a crucial role in the immune response and are crucial mediators of tissue homeostasis and remodeling [7,62]. In this study, a small number of Mn-doped surfaces exhibited good biocompatibility with Raw264.7 cells and could stimulate the polarization of Raw264.7 cells towards the M1 phenotype by stimulating the secretion of iNOS (M1 marker) and upregulating the mRNA expressions of pro-inflammatory cytokines (TNF-α) and chemokines (CCL3, CCL5, CSF2) under mono-culture conditions. Chemokines are found to exert a great effect on modulating the recruitment of immune cells, such as macrophages, which can clear cellular debris and promote healing [63]. For example, Mn can promote the recruitment and activation of monocytes to macrophages by increasing the gene expression of CCL3 [64,65]. Polarization and secreted cytokines of macrophages have been identified as important factors regulating tissue vascularization, which plays a key role in the process of bone development and remodeling [66]. Elevated exogenous Mn ions were proposed to increase the activity of Mn-SOD, which can help alleviate the oxidative stress state of endothelial cells and restore impaired angiogenic function [67]. The Mn-induced upregulation of specific chemokines, such as CSF2, can foster blood vessel formation around the implant, providing a source of oxygen and nutrients for the cells responsible for bone formation [68]. Additionally, the Mn-induced inflammation may cause a remodeling of the lymphatic network, which can form a versatile transport network to promote cellular osteogenic activities [66]. Therefore, inflammation at a low-grade level may be beneficial for osteogenic activities.

Different from the pro-inflammatory immune microenvironment constructed in mono-culture conditions, the co-cultured macrophages on Mn-implanted surfaces exhibited a more prominent M2 phenotype compared to those on the Ti surface (Figure 6c). Although BMSCs have been revealed to transform macrophages to the M2 phenotype by their interaction via direct and indirect contact, Mn-implanted surfaces significantly enhanced the mRNA expression of anti-inflammation markers including IL-4, IL-10, CD206, and Arg-1 while downregulating pro-inflammatory marker TNF-α and chemokines such as CCL3, CCL5, and CSF2 (Figure 6c). Comparatively speaking, macrophages were more sensitive to Mn at a low concentration (0–10 ppb) than mBMSCs. After co-culture, the presence of mBMSCs and manganese in the environment together promoted the transformation of macrophages from the M1 to the M2 phenotype. This work shows an enhanced osteogenic differentiation of mBMSCs during a 10-day culture period. Nonetheless, since bone remodeling is a long-term process and the phenotypic shift of mBMSCs and macrophages occurs during that process, it may be necessary to conduct experiments over shorter or longer periods to refine the mechanism of the crosstalk between the two types of cells [4]. Moreover, the in vitro result also needs to be more rigorously validated by in vivo animal experiments in the future. Given these findings, it is believed that there is further potential for the application of Mn-doped orthopedic implants.

5. Conclusions

In this study, we utilized the PIII&D technique to successfully introduce Mn onto the surface of titanium plates and characterized the changes in the physicochemical properties. Mn-implanted surfaces, especially Mn120, exhibited elevated surface zeta potentials (pH = 7.4) and enhanced surface nano-hardness (within 110 nm). Although Mn implantation stimulated M1 phenotypes of macrophages and showed little effect on mBMSCs under mono-culture conditions, it can improve cellular osteogenic differentiation by mediating the crosstalk between mBMSCs and macrophages under co-culture conditions. Therefore, Mn-modified materials showed the potential to promote bone formation through immunomodulation.

Supplementary Materials: The following supporting information can be downloaded at https://www.mdpi.com/article/10.3390/jfb14090456/s1, Figure S1: Schematic of the co-culture experiment (Raw264.7 cells with mBMSCs); Table S1: Primers used in RT-qPCR assays.

Author Contributions: K.Y.: investigation, writing—original draft, and writing—review and editing. X.Z.: investigation. L.S.: investigation. X.L.: investigation. X.N.: investigation. Y.Q.: supervision, writing—review and editing, and funding acquisition. All authors have read and agreed to the published version of this manuscript.

Funding: This work was financially supported by the National Natural Science Foundation of China (31870944), Science and Technology Commission of Shanghai Municipality (20S31903300).

Institutional Review Board Statement: Not applicable.

Informed Consent Statement: Not applicable.

Data Availability Statement: Not applicable.

Conflicts of Interest: The authors declare no conflict of interest.

References

1. Shah, F.A.; Thomsen, P.; Palmquist, A. Osseointegration and current interpretations of the bone-implant interface. *Acta Biomater.* **2019**, *84*, 1–15. [CrossRef]
2. Loi, F.; Córdova, L.A.; Pajarinen, J.; Lin, T.-h.; Yao, Z.; Goodman, S.B. Inflammation, fracture and bone repair. *Bone* **2016**, *86*, 119–130. [CrossRef] [PubMed]
3. Allard, C. Treating inflammatory bone loss. *Nat. Rev. Mater.* **2022**, *7*, 932. [CrossRef]
4. Wynn, T.A.; Vannella, K.M. Macrophages in tissue repair, regeneration, and fibrosis. *Immunity* **2016**, *44*, 450–462. [CrossRef] [PubMed]
5. Baht, G.S.; Vi, L.; Alman, B.A. The role of the immune cells in fracture healing. *Curr. Osteoporos. Rep.* **2018**, *16*, 138–145. [CrossRef]
6. Wculek, S.K.; Dunphy, G.; Heras-Murillo, I.; Mastrangelo, A.; Sancho, D. Metabolism of tissue macrophages in homeostasis and pathology. *Cell. Mol. Immunol.* **2022**, *19*, 384–408. [CrossRef] [PubMed]
7. Schlundt, C.; Fischer, H.; Bucher, C.H.; Rendenbach, C.; Duda, G.N.; Schmidt-Bleek, K. The multifaceted roles of macrophages in bone regeneration: A story of polarization, activation and time. *Acta Biomater.* **2021**, *133*, 46–57. [CrossRef] [PubMed]
8. Orecchioni, M.; Ghosheh, Y.; Pramod, A.B.; Ley, K. Macrophage polarization: Different gene signatures in M1(LPS+) vs. classically and M2(LPS−) vs. alternatively activated macrophages. *Front. Immunol.* **2019**, *10*, 1084. [CrossRef]
9. Chen, S.; Saeed, A.; Liu, Q.; Jiang, Q.; Xu, H.; Xiao, G.G.; Rao, L.; Duo, Y. Macrophages in immunoregulation and therapeutics. *Signal Transduct. Target. Ther.* **2023**, *8*, 207. [CrossRef]
10. Tu, Z.X.; Zhong, Y.L.; Hu, H.Z.; Shao, D.; Haag, R.N.; Schirner, M.; Lee, J.; Sullenger, B.; Leong, K.W. Design of therapeutic biomaterials to control inflammation. *Nat. Rev. Mater.* **2022**, *7*, 557–574. [CrossRef]
11. Wang, T.; Bai, J.X.; Lu, M.; Huang, C.L.; Geng, D.C.; Chen, G.; Wang, L.; Qi, J.; Cui, W.G.; Deng, L.F. Engineering immunomodulatory and osteoinductive implant surfaces via mussel adhesion-mediated ion coordination and molecular clicking. *Nat. Commun.* **2022**, *13*, 160. [CrossRef]
12. Baseri, M.; Radmand, F.; Hamedi, R.; Yousefi, M.; Kafil, H.S. Immunological aspects of dental implant rejection. *Biomed. Res. Int.* **2020**, *2020*, 7279509. [CrossRef] [PubMed]
13. Zhou, P.; Xia, D.; Ni, Z.; Ou, T.; Wang, Y.; Zhang, H.; Mao, L.; Lin, K.; Xu, S.; Liu, J. Calcium silicate bioactive ceramics induce osteogenesis through oncostatin M. *Bioact. Mater.* **2021**, *6*, 810–822. [CrossRef] [PubMed]
14. Lu, L.Y.; Loi, F.; Nathan, K.; Lin, T.-h.; Pajarinen, J.; Gibon, E.; Nabeshima, A.; Cordova, L.; Jämsen, E.; Yao, Z.; et al. Pro-inflammatory M1 macrophages promote Osteogenesis by mesenchymal stem cells via the COX-2-prostaglandin E2 pathway. *J. Orthop. Res.* **2017**, *35*, 2378–2385. [CrossRef] [PubMed]
15. Zhang, Y.; Böse, T.; Unger, R.E.; Jansen, J.A.; Kirkpatrick, C.J.; van den Beucken, J.J.J.P. Macrophage type modulates osteogenic differentiation of adipose tissue MSCs. *Cell Tissue Res.* **2017**, *369*, 273–286. [CrossRef]
16. Rai, B.; Lin, J.L.; Lim, Z.X.H.; Guldberg, R.E.; Hutmacher, D.W.; Cool, S.M. Differences between in vitro viability and differentiation and in vivo bone-forming efficacy of human mesenchymal stem cells cultured on PCL-TCP scaffolds. *Biomaterials* **2010**, *31*, 7960–7970. [CrossRef] [PubMed]
17. Liu, W.; Li, J.H.; Cheng, M.Q.; Wang, Q.J.; Qian, Y.B.; Yeung, K.W.K.; Chu, P.K.; Zhang, X.L. A surface-engineered polyetheretherketone biomaterial implant with direct and immunoregulatory antibacterial activity against methicillin-resistant Staphylococcus aureus. *Biomaterials* **2019**, *208*, 8–20. [CrossRef] [PubMed]
18. Murdoch, C.C.; Skaar, E.P. Nutritional immunity: The battle for nutrient metals at the host-pathogen interface. *Nat. Rev. Microbiol.* **2022**, *20*, 657–670. [CrossRef] [PubMed]
19. Li, S.R.; Cui, Y.T.; Liu, H.; Tian, Y.H.; Wang, G.; Fan, Y.; Wang, J.W.; Wu, D.K.; Wang, Y.B. Application of bioactive metal ions in the treatment of bone defects. *J. Mater. Chem. B* **2022**, *10*, 9369–9388. [CrossRef] [PubMed]
20. Knoell, D.; Rink, L. Metal ions in immune function and host defense. *Semin. Cell Dev. Biol.* **2021**, *115*, 1–2. [CrossRef] [PubMed]
21. Jacobs, A.; Renaudin, G.; Forestier, C.; Nedelec, J.M.; Descamps, S. Biological properties of copper-doped biomaterials for orthopedic applications: A review of antibacterial, angiogenic and osteogenic aspects. *Acta Biomater.* **2020**, *117*, 21–39. [CrossRef] [PubMed]
22. Wang, P.; Yuan, Y.H.; Xu, K.; Zhong, H.S.; Yang, Y.H.; Jin, S.Y.; Yang, K.; Qi, X. Biological applications of copper-containing materials. *Bioact. Mater.* **2021**, *6*, 916–927. [CrossRef]
23. Huang, Q.L.; Ouyang, Z.X.; Tan, Y.N.; Wu, H.; Liu, Y. Activating macrophages for enhanced osteogenic and bactericidal performance by Cu ion release from micro/nano-topographical coating on a titanium substrate. *Acta Biomater.* **2019**, *100*, 415–426. [CrossRef]
24. Kitada, M.; Xu, J.; Ogura, Y.; Monno, I.; Koya, D. Manganese superoxide dismutase dysfunction and the pathogenesis of kidney disease. *Front. Physiol.* **2020**, *11*, 755. [CrossRef]
25. Wang, C.; Zhang, R.; Wei, X.; Lv, M.; Jiang, Z. Metalloimmunology: The metal ion-controlled immunity. *Adv. Immunol.* **2020**, *145*, 187–241. [CrossRef]
26. Wang, C.; Guan, Y.; Lv, M.; Zhang, R.; Guo, Z.; Wei, X.; Du, X.; Yang, J.; Li, T.; Wan, Y.; et al. Manganese increases the sensitivity of the cGAS-STING pathway for double-stranded DNA and is required for the host defense against DNA viruses. *Immunity* **2018**, *48*, 675–687.e7. [CrossRef]

27. Prasadh, S.; Gupta, M.; Wong, R. In vitro cytotoxicity and osteogenic potential of quaternary Mg-2Zn-1Ca/X-Mn alloys fo craniofacial reconstruction. *Sci. Rep.* **2022**, *12*, 8259. [CrossRef] [PubMed]
28. Wu, T.; Shi, H.; Liang, Y.; Lu, T.; Lin, Z.; Ye, J. Improving osteogenesis of calcium phosphate bone cement by incorporating with manganese doped β-tricalcium phosphate. *Mater. Sci. Eng. C* **2020**, *109*, 110481. [CrossRef] [PubMed]
29. Westhauser, F.; Wilkesmann, S.; Nawaz, Q.; Hohenbild, F.; Rehder, F.; Saur, M.; Fellenberg, J.; Moghaddam, A.; Ali, M.S.; Peukert W.; et al. Effect of manganese, zinc, and copper on the biological and osteogenic properties of mesoporous bioactive glass nanoparticles. *J. Biomed. Mater. Res. Part A* **2021**, *109*, 1457–1467. [CrossRef]
30. Chu, P.K.; Chen, J.Y.; Wang, L.P.; Huang, N. Plasma-surface modification of biomaterials. *Mater. Sci. Eng. R Rep.* **2002**, *36*, 143–206 [CrossRef]
31. Kong, L.; Smith, W.; Hao, D. Overview of RAW264.7 for osteoclastogensis study: Phenotype and stimuli. *J. Cell. Mol. Med.* **2019**, *23*, 3077–3087. [CrossRef]
32. Berghaus, L.J.; Moore, J.N.; Hurley, D.J.; Vandenplas, M.L.; Fortes, B.P.; Wolfert, M.A.; Boons, G.-J. Innate immune responses of primary murine macrophage-lineage cells and RAW 264.7 cells to ligands of Toll-like receptors 2, 3, and 4. *Comp. Immunol. Microbiol. Infect. Dis.* **2010**, *33*, 443–454. [CrossRef]
33. Kershner, R.J.; Bullard, J.W.; Cima, M.J. Zeta potential orientation dependence of sapphire substrates. *Langmuir* **2004**, *20*, 4101–4108. [CrossRef]
34. Chen, L.; Wang, D.; Qiu, J.; Zhang, X.; Liu, X.; Qiao, Y.; Liu, X. Synergistic effects of immunoregulation and osteoinduction of ds-block elements on titanium surface. *Bioact. Mater.* **2021**, *6*, 191–207. [CrossRef] [PubMed]
35. Li, K.; Xia, C.; Qiao, Y.; Liu, X. Dose-response relationships between copper and its biocompatibility/antibacterial activities. *J. Trace Elem. Med. Biol.* **2019**, *55*, 127–135. [CrossRef] [PubMed]
36. Li, G.; Fu, H.; Chen, X.; Gong, P.; Chen, G.; Xia, L.; Wang, H.; You, J.; Wu, Y. Facile and sensitive fluorescence sensing of alkaline phosphatase activity with photoluminescent carbon dots based on inner filter effect. *Anal. Chem.* **2016**, *88*, 2720–2726. [CrossRef]
37. Du, H.; Wang, Y.; Arandiyan, H.; Younis, A.; Scott, J.; Qu, B.; Wan, T.; Lin, X.; Chen, J.; Chu, D. Design and synthesis of CeO_2 nanowire/MnO_2 nanosheet heterogeneous structure for enhanced catalytic properties. *Mater. Today Commun.* **2017**, *11*, 103–111 [CrossRef]
38. Chang, S.-L.; Anderegg, J.; Thiel, P. Surface oxidation of an Al Pd Mn quasicrystal, characterized by X-ray photoelectron spectroscopy. *J. Non-Cryst. Solids* **1996**, *195*, 95–101. [CrossRef]
39. Pandey, J.; Hua, B.; Ng, W.; Yang, Y.; van der Veen, K.; Chen, J.; Geels, N.J.; Luo, J.-L.; Rothenberg, G.; Yan, N. Developing hierarchically porous MnOx/NC hybrid nanorods for oxygen reduction and evolution catalysis. *Green Chem.* **2017**, *19*, 2793–2797 [CrossRef]
40. Ruytinx, P.; Proost, P.; Van Damme, J.; Struyf, E. Chemokine-Induced Macrophage Polarization in Inflammatory Conditions. *Front. Immunol.* **2018**, *9*, 1930. [CrossRef]
41. Hamilton, J.A. GM-CSF in inflammation. *J. Exp. Med.* **2020**, *217*, e20190945. [CrossRef] [PubMed]
42. Sorgi, C.A.; Rose, S.; Court, N.; Carlos, D.; Paula-Silva, F.W.G.; Assis, P.A.; Frantz, F.G.; Ryffel, B.; Quesniaux, V.; Faccioli, L.H. GM-CSF Priming Drives Bone Marrow-Derived Macrophages to a Pro-Inflammatory Pattern and Downmodulates PGE(2) in Response to TLR2 Ligands. *PLoS ONE* **2012**, *7*, e40523. [CrossRef] [PubMed]
43. Liu, C.; Yao, Z.Y.; Wang, J.N.; Zhang, W.; Yang, Y.; Zhang, Y.; Qu, X.L.; Zhu, Y.B.; Zou, J.J.; Peng, S.S.; et al. Macrophage-derived CCL5 facilitates immune escape of colorectal cancer cells via the p65/STAT3-CSN5-PD-L1 pathway. *Cell Death Differ.* **2020**, *27*, 1765–1781. [CrossRef] [PubMed]
44. Saha, K.; Keung, A.J.; Irwin, E.F.; Li, Y.; Little, L.; Schaffer, D.V.; Healy, K.E. Substrate modulus directs neural stem cell behavior. *Biophys. J.* **2008**, *95*, 4426–4438. [CrossRef]
45. Kim, J.M.; Yang, Y.S.; Park, K.H.; Ge, X.P.; Xu, R.; Li, N.; Song, M.Y.; Chun, H.; Bok, S.; Charles, J.F.; et al. A RUNX2 stabilization pathway mediates physiologic and pathologic bone formation. *Nat. Commun.* **2020**, *11*, 2289. [CrossRef]
46. Yang, W.; Li, H.Y.; Wu, Y.F.; Mi, R.J.; Liu, W.Z.; Shen, X.; Lu, Y.X.; Jiang, Y.H.; Ma, M.J.; Shen, H.Y. ac4C acetylation of RUNX2 catalyzed by NAT10 spurs osteogenesis of BMSCs and prevents ovariectomy-induced bone loss. *Mol. Ther.-Nucl. Acids* **2021**, *26*, 135–147. [CrossRef]
47. Haase, H. Innate immune cells speak manganese. *Immunity* **2018**, *48*, 616–618. [CrossRef]
48. Gao, M.; Xie, Y.Q.; Lei, K.; Zhao, Y.; Kurum, A.; Van Herck, S.; Guo, Y.; Hu, X.; Tang, L. A manganese phosphate nanocluster activates the cGAS-STING pathway for enhanced cancer immunotherapy. *Adv. Ther.* **2021**, *4*, 2100065. [CrossRef]
49. Sridharan, R.; Cameron, A.R.; Kelly, D.J.; Kearney, C.J.; O'Brien, F.J. Biomaterial based modulation of macrophage polarization: A review and suggested design principles. *Mater. Today* **2015**, *18*, 313–325. [CrossRef]
50. Abaricia, J.O.; Farzad, N.; Heath, T.J.; Simmons, J.; Morandini, L.; Olivares-Navarrete, R. Control of innate immune response by biomaterial surface topography, energy, and stiffness. *Acta Biomater.* **2021**, *133*, 58–73. [CrossRef] [PubMed]
51. Murray, J.L. The Mn-Ti (Manganese-Titanium) system. *Bull. Alloy Phase Diagr.* **1981**, *2*, 334–343. [CrossRef]
52. Sharkeev, Y.P.; Gritsenko, B.P.; Fortuna, S.V.; Perry, A.J. Modification of metallic materials and hard coatings using metal ion implantation. *Vacuum* **1999**, *52*, 247–254. [CrossRef]
53. Jones, E.C.; En, W.; Ogawa, S.; Fraser, D.B.; Cheung, N.W. Anomalous behavior of shallow BF3 plasma immersion ion implantation. *J. Vac. Sci. Technol. B Microelectron. Nanometer Struct. Process. Meas. Phenom.* **1994**, *12*, 956–961. [CrossRef]

54. Thorwarth, G.; Hammerl, C.; Kuhn, M.; Assmann, W.; Schey, B.; Stritzker, B. Investigation of DLC synthesized by plasma immersion ion implantation and deposition. *Surf. Coat. Technol.* **2005**, *193*, 206–212. [CrossRef]
55. Yan, C.; Zeng, Q.F.; He, W.J.; Zhu, J.N. Enhanced surface hardness and tribocorrosion performance of 60NiTi by boron ion implantation and post-annealing. *Tribol. Int.* **2021**, *155*, 106816. [CrossRef]
56. Maremonti, M.I.; Panzetta, V.; Dannhauser, D.; Netti, P.A.; Causa, F. Wide-range viscoelastic compression forces in microfluidics to probe cell-dependent nuclear structural and mechanobiological responses. *J. R. Soc. Interface* **2022**, *19*, 20210880. [CrossRef]
57. Yi, B.; Xu, Q.; Liu, W. An overview of substrate stiffness guided cellular response and its applications in tissue regeneration. *Bioact. Mater.* **2022**, *15*, 82–102. [CrossRef]
58. Kubiak-Ossowska, K.; Jachimska, B.; Al Qaraghuli, M.; Mulheran, P.A. Protein interactions with negatively charged inorganic surfaces. *Curr. Opin. Colloid. Interface Sci.* **2019**, *41*, 104–117. [CrossRef]
59. Wu, Q.; Mu, Q.; Xia, Z.; Min, J.; Wang, F. Manganese homeostasis at the host-pathogen interface and in the host immune system. *Semin. Cell Dev. Biol.* **2021**, *115*, 45–53. [CrossRef]
60. Li, Y.; Pan, Q.; Xu, J.; He, X.; Li, H.A.; Oldridge, D.A.; Li, G.; Qin, L. Overview of methods for enhancing bone regeneration in distraction osteogenesis: Potential roles of biometals. *J. Orthop. Transl.* **2021**, *27*, 110–118. [CrossRef]
61. Rondanelli, M.; Faliva, M.A.; Peroni, G.; Infantino, V.; Gasparri, C.; Iannello, G.; Perna, S.; Riva, A.; Petrangolini, G.; Tartara, A. Essentiality of manganese for bone health: An overview and update. *Nat. Prod. Commun.* **2021**, *16*, 1934578X211016649. [CrossRef]
62. Bozec, A.; Soulat, D. Latest perspectives on macrophages in bone homeostasis. *Pflug. Arch.-Eur. J. Physiol.* **2017**, *469*, 517–525. [CrossRef] [PubMed]
63. Tang, P.; Wang, J.M. Chemokines: The past, the present and the future. *Cell. Mol. Immunol.* **2018**, *15*, 295–298. [CrossRef] [PubMed]
64. Shi, C.; Pamer, E.G. Monocyte recruitment during infection and inflammation. *Nat. Rev. Immunol.* **2011**, *11*, 762–774. [CrossRef] [PubMed]
65. Zhang, R.; Wang, C.; Guan, Y.; Wei, X.; Sha, M.; Yi, M.; Jing, M.; Lv, M.; Guo, W.; Xu, J.; et al. Manganese salts function as potent adjuvants. *Cell Mol. Immunol.* **2021**, *18*, 1222–1234. [CrossRef]
66. Biswas, L.; Chen, J.; De Angelis, J.; Singh, A.; Owen-Woods, C.; Ding, Z.; Pujol, J.M.; Kumar, N.; Zeng, F.; Ramasamy, S.K.; et al. Lymphatic vessels in bone support regeneration after injury. *Cell* **2023**, *186*, 382–397.e24. [CrossRef]
67. Wang, X.R.; Zhang, M.W.; Chen, D.D.; Zhang, Y.; Chen, A.F. AMP-activated protein kinase rescues the angiogenic functions of endothelial progenitor cells via manganese superoxide dismutase induction in type 1 diabetes. *Am. J. Physiol.-Endocrinol. Metab.* **2011**, *300*, E1135–E1145. [CrossRef]
68. Sielska, M.; Przanowski, P.; Pasierbińska, M.; Wojnicki, K.; Poleszak, K.; Wojtas, B.; Grzeganek, D.; Ellert-Miklaszewska, A.; Ku, M.-C.; Kettenmann, H.; et al. Tumour-derived CSF2/granulocyte macrophage colony stimulating factor controls myeloid cell accumulation and progression of gliomas. *Br. J. Cancer* **2020**, *123*, 438–448. [CrossRef]

Journal of
Functional
Biomaterials

Review

Musculoskeletal Biomaterials: Stimulated and Synergized with Low Intensity Pulsed Ultrasound

Wanru Jia [1], Zifei Zhou [2,*] and Weiwei Zhan [1,*]

[1] Department of Ultrasound, Ruijin Hospital, Shanghai Jiao Tong University School of Medicine, Shanghai 200025, China; jiawanru@126.com

[2] Department of Orthopedics, Shanghai Tenth People's Hospital, Tongji University School of Medicine, Shanghai 200072, China

* Correspondence: zhouzf@tongji.edu.cn (Z.Z.); shanghairuijin@126.com (W.Z.)

Abstract: Clinical biophysical stimulating strategies, which have significant effects on improving the function of organs or treating diseases by causing the salutary response of body, have shown many advantages, such as non-invasiveness, few side effects, and controllable treatment process. As a critical technique for stimulation, the low intensity pulsed ultrasound (LIPUS) has been explored in regulating osteogenesis, which has presented great promise in bone repair by delivering a combined effect with biomaterials. This review summarizes the musculoskeletal biomaterials that can be synergized with LIPUS for enhanced biomedical application, including bone regeneration, spinal fusion, osteonecrosis/osteolysis, cartilage repair, and nerve regeneration. Different types of biomaterials are categorized for summary and evaluation. In each subtype, the verified biological mechanisms are listed in a table or graphs to prove how LIPUS was effective in improving musculoskeletal tissue regeneration. Meanwhile, the acoustic excitation parameters of LIPUS that were promising to be effective for further musculoskeletal tissue engineering are discussed, as well as their limitations and some perspectives for future research. Overall, coupled with biomimetic scaffolds and platforms, LIPUS may be a powerful therapeutic approach to accelerate musculoskeletal tissue repair and even in other regenerative medicine applications.

Keywords: low intensity pulsed ultrasound; bone repair; musculoskeletal biomaterials; tissue engineering; regenerative medicine

1. Introduction

Designers of musculoskeletal materials commonly try to recreate hierarchical structures or to offer bone components such as bioactive proteins, minerals, and cells to facilitate and assist the formation of new bone tissue and restore its function [1–3]. In addition, the typical fracture healing process inspires novel approaches for bone tissue repair. The majority of bone scaffolds are fabricated by bioceramic, polymer, metal, or hybrid, which typically serve as a mechanical support and a 3D environment for cell adhesion, proliferation, and ingrowth [2,4–6]. Musculoskeletal biomaterials include not only transitional bone tissue engineering scaffolds for bone healing, but also biomaterials for spine fusion, bone–tendon, bone–ligament, and cartilage repair. Notably, neurogenesis and neovascularization are also crucial factors which indirectly benefit bone repair [7–10].

Although it is of potential value to incorporate biological components into bone tissue engineering strategies, the costs of such components limit their further application. Material-only approaches, which exclude biological components and instead rely on the body's own cells to promote bone regeneration, are highly beneficial for preclinical research and clinical translation [2]. Biophysical stimulating techniques, such as low intensity pulsed ultrasound (LIPUS) stimulation, have shown promise by delivering a combined effect with bioactive materials at a lower cost and a shorter cycle time [11,12]. Generally,

LIPUS is a specific type of ultrasound with a frequency of 0.045–3 MHz and intensity of 0.02–1 W/cm^2 [13–17]. The widely used parameters are 1.5 MHz frequency, 30 mW/cm^2 intensity, 200 μs pulse duration, and 20 min/d exposure time [15,18]. LIPUS has been proven to promote cellular viability, proliferation, differentiation, and migration [13,19,20]. Moreover, LIPUS has shown favorable outcomes in promoting bone fracture healing by inducing molecular, biological, and biomechanical alterations in the fracture vicinity. In addition, LIPUS has been scientifically validated to accelerate bone regeneration in cases of fresh fractures, delayed unions, non-unions, distraction osteogenesis, and musculoskeletal soft tissue injuries [17,18,21–26].

According to a meta-analysis of randomized clinical trials, LIPUS therapy may shorten the overall treatment period (mean difference = −15.236 d/cm, 95% confidence interval = −19.902 to −10.569 d/cm) for tibial distraction osteogenesis [27]. Several studies have been conducted in order to improve the growth of osteogenic cells utilizing LIPUS and osteoconductive materials [11,28–30]. In certain studies, LIPUS has been shown to improve bone growth and local blood flow in an animal model of fracture repair [31,32]. Meanwhile, it could also contribute to cell spreading, either seeding cells into the scaffold system before implantation or recruiting from graft-surrounded native tissue after implantation in vivo [33,34]. These findings accelerate the development of a clinically applicable LIPUS therapy for bone defects, allowing for the transdermal application of mechanical stress to bone defects without physically destabilizing the defect site.

The mechanisms of ultrasound (US)-responsive nanomaterials include cavitation, acoustic radiation force, acoustic droplet vaporization, hyperthermia, and free radical generation. At least one of these mechanisms can be employed by the nanomaterials [35–40]. Acoustic cavitation refers to formation of gas bubbles, pockets, caused by interaction between ultrasonicated materials with acoustic waves [39,41]. Acoustic radiation force is defined as a mechanical force generated by transferring momentum from the ultrasound wave to the medium [38]. Any particles suspended in the fluid will drift, form clusters, and attract or repel one another due to the radiation force [35]. Acoustic droplet vaporization is a process that converts superheated liquid droplets of micron-sized to gaseous microbubbles 5–6 times larger [37]. When focused US beams are targeted at certain tissue, local hyperthermia will occur by absorbing the acoustic energy [40]. Moreover, when US interacts with certain components in a water-based medium, free radical molecules are created for both therapeutic and diagnostic purposes [36,42]. In summary, all these mechanisms may lead to either positive or negative impacts on living tissues.

In this review, an overview of the enhanced or synergistic effects of LIPUS with musculoskeletal biomaterials will be provided. First, the materials that are designed to promote LIPUS-enhanced regeneration for musculoskeletal injuries are described, including bone, spinal fusion, cartilage, bone–tendon, bone–ligament, and nerve (Figure 1). Furthermore, the most promising LIPUS stimulation techniques will be discussed, including their conducting parameters during stimulus delivery. Finally, a new perspective for LIPUS application in musculoskeletal tissue engineering in the future will be proposed.

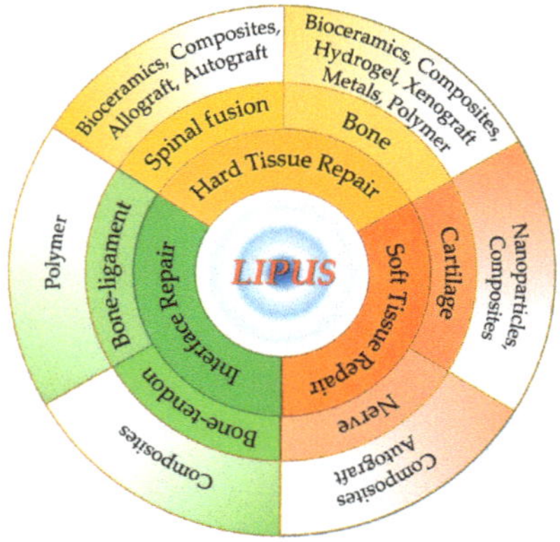

Figure 1. The application of different biomaterials in hard tissue, soft tissue, and interface repair synergized with LIPUS.

2. Bone Regeneration

Many researchers focused on the investigation of the effects of LIPUS on bone healing. Osteo-inductive biomaterials are prone to promoting bone regeneration when coupled and modulated with LIPUS appropriately. In this part, we discuss the combined effect of LIPUS with several kinds of biomaterials in promoting bone repair (Table 1).

2.1. Bioceramics

Bioceramic materials are intriguing for the fabrication of bone scaffold, since they are desirable alternatives to autogenous or heterogeneous bone grafts. Bioceramics-based scaffolds are good for osteoinduction, osteoconduction, osseointegration, and vascularization [43–47]. In a 1.5 cm rabbit ulna defect model, 20 min of LIPUS stimulation with β-tricalcium phosphate (β-TCP) bone graft increased bone formation at 4 and 12 weeks. Statistically significant differences were found in bone mineral density at 4 weeks, and in new woven bone formation at 4 and 12 weeks. VEGF expression was increased with LIPUS treatment at 4 weeks and remained elevated at 12 weeks compared with controls, while RUNX2 expression levels were elevated with LIPUS treatment at both time points [48]. Wang et al. tried to study the concordant effect of cell/scaffold and LIPUS. Rat bone mesenchymal stem cells (BMSCs) were co-cultured with β-TCP for 2 weeks to form a composite, and then such composite was subcutaneously implanted into rats for further LIPUS treatment [49]. Tests of harvested composites at 5, 10, 25, 50 days showed elevated compressive strength, increased numbers of the vessels, and upregulated expression of CD31 and OCN, denoting that LIPUS stimulation could promote osteogenesis and angiogenesis in the rat BMSCs/TCP composites [49].

The application of LIPUS did not weaken the mechanical property of porous ceramic in vitro. In vivo rabbit experiments revealed that LIPUS treatment for 2 weeks significantly increased osteoblast numbers and bone area, while LIPUS for 3 weeks significantly increased mineralized tissue volume and mineral content in the porous HA ceramic. LIPUS application increased cell migration of MC3T3-E1. It may be a good choice to fill large bone defects in a preclinical model by combining a porous inorganic scaffold with LIPUS [50].

Nagasaki et al. tried to investigate the potential synergistic effects of LIPUS and nanohydroxyapatite in the osteogenic differentiation of human adipose-derived stem cells (hADSCs) [51]. ADSCs isolated from human extirpated buccal fat pad (BFP) were mixed with porcine atelocollagen, with or without nanohydroxyapatite. Then, the mixture was transplanted into the bone defects area of mice calvarium. Experiments in vitro and in vivo revealed the combinational effects of LIPUS and nanocrystalline hydroxyapatite (nHA) in inducing the osteogenic differentiation of ADSCs into osteoblasts, and bone regeneration.

The new bone formation only occurred in the defect margin, which can be explained by the heterogeneous sources of cells and organic scaffolds; however, this research provided a novel strategy in autologous sources of ADSCs in combined application with LIPUS [51]. Using stereolithography 3D printer, PEGDA-RGDS-nHA scaffolds (polyethylene glycol diacrylate bioinks containing RGDS peptide and nHA) were fabricated, which could greatly promote hBMSC proliferation rate, filopodia growth, ALP activity, and calcium deposition under LIPUS stimulation [52].

LIPUS has been demonstrated to facilitate the cellular ingrowth in a silicon carbide porous ceramic scaffold, and enhance the proliferation and early osteogenic differentiation of MC3T3-E1 cells [53]. Another study used commercial OsteoBone^{TM} scaffold to find the potential osteogenic capacity of dental follicle cells (DFCs) after LIPUS application. The expression of osteoblast gene markers and formation of mineralized nodules and blood vessels of the DFCs/OsteoBone/LIPUS group increased in vivo. However, the subcutaneous transplantation mouse model limited the observation of potential osteogenesis effect in vivo [29].

2.2. Metals

It has been demonstrated that more bone formation was induced in a rabbit nasal bone defect model by porous titanium mesh with high density (10 holes/cm^2) than that with low density (5 holes/cm^2), and the application of LIPUS with high density titanium mesh induced a significant augmentation of new bone formation than titanium mesh only [54]. Moreover, the osseointegration of titanium implants in the rabbit metaphyseal area occurred earlier and more adequately in the LIPUS-treated group than in the control group [55]. LIPUS could also promote cell proliferation and migration on a pure titanium plate. In a rabbit study, it accelerated blood flow and maturation of type I collagen around titanium screws, and then promoted bone formation [56].

Electron beam melting (EBM)–microarc oxidation (MAO)-modified porous titanium-6aluminum-4vanadium (Ti6Al4V) scaffolds facilitated cellular filopodia/lamellipodia of MG63 cells, indicated good spreading ability. It was found in vitro that cell proliferation, attachment, and osteogenesis differentiation cultured on these scaffolds were also improved by LIPUS [33]. LIPUS also promoted ALP activity and osteocalcin levels of MC3T3-E1 cultured on porous Ti6Al4V alloy scaffolds, with neither inhibited nor stimulated effect on proliferation or attaching. Moreover, bone ingrowth, bone formation, and maturity were also enhanced in a bony defect model of rabbit mandibles [57,58].

Barium titanate ($BaTiO_3$)-coated Ti6Al4V scaffold ($BaTiO_3$/Ti6Al4V) improved the surface hydrophilicity and roughness, and showed better cellular attachment, proliferation, and osteoblast differentiation of rabbit BMSCs (rbBMSCs), which could be caused by the LIPUS-triggered piezoelectric effect of $BaTiO_3$. Enhanced osteoinduction and osseointegration were found in rabbit radius defects after scaffold implantation and LIPUS application for 6 and 12 weeks [59]. Another study of $BaTiO_3$/Ti6Al4V + LIPUS using rat BMSCs (raBMSCs) and sheep femur bone defect model verified such osteogenesis and osseointegration property [60]. However, no experiment was conducted to detect the potential current induced by LIPUS on $BaTiO_3$ in these two studies. Another study found that an induced current of 10–17.5 μA was generated by application of LIPUS on $BaTiO_3$/Ti6Al4V scaffold. The microcurrent could activate mitochondria, which might be the reason for this piezoelectric effect on cell behaviors, including better viability and adhesion. LIPUS on day 1 caused little damage to cell survival, but the piezodynamic effect weakened the damaged apoptosis and promoted cell proliferation after 4 days' application. Interestingly, continuous electric cues could be observed even 24 h after intermittent LIPUS stimulation. Thus, there was adequate current to upregulate the expression of osteogenic-related genes [28]. In a subcutaneous implantation rat model, tissues surrounding the poled $BaTiO_3$/Ti6Al4V scaffolds showed a high proportion of $CD68^+$ $CD206^+$ M2 macrophages under LIPUS stimulation. Improvements in macrophage M2 polarization and bone repair were also observed in a sheep cervical corpectomy model. The piezoelectric poled $BaTiO_3$/Ti6Al4V scaffold can regulate the immune microenvironment to enhance bone regeneration. This is achieved by inhibiting the inflammatory MAPK/JNK signal pathway and activating oxidative phosphorylation and adenosine triphosphate synthesis in macrophages (Figure 2) [61]. In conclusion, LIPUS might induce electrical signal on electroactive material. Uniform nanosphere-shaped $BaTiO_3$ piezoelectric ceramic was coated on the surface of a TC4 titanium alloy to synthesize a $BaTiO_3$/TC4 material. Microcurrent ($\approx$10 μA/cm^2) could be detected when LIPUS was applied on the $BaTiO_3$/TC4 disks. Meanwhile, the concentration of intracellular calcium ion and the Ca_V1.2 protein expression increased. All these mechanisms could introduce synergies of accelerating cell behaviors, cell attachment, migration, proliferation, and osteoblastic differentiation [34].

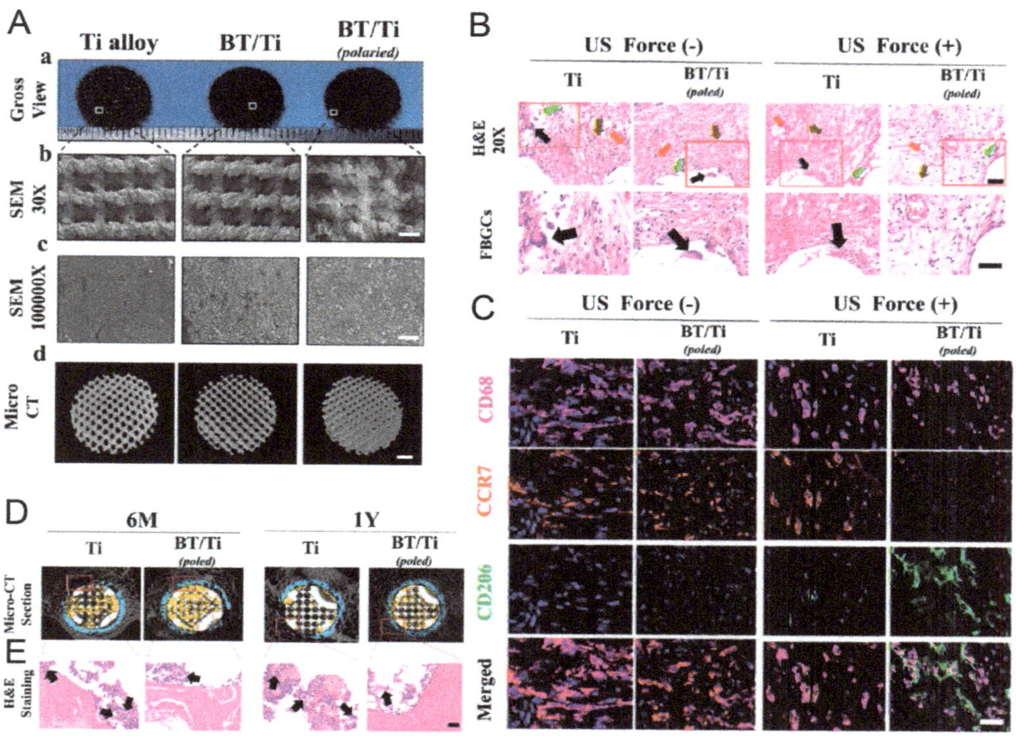

Figure 2. Morphologic views (**A**) and surface observation by SEM (**B**,**C**), and micro-CT 3D reconstruction (**D**,**E**). (**B**) Representative H&E staining images of tissues and FBGCs around scaffolds after subcutaneous implantation for 14 days. (**C**) Representative immunofluorescence images of CD68$^+$ macrophages (purple), CCR7$^+$ M1 (red), and CD206$^+$ M2 (green) macrophages in the tissues after subcutaneous implantation for 14 days. (**D**) Micro-CT section of the osteointegration around the artificial scaffolds. (**E**) Representative H&E staining images of FBGCs in bone tissues around artificial vertebral scaffolds. Reprinted with permission from Ref. [61]. Copyright 2023 Elsevier.

2.3. *Sponge- or Hydrogel-Based Composites*

Hydrogel or sponge type scaffolds were used as a platform for tissue regeneration. Collagen, especially type I collagen, is the most widely used substrate for 3D cell culture [62–64]. Hydrogel or sponge scaffolds made of collagen were widely checked for further application in LIPUS-enhanced bone regeneration [65–67]. With the help of atelocollagen sponge, LIPUS alone or simvastatin alone can promote bone regeneration; however, the combination of LIPUS and simvastatin does not induce acceleration in bone formation than LIPUS alone or simvastatin alone [65]. In rat femoral segmental defects, LIPUS enhanced radiographic healing and increased bone volume of rhBMP-2 loaded absorbable collagen sponges. Lower doses (1.2 and 6 μg) of rhBMP-2 delivery induced bone formation, while higher dose (12 μg) induced callus maturation [66]. In addition, LIPUS exposure of seeded hMSCs on magnesium-HA/collagen I hybrid could improve cell colonization and osteogenic differentiation [67].

The treatment of LIPUS showed no influence on cell proliferation within collagen I hydrogel. The elevated gene expression of ALP and osteocalcin denoted that the application of LIPUS could enhance the osteogenetic differentiation of collagen I-encapsulated MC3T3 preosteoblasts. After the BMSCs were localized in type I collagen hydrogels, LIPUS could induce undifferentiated BMSCs to the osteoblastic lineage, and also in vivo fracture healing [68]. Notwithstanding, stiffer collagen I hydrogels could reduce or reverse such osteoblastic response [69]. Either encapsulating cells within a flexible hydrogel or LIPUS exposure could induce high cyclooxygenase 2 (COX-2) and prostaglandin E2 (PGE2) expression. The cumulative higher expression of COX-2 and PGE2 could be observed after combining the two distinct conditions [70].

Moreover, the positive synergistic effect of LIPUS and RGD on the enhancement of proliferation and differentiation of hMSCs was observed. With the hybrid use of LIPUS with RGD, a significant increase was obtained in cell numbers, ALP activity, and mineralized nodule formation assay. With LIPUS, RGD-grafted oxidized sodium alginate/N-succinyl chitosan (RGD–OSA/NSC) hydrogel presented good biological properties in the attachment, proliferation, and osteogenic differentiation of human BMSCs (hBMSCs), suggesting that by combining RGD modification with LIPUS, a high level of bone formation and vascularization would be achieved [71].

2.4. GBR/GTR and Xenograft

Guided tissue regeneration (GTR) and guided bone regeneration (GBR) procedures were initially applied to regenerate periodontal tissue, and further used in bone tissue engineering. Generally, GTR and GBR adopted scaffolds or membranes to prevent growth of epithelial and connective tissues into the bone defect, so as to facilitate bone reconstruction. It has been found that LIPUS + collagen barrier membranes can facilitate the osteoblastic differentiation of dog periodontal ligament cells in vitro and promote new alveolar bone formation in vivo [72]. New bone maturation can be accelerated by LIPUS after the implantation of polytetrafluoro ethylene membranes on the surface of a bone defect [73]. The asymmetrically porous membrane was another choice for guided bone regeneration. The cross section of asymmetrically porous polycaprolactone/pluronic F127 membrane showed a column-shaped pore structure. The exterior surface had nanosized pores (≈100 nm) to prevent the infiltration of dense connective tissue but benefit from the permeation of nutrients, while the interior surface had microsized pores (≈100 μm) to improve ingrowth of new bone tissue. The selective permeability, hydrophilicity, and osteoconductivity allowed this membrane to achieve a favorable induction of osteogenesis [74]. In addition, metal membranes showed better induction of new bone than polymer membranes. In vivo, combined with LIPUS, more new bone was observed in rat calvarium defects with a cover of titanium membranes than in those with a cover of GC membrane [75]. LIPUS promoted the repair of periodontal bone defects in beagle dogs, where the bone defect was transplanted with Bio Gide® collagen membrane + autogenous bone graft [76].

Before implantation of a cell-seeded scaffold, pre-treatment of such scaffold with LIPUS could probably facilitate cell ingrowth and thus accelerate fracture healing and tissue regeneration. For MC3T3 cells in 3D trabecular bone scaffold, LIPUS treatment yielded enhanced calcific deposition, but reduced proliferation [77].

2.5. Polymers or Microbubbles

The mechanical effects of LIPUS could be amplified by adding microbubbles into cell culture, such as local shear forces and controllable mechanical stress in cells. Yao et al.

reported cyclic arginine-glycine-aspartic acid-modified nanobubbles (cRGD-NBs), which could target BMSCs mediated by integrin receptors. LIPUS/cRGD-NBs could promote the osteogenic differentiation of BMSCs induced by polymerization of actin microfilaments, TRPM7 regulation (Figure 3), and extracellular Ca^{2+} influx [78]. Microbubbles coated by a monolayer of lipids have been approved by the FDA for contrast-enhanced ultrasound imaging [79]. Integrating LIPUS and lipid mixture-coated microbubbles was proven to be effective in boosting cell proliferation and osteogenic differentiation of hBMSCs which were cultured on 3D printed porous poly(lactic acid) scaffolds. The microbubbles could maintain stable structure during LIPUS exposure. Sustained oscillations by LIPUS demonstrably contributed to the transmission of ultrasound energy toward surrounding cells of microbubbles [80]. Specific three-dimensional architectures of 3D-printed scaffolds are good for the acoustic wave transmission of LIPUS. Cells attached on the 3D-printed scaffolds receive more sonic stimulation, and thus tend to exhibit active cellular activity.

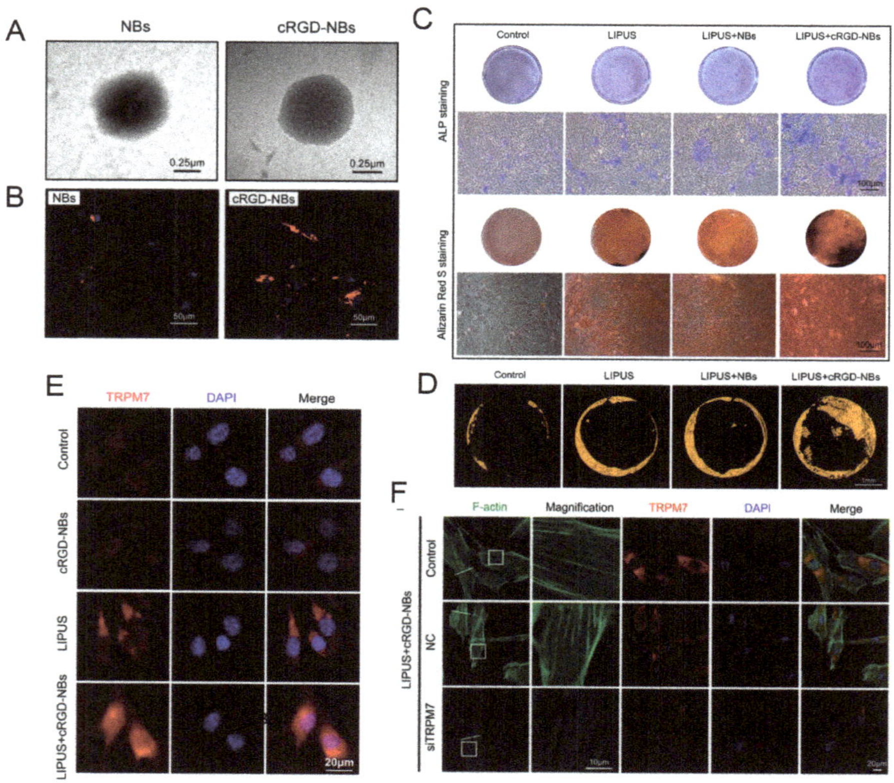

Figure 3. (**A**) Representative TEM images of NBs and cRGD-NBs. (**B**) Representative images of BMSCs co-incubated with Dil-labeled NBs and cRGD-NBs. (**C**) Alkaline phosphatase staining on day 7 and Alizarin Red S staining on day 21. (**D**) Micro-CT images of mouse calvarial bone regeneration after 4 weeks of treatment. (**E**) Representative immunofluorescence images of BMSCs co-incubated with cRGD-NBs and LIPUS, TRPM7 (red), and DAPI (blue). (**F**) Representative immunofluorescence images of BMSCs co-incubated with cRGD-NBs and LIPUS treatment for 4 h, TRPM7 (red), F-actin (green), and DAPI (blue). Reprinted from Ref. [78].

Huang demonstrated the ability of poly-L-lactic acid (PLLA) electrospun nanofibrous membrane coupled with LIPUS in enhancing the development of nascent bone using the rabbit tibia defect model [81]. Moreover, combining LIPUS with lipid microbubbles on poly (lactic-glycolic acid copolymer) (PLGA)/α-tricalcium phosphate (TCP) 3D-printed scaffolds can also enhance the growth and osteogenesis of BMSCs [82]. Ramie-based carboxymethyl cellulose (CMC) displayed cytocompatity of MC3T3-E1 cells, and synergistic effects caused by LIPUS and CMC further promoted cellular proliferation and osteogenic differentiation [83].

Table 1. LIPUS + biomaterials for bone repair.

No	Biomaterials	Constituent	Evidence In Vitro						Evidence In Vivo					Ref.
			Cell	Prolif	Adhes	Migra	Osteog Differ		Animal	Osteo-ind	Osteo-cond	Osseo-int	Angio-ge	
1	Bioceramics	TCP	/	/	/	/	/	/	Rabbit	+	/	/	+	[48]
2	Bioceramics	Hydroxyapatite	MC3T3-E1	/	/	+	+	/	Rabbit	+	+	/	/	[50]
3	Metals	Titanium	/	/	/	/	/	/	Rabbit	/	/	+	/	[55]
4	Polymers	ePTFE	/	/	/	/	/	/	Dog	+	/	/	/	[73]
5	Metals	Titanium	/	/	/	/	/	/	Rat	+	/	/	/	[75]
6	Polymers	PCL/F127	/	/	/	/	/	/	Rat	+	+	/	/	[74]
7	Metals	Titanium	MG63	+	+	/	+	/	Rabbit	+	/	+	/	[56]
8	Composites	Bio Gide®/autogenous bone graft	/	/	/	/	/	/	Dog	+	/	/	/	[76]
9	Composites	raBMSCs/TCP	/	/	/	/	/	/	Rat	+	/	/	+	[49]
10	Composites	Sim@ACS	/	/	/	/	/	/	Rabbit	+	/	/	/	[65]
11	Hydrogel	RGD–OSA/NSC	hBMSCs	+	+	/	+	Vascul	/	/	/	/	/	[71]
12	Composites	rhBMP-2@ACS	/	/	/	/	/	/	Rat	+	/	/	/	[66]
13	Composites	Atelocollagen/ nanohydroxyapatite	hADSCs	/	/	/	+	/	Mouse	+	/	/	/	[51]
14	Bioceramics	Silicon carbide	MC3T3-E1	+	+	+	+	/	/	/	/	/	/	[53]
15	Composites	PEGDA-RGDS-nHA	hBMSCs	+	+	/	+	/	/	/	/	/	/	[52]
16	Hydrogel	Collagen I	MC3T3	-	/	/	+	/	/	/	/	/	/	[69]
17	Metals	Titanium	/	/	/	/	/	/	Rabbit	+	/	/	/	[54]
18	Composites	MgHA/Col I	hMSCs	+	/	/	+	/	/	/	/	/	/	[67]
19	Metals	Ti6Al4V	MC3T3-E1	-	-	/	+	/	Rabbit	+	+	/	/	[57]
20	Polymer	PLLA	/	/	/	/	/	/	Rabbit	+	/	/	/	[81]
21	Bioceramics	OsteoBone™	DFCs	/	+	/	+	/	Mouse	/	/	/	+	[29]
22	Xenograft	Trabecular bone	MC3T3	-	/	/	+	/	/	/	/	/	/	[77]
23	Composites	Collagen	dPDLCS	/	/	/	+	/	Dog	+	/	/	/	[72]
24	Composites	LMBs + PLA	hBMSCs	+	/	/	+	/	/	/	/	/	/	[80]
25	Metals	Ti6Al4V	MG63	+	+	/	+	/	/	/	/	/	/	[33]
26	Metals	Ti6Al4V	MC3T3-E1	-	/	/	+	/	Rabbit	+	+	/	/	[58]
27	Metals	$BaTiO_3$/Ti6Al4V	rbBMSCs	+	+	/	+	/	Rabbit	+	/	+	/	[59]
28	Metals	$BaTiO_3$/Ti6Al4V	raBMSCs	/	/	/	+	/	Sheep	+	+	+	/	[60]
29	Metals	$BaTiO_3$/Ti6Al4V	raBMSCs	+	+	/	+	/	/	/	/	/	/	[28]
30	Metals	$BaTiO_3$/TC4	MC3T3-E1	+	+	+	+	/	/	/	/	/	/	[34]
31	Metals	$BaTiO_3$/Ti6Al4V	RAW264.7 MC-3T3	/	/	+	+	Polari	Rat Sheep	+	+	+	/	[61]
32	Composites	Carboxymethyl cellulose	MC3T3-E1	+	/	+	+	/	/	/	/	/	/	[83]
33	Composites	cRGD-NBs	mBMSCs	/	/	/	+	/	/	/	/	/	/	[78]
34	Composites	PLGA/TCP	raBMSCs	+	+	/	+	/	/	/	/	/	/	[82]
35	Hydrogel	Collagen I	/	/	/	/	/	/	Rat	+	/	/	/	[68]

Prolif: proliferation, Adhes: adhesion, Migra: migration, Osteog Differ: osteogenic differentiation, Vascul: vascularization, Polari: polarization, Osteoind: osteoinduction, Osteocond: osteoconduction, Osseoint: osseointegration, Angioge: angiogenesis. hBMSCs: human bone mesenchymal stem cells, hADSCs: human adipose-derived stem cells, DFCs: dental follicle cells, dPDLCS: dog periodontal ligament cells, rbBMSCs: rabbit bone mesenchymal stem cells, raBMSCs: rat bone mesenchymal stem cells. TCP: tricalcium phosphate, ePTFE: expanded polytetrafluoro ethylene, PCL/F127: polycaprolactone/pluronic F127, Sim@ACS: simvastatin loaded atelocollagen sponge, RGD: arginine-glycine-aspartic acid, RGD–OSA/NSC: RGD-grafted oxidized sodium alginate/N-succinyl chitosan, rhBMP-2@ACS: rhBMP-2 loaded absorbable collagen sponges, PEGDA: polyethylene (glycol) diacrylate, nHA: nanocrystalline hydroxyapatite, MgHA/Col I: magnesium-HA/collagen I, Ti6Al4V: titanium-6aluminum-4vanadium, PLLA: poly(L-lactic acid), DFBA: demineralized freeze-dried bone allograft, LMBs + PLA: lipid-coated microbubbles + poly(lactic acid) porous scaffolds, cRGD-NBs: cyclic arginine-glycine-aspartic acid-modified nanobubbles, PLGA/TCP: poly(lactic-glycolic acid copolymer)/α-tricalcium phosphate. +: Positive, -: Invalid or negative, /: not tested.

3. Spinal Fusion

In addition to the effective promotion of fracture repair, including new fractures, the method may also be of use in delayed union or nonunion and bone defects [84–87]. Meanwhile, LIPUS therapy may be a useful means to ensure successful spine fusion (Table 2). Several studies focused on spinal fusion using LIPUS and autologous bone graft. LIPUS treatment improved the lumber fusion rate of autologous iliac bone graft after 12 weeks of implantation.

The ABG + LIPUS group achieved 100% fusion rate, both in radiographic and histologic fusion, while the ABG group achieved 78% radiographic fusion and 44% histologic fusion [88]. Another study verified the conclusion in a rabbit model of posterolateral intertransverse process spine arthrodesis using muscle-pediculated bone grafts [89]. In a rabbit lumbar posterolateral fusion model, augmented by LIPUS, stiffer fusion mass and an analogous fusion rate can be achieved by laminectomy chip bone graft (LCBG) than those of an AIBG [90]. In nicotine-administrated rabbit, LIPUS could not promote fusion rate without any implantation, and remained at 0% in the control group with no implantation and LIPUS. However, LIPUS could increase the fusion rate from 29% to 57% in the AIBG-implanted rabbits [91].

Hui et al. established a posterior spinal fusion model in New Zealand white rabbit, to evaluate the synergistic effects of LIPUS by implanting porous TCP bioceramic scaffolds and mesenchymal stem cells (MSCs) [92]. They found that LIPUS could enhance endochondral ossification at the fusion site and bone formation with porous TCP scaffold which was impregnated with MSCs. Thus, it was LIPUS that could achieve better osseointegration between the host bone and implanted composites [92]. LIPUS can increase rabbit spinal posterolateral fusion, bone density, trabecular bone formation, and accelerate bone ingrowth into hydroxyapatite ceramics [93]. Interestingly, in a rabbit model of posterolateral lumbar fusion, there was no significant difference in the number of chondrocytes and relative gray-scale between the hydroxyapatite and the AIBG [94].

Generally, demineralized freeze-dried bone allograft (DFDBA) was usually used for studies on LIPUS-accelerated spinal fusion. Stimulated by LIPUS, a number of type H vessels could be observed in the fusion mass of rat spinal fusion model, and more osteoblasts were located on the bone callus of the allograft and were enclosed by type H vessels (Figure 4) [95]. Further study indicated that LIPUS could promote not only osteoblast differentiation but also cell migration of osteoblast-like MG63 cells, which contributed to DFDBA-induced spinal fusion. The upregulated sonic hedgehog (Shh) signal pathway was involved in those cell behaviors. In contrast, inhibited Shh signaling reduced the migratory and proliferative ability of MG63 cells and impeded the efficacy of LIPUS treatment [96]. Cell experiments of Raw264.7 cells and bone marrow-derived macrophages (BMDM) indicated that the polarization changes of macrophages were found from inflammatory type M1 to resident type M2 after LIPUS application. The authors deemed that the macrophages' earlier polarization transition might be one cause of the confirmed effect of DFDBA + LIPUS on spinal fusion [97]. Overall, multiple factors, including vessel formation, Shh signaling, and polarization transition, may be involved in the LIPUS-enhanced spinal fusion of DFDBA.

During callus formation or bone remodeling, calcitonin gene-related peptide (CGRP)-positive sensory nerve fibers proliferated rapidly and may play an important role in bone repair [98,99]. However, sensory innervation decreased or even disappeared when the delayed fracture or non-union existed [100,101]. The rhBMP-4-loaded porous poly-D, L-lactic acid blocks were implanted in the rabbit under bilateral posterolateral intertransverse process fusion. After LIPUS treatment, the number and density of CGRP-positive nerve fibers were higher in newly formed cartilage and bone tissue. Thus, LIPUS promoted the growth of CGRP sensory nerves into heterotopic bone, thereby contributing to the promotion of LIPUS on ectopic ossification [102]. Another study from a rat spinal fusion model confirmed that CGRP innervation located closely surrounding the demineralized freeze-dried bone allograft and newly formed cartilage [103].

Table 2. LIPUS + biomaterials for spinal fusion.

No	Biomaterials	Constituent	Evidence In Vitro				Evidence In Vivo							Ref.
			Cell	Prolif	Migra	Polari	Animal	CGRP Innerv	Osteo-ind	Osteo-cond	Osseo-int	Angio-ge	
1	Autograft	AIBG	/	/	/	/	Dog	/	+	/	/	/	[88]
2	Autograft	MPBG	/	/	/	/	Rabbit	/	+	/	/	/	[89]
3	Bioceramics	HA	/	/	/	/	Rabbit	/	+	+	/	/	[93]
4	Composites	BMP4/PDLA	/	/	/	/	Rabbit	+	+	/	/	/	[102]
5	Composites	DFDBA	/	/	/	/	Rat	+	+	/	/	/	[103]

Table 2. *Cont.*

No	Biomaterials	Constituent	Evidence In Vitro				Evidence In Vivo						Ref.
			Cell	Prolif	Migra	Polari	Animal	CGRP Innerv	Osteo-ind	Osteo-cond	Osseo-int	Angio-ge	
6	Bioceramics	TCP	/	/	/	/	Rabbit	/	+	+	+	/	[92]
7	Autograft	LCBG	/	/	/	/	Rabbit	/	+	/	/	/	[90]
8	Autograft	AIBG	/	/	/	/	Rabbit	/	+	/	/	/	[94]
9	Autograft	AIBG	/	/	/	/	Rabbit #	/	+	/	/	/	[91]
10	Allograft	DFDBA	/	/	/	/	Rat	/	+	/	/	+	[95]
11	Allograft	DFDBA	MG63	+	+	/	Rat	/	+	/	/	/	[96]
12	Allograft	DFDBA	Raw264.7, BMDM	/	/	+	Rat	/	+	/	/	/	[97]

Prolif: proliferation, Adhes: adhesion, Migra: migration, Chondr Differ: chondrogenic differentiation, Polari: polarization, Innerv: innervation, Chondrogen: chondrogenesis, Vascul: vascularization, CGRP: calcitonin gene-related peptide, Osteoind: osteoinduction, Osteocond: osteoconduction, Osseoint: osseointegration, Angioge: angiogenesis. AIBG: autologous iliac bone graft, MPBG: muscle-pediculated bone grafts, HA: hydroxyapatite, BMP4/PDLA: BMP4-loaded poly-D,L-lactic acid, DFDBA: demineralized freeze-dried bone allograft, TCP: tricalcium phosphate, LCBG: laminectomy chip bone graft. +: Positive, /: not tested, #: nicotine-administered.

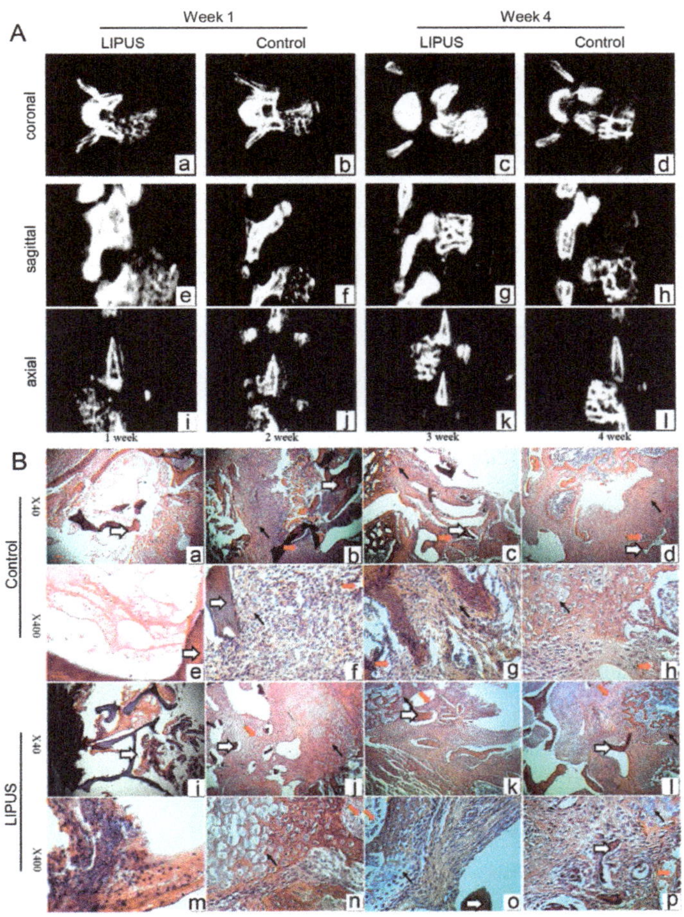

Figure 4. (**A**) Micro-CT scan views at 1 and 4 weeks post-surgery in the LIPUS group and control group. (**B**) H&E photomicrographs of the fusion mass in control and LIPUS-treated rats at 1–4 weeks. White arrows denotes allografts. Fat red arrows denotes microvessels. Thin red arrows denotes chondrocytes. Reprinted from Ref. [95].

4. Osteonecrosis/Osteolysis

Non-medical LIPUS has been used in common musculoskeletal disorders, including osteoporosis, osteonecrosis, and osteolysis [104,105]. By increasing bone formation and decreasing bone resorption, LIPUS was found to counteract the bone loss effects induced by spinal cord injury. The osteogenic effects of LIPUS lay in partially restoring endochondral ossification during callus formation, which would finally result in newly formed tissue with enhanced microarchitecture and mechanical integrity [106].

In a steroid-associated osteonecrosis rabbit model, LIPUS was proven to promote bone regeneration by increasing osteogenesis and neovascularization [32]. The potential biomechanical mechanism of LIPUS in the treatment of disuse osteoporosis may be the mechanical micro-environment improvement of trabecular bone and osteoblasts [107]. Pilot studies have been conducted regarding the combined effect of biomaterials and LIPUS (Figure 5) [108,109].

Periprosthetic osteolysis was the leading cause of polyethylene artificial joint invalidation [110,111]. Yan et al. reported that LIPUS could prevent or delay the polyethylene debris-caused osteolysis. The changes of shear strength, bone mineral density (BMD) and histopathology indicated that LIPUS-induced bone growth reversed the polyethylene-caused periprosthetic osteolysis [108], and the underlying mechanism may lie in the stimulation of bone tissue growth and inhibition of fibroblast growth. Further studies are needed to determine whether osteoclasts played an essential role in the LIPUS treatment of periprosthetic osteolysis [108].

Corticosteroid use is one of the major risks of osteonecrotic lesions; LIPUS treatment alone was validated to contribute to the alleviation of osteonecrosis [112,113], so it was a promising strategy to take advantage of the biomaterials-enhanced effect after LIPUS intervention. In another steroid-induced osteonecrosis model, bone morphogenetic protein-2 (BMP-2)-loaded poly-L-lactic acid/polylactic-co-glycolic acid/poly-ε-caprolactone (PLLA/PLGA/PCL) composite scaffolds stimulated by LIPUS could facilitate osteoblast differentiation, vascularization, and bone formation [109].

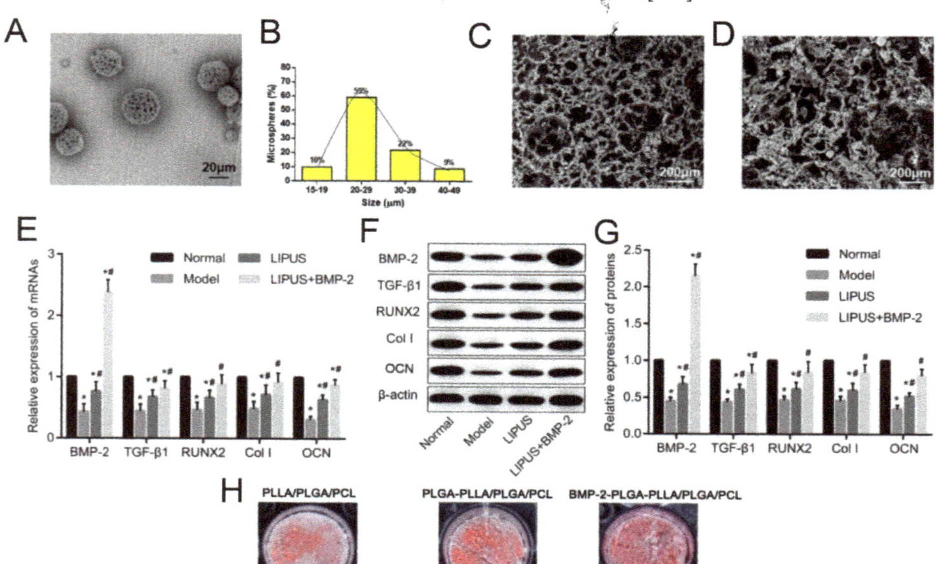

Figure 5. (**A**) TEM images of BMP-2-loaded PLGA microspheres. (**B**) Particle size distribution of BMP-2-loaded PLGA nanoparticles. (**C**) Surface morphology of the PLLA/PLGA/PCL scaffold. (**D**) Morphology of the BMP-2-loaded PLGA-PLLA/PLGA/PCL composite scaffold. (**E**) Measurement

of BMP 2, TGF β1, RUNX 2, and Col I and OCN mRNA expression in tissues by RT-PCR. (**F**,**G**) Measurement of BMP 2, TGF β1, RUNX 2, Col I, and OCN protein expression in tissues by Western blotting. * denotes $p < 0.05$ compared with Normal group. # denotes $p < 0.05$ compared with Model group. (**H**) Measurement of calcium deposition by alizarin red staining in MC3T3 E1 cell lines. Reprinted from Ref. [109].

5. Cartilage Repair

Chen et al. designed a four-layer scaffold, including layer 1: for cartilage repair, layer 2: for cartilage calcification, layer 3: for spatial distribution restriction of cells, and layer 4: for bone repair [114]. The hybrid use of multiple growth factors and LIPUS treatment exhibited good potential in facilitating vascularization and osteochondral repair [114] (Table 3).

In rabbit articular cartilage defects, LIPUS promoted the hyaline chondroid tissue formation after transplantation of allogeneic chondrocytes–calcium alginate gel composite (C-CAG). The smooth surface and integration degree optimized such effect in the LIPUS/C-CAG group. In line with the gross appearance, histological observation found that the collagen II positive area in the LIPUS/C-CAG group was larger than that in the model group and C-CAG group [115].

Combined with LIPUS, liposome-encapsulated rapamycin (L-rapa) not only increased proteoglycan production in human normal chondrocytes, but also improved type II collagen production. Moreover, in human osteoarthritis (OA) chondrocytes (HOACs), L-rapa + LIPUS upregulated mRNA expression or synthesis of aggrecan, type II collagen, and proteoglycan, while inhibiting the expression of MMP-13 and IL-6. Immunohistochemical findings from spontaneous OA Dunkin-Hartley guinea pig models proved significant enhancement of glycosaminoglycans and type II collagen in articular cartilage in the L-rapa + LIPUS group. Moreover, decreased expression of MMP-13 was principally consistent with those found in HOACs in vitro. The verified results in vitro and in vivo evidently ascertained that the L-rapa combined with LIPUS showed promising anabolic and anti-catabolic activities against OA [116].

However, one study reported that LIPUS had limited potential in stimulating the synthesis of sulphated glycosaminoglycan from bovine articular chondrocytes which were cultured in monolayer or agarose constructs [117]. Composites of bovine chondrocyte-fibrinogen exhibited no significant difference in neocartilage formation between LIPUS-treated and sham-treated groups [118]. Subcutaneously implanted composites in the backs of nude mice might account for the similar cartilage maturation and regeneration stability.

The eradication of reactive oxidative stress (ROS) could relieve chondrocyte apoptosis and extracellular matrix (ECM) degradation, which were pathological changes in cartilage suffering from osteoarthritis. Prussian blue nanoparticles (PBNPs) + LIPUS application drastically reversed lipopolysaccharide (LPS)-induced cellular ROS level and apoptosis rate by activating the PI3K/Akt/mTOR pathway. Meanwhile, PBNPs/LIPUS combination treatment resulted in the inhibition of IL-1β and MMPs by the suppression of JNK/c-Jun signal pathway in LPS-incubated chondrocytes. The anterior cruciate ligament was transected to construct a knee osteoarthritis rabbit model. Consistent with the result from chondrocytes, PBNPs + LIPUS application could activate the PI3K/Akt/mTOR signaling and suppress the JNK/c-Jun axis, leading to reverse cellular apoptosis and ECM degradation, which in turn provided an exciting repair of femoral condylar cartilage (Figure 6) [119].

Microbubbles, one kind of clinically approved agent for contrast-enhanced ultrasound imaging, could be combined with LIPUS for further application in bone tissue engineering. Along with LIPUS, lipid-coated, perfluorobutane-filled microbubbles could boost the proliferation properties and chondrogenic differentiation of human mesenchymal stem cells (hMSCs) which were cultured on 3D printed poly-(ethylene glycol)-diacrylate (PEG-DA) hydrogel scaffold (LPM + PEG-DA). The hMSCs produced more glycosaminoglycan (GAG) of 17% and type II collagen of 78% in the LIPUS+ microbubbles + LPM + PEG-DA group, whereas in the LIPUS group, they were 5% and 44% [120].

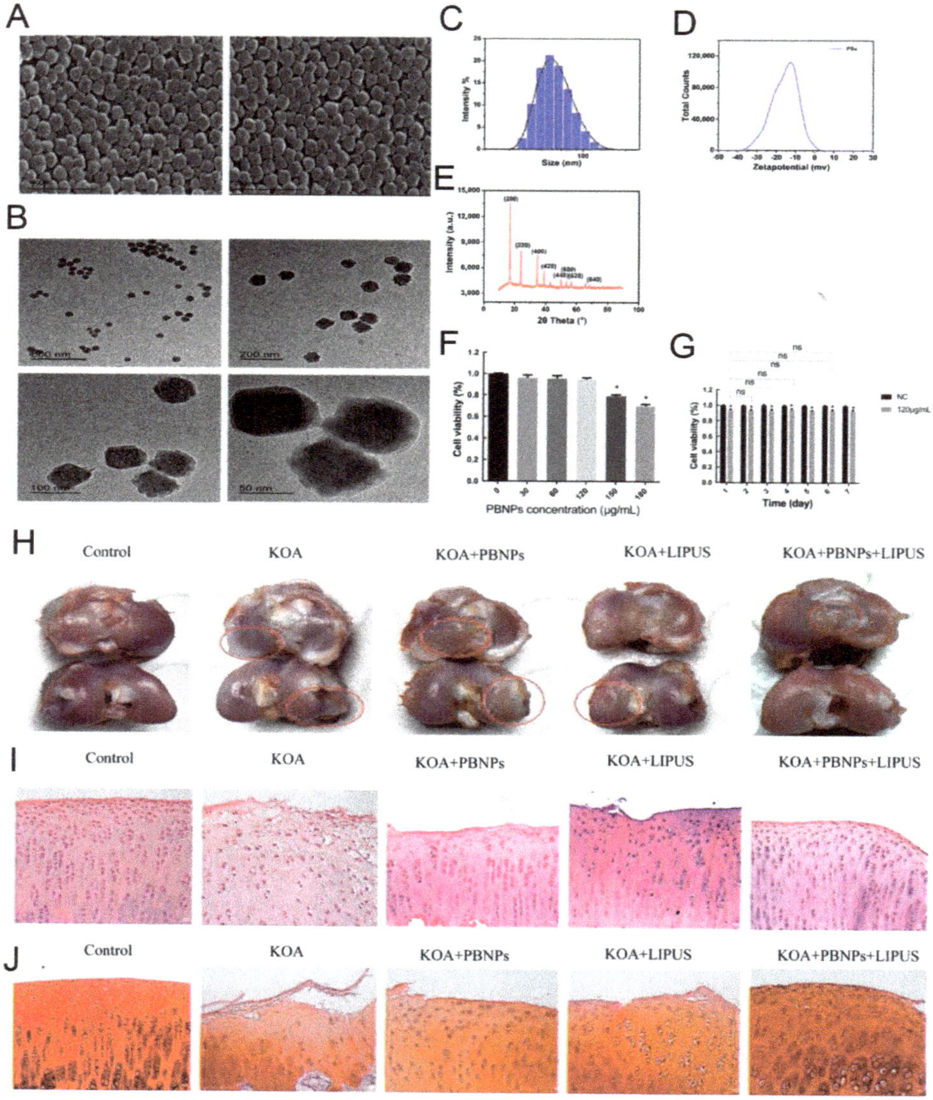

Figure 6. SEM image (**A**), TEM image (**B**), size distribution profiles (**C**), Zeta potential (**D**), and X-ray diffractometer result (**E**) of PBNPs. (**F**) Cell viability of various PBNP concentrations. (**G**) Effect of PBNPs (120 μg/mL) on cells viability for 7 d. Macroscopic observation (**H**) and H&E staining (**I**) and Safranin O staining (**J**) of effects of PBNPs/LIPUS on repair of the femoral condylar cartilage. * $p < 0.05$ vs. 0 μg/mL PBNPs group. ns: no significance. Reprinted from Ref. [119].

Table 3. LIPUS + biomaterials for chondral/osteochondral repair.

No	Biomaterials	Constituent	Evidence In Vitro			Evidence In Vivo				Ref.
			Cell	Prolif	Chondr Differ	Animal	Chondrogen	Osteo-ind	Angio-ge	
1	Composites	C-CAG	/	/	/	Rabbit	+	/	+	[115]
2	Composites	LPMBs + PEG-DA	hMSCs	+	+	/	/	/	/	[120]
3	Composites	Four-layers scaffold	/	/	/	Rabbit	+	+	+	[114]
4	Composites	L-rapa	Human Chondrocyte	-	+	Pig	+	/	/	[116]
5	Nanoparticles	PBNPs	Rabbit Chondrocyte	#	/	Rabbit	+	/	/	[119]

Prolif: proliferation, Chondr Differ: chondrogenic differentiation, Chondrogen: chondrogenesis, Osteo-ind: osteoinduction, Angio-ge: angiogenesis. hMSCs: human mesenchymal stem cells. C-CAG: chondrocytes-calcium alginate gel, LPMBs + PEG-DA: lipid-coated, perfluorobutane-filled microbubbles and poly-(ethylene glycol)-diacrylate (PEG-DA) hydrogel scaffold, Four-layers scaffold: first layer: hydrogel of oxidized sodium alginate and N-succinyl chitosan (OSA/NSC) loaded with FGF-2, BMP-2, and TGF-1; second layer: hydrogel of OSA/NSC loaded with micro hydroxyapatite (μHA) and wnt/β-catenin; third layer: PCL/PEG electrospun fiber membrane; fourth layer: porous composite of SA/nano HA/BMP-2-loaded coaxial short fibers, L-rapa: liposome-encapsulated rapamycin, PBNPs: Prussian blue nanoparticles. +: Positive, -: Invalid or negative, /: not tested, #: Reverse LPS-induced apoptosis.

6. Bone–Ligament or Bone–Tendon Repair

Although artificial ligament grafts were a viable strategy for replacing autologous grafts, single artificial ligament grafts frequently resulted in poor integration. It was vital to design a strategy that was both effective and quickly useful for promoting graft-bone healing of artificial ligaments (Table 4). Liu et al. recently organized a study to investigate the effects of LIPUS on polyethylene terephthalate (PET) artificial ligament concerning cell behaviors in vitro and osseointegration in the extra-articular graft-bone healing model (Figure 7) [121]. LIPUS promoted the cell proliferation, adherence, and osteoblastic differentiation of MC3T3-E1 preosteoblasts seeded on PET sheets. Meanwhile, in vivo study of rabbits confirmed the promoting effect of LIPUS on bone formation, and enhanced effect on graft-bone healing, such as less fibrous tissue, narrower interface, direct contact and higher ultimate failure load [121].

LIPUS treatment appeared to accelerate bone–tendon interface healing after liposomal clodronate or liposomes injection. Additionally, liposomes + LIPUS exhibited significantly more fibrocartilage than liposomal clodronate + LIPUS. Biomechanical tests of mouse supraspinatus muscle–supraspinatus tendon–humerus structure was in line with the histological results [122].

LIPUS could facilitate the osteogenesis and microvascular formation of periodontal ligament stem cells [123,124]. Stimulated by LIPUS, autologous ADSC transplantation with fibrin can lead to superior bone–tendon healing quality in the patella–patellar tendon junctions when compared with LIPUS or ADSCs alone. Compared with other groups at postoperative 8 and 16 weeks, the LIPUS + ADSCs group showed more regeneration and maturity both in fibrocartilage layer and new bone histologically, and significantly higher ultimate failure load and stiffness biomechanically [125].

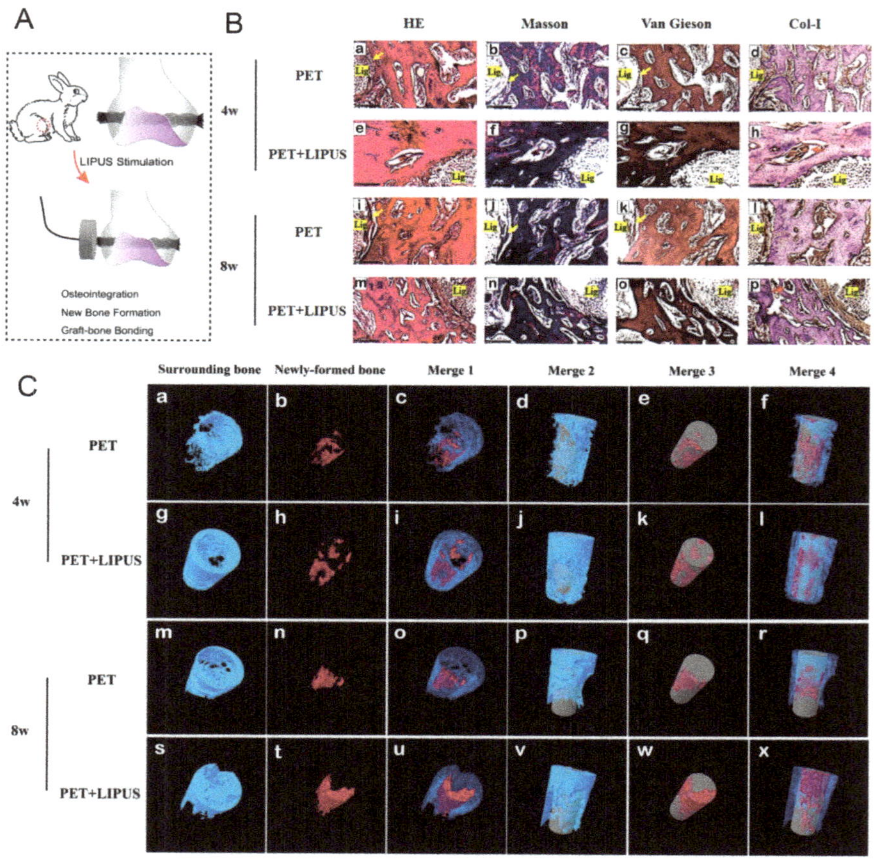

Figure 7. (**A**) Schematic illustration for the in vivo evaluation of LIPUS on the graft-bone healing of PET artificial ligaments. (**B**) HE, Masson, and van Gieson staining and Col-I immunohistochemical staining for the surrounding bone at 4 and 8 weeks postoperatively. Fibrous tissue (yellow arrow), Col-I (red arrow). (**C**) Micro-CT analysis of the surrounding bone and newly formed bone at 4 weeks and 8 weeks postoperatively. Blue, surrounding bone. Red, newly formed bone. Gray, PET graft. Reprinted with permission from Ref. [121]. Copyright 2022 American Orthopaedic Society for Sports Medicine.

Table 4. LIPUS + biomaterials for bone–ligament/bone–tendon repair.

No		Biomaterials	Constituent	Evidence In Vitro				Evidence In Vivo				Ref.
				Cell	Prolif	Adhes	Osteog Differ	Animal	Osteoind	Bone Rem	Interf Heal	
1	Bone-Ligament	Polymer	PET	MC3T3-E1	+	+	+	Rabbit	+	/	+	[121]
2	Bone-Tendon	Lipidosome	Lipo clodro, Lipo,	/	/	/	/	Mouse	+	/	+	[122]
3	Bone-Tendon	Fibrin	ADSCs@Fib	/	/	/	/	Rabbit	+	+	+	[125]

Prolif: proliferation, Adhes: adhesion, Osteog Differ: osteogenic differentiation, Osteoind: osteoinduction, Bone Rem: bone remodeling, Interf heal: interface healing. PET: polyethylene terephthalate, Lipo clodro: liposomal clodronate, ADSCs@Fib: adipose-derived stromal cells loaded fibrin. +: Positive, /: not tested.

7. Nerve Repair

LIPUS affected the proliferation and myelinating activity of Schwann cells, in a both time- and duty ratio-dependent manner. Thus, LIPUS can be used to repair peripheral nerve injury and peripheral neuropathies [126,127]. The current knowledge about the influence of LIPUS on animal and human models revealed that LIPUS may have an impact on nerve regeneration and axonal alterations in the situation of carpal tunnel syndrome, transected nerve, dementia, and neurogenic erectile dysfunction [128,129]. Meanwhile, a growing number of studies are being conducted to investigate the function of LIPUS in materials-induced neuron regeneration (Table 5). LIPUS could accelerate autografting the sciatic nerve, and low-intensity US (250 mW/cm^2) showed faster regeneration than higher intensity (500 and 750 mW/cm^2) [130]. Notably, several studies reported that LIPUS could induce CGRP innervation, and then indirectly promote bone formation and spinal fusion [98,99,102,103].

Other studies focused on the potential accelerated effect of LIPUS on nerve conduit. L-ornithine was coated in the internal wall of PLGA conduits to promote cell adherence of the Schwann cells. In vitro, such conduit combined with LIPUS could promote cell proliferation of Schwann cells [131]. In vivo, the LIPUS stimulated the seeded Schwann cells to form regenerated nerves while inducing retarded axon regeneration in the silicone conduit [131,132].

After being blended with Matrigel solution, induced pluripotent stem cell-derived neural crest stem cells (iPSCs-NCSCs) were filled into the center of PLLA nanofiber nerve conduit, which acted as a scaffold in the rat transected sciatic nerve model. After LIPUS treatment, the neurophysiological parameters of the rat sciatic nerve were significantly improved. Staining of tissue sections revealed increased new blood vessels and neurofilaments, and increased expression of the neural marker Tuj1. Above all, the combination of LIPUS and iPSCs-NCSCs/PLLA promoted the regeneration and reconstruction of the rat sciatic nerve [133]. The activation of FAK-ERK1/2 signaling in iPSCs-NCSCs might contribute to the promotion effect of LIPUS on nerve regrowth. Based on this finding, this team fabricated an allogeneic decellularized nerve conduit containing iPSCs-NCSCs, perfluorotributylamine (PFTBA), and growth differentiation factor 5 (GDF5). The addition of PFTBA and GDF5 could provide an advantageous microenvironment for nerve regeneration, because PFTBA could supply enough oxygen and the addition of GDF5 could promote neural differentiation. This conduit showed rather good influence on the repair of rat transected sciatic nerves (Figure 8) [134].

Polycaprolactone/Pluronic F127 membrane (PCL/F127) was used to conduct a nerve guide conduit (NGC). The favorable permeability, hydrophilicity, and structural stability of PCL/F127 was good for the permeation of nutrients from the whole surroundings, whereas no nutrition could permeate the hydrophobic PLGA tube. Thus, this PCL/F127 conduit could favor nerve regeneration in the rat sciatic nerve defect model [135]. A further study used nerve growth factor (NGF) and LIPUS as double biophysical stimulation; by combining them, the NGF@PCL/F127 + LIPUS system could provide a synergetic effect on peripheral nerve repair, potentially for the repair of delayed and malfunctioned peripheral nerve [136].

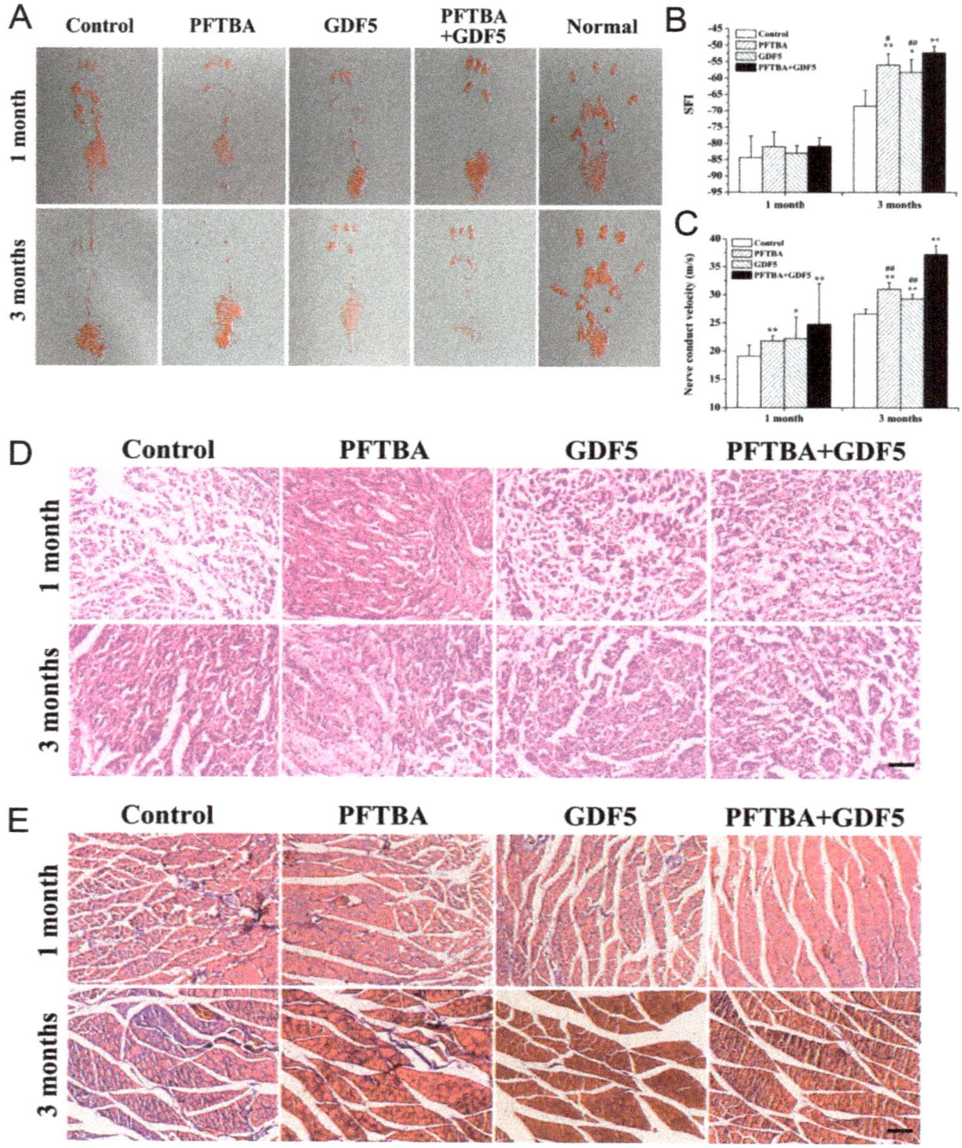

Figure 8. Promotion of LIPUS with allogeneic decellularized nerve conduit containing PFTBA and GDF5 on the repair of rat sciatic nerve injury. The footprint images (**A**), the SFI (**B**), and NCV (**C**) at 1 and 3 months post operation. * denotes $p < 0.05$ and ** denotes $p < 0.01$ vs. control group. # denotes $p < 0.05$ and ## denotes $p < 0.01$ vs. PFTBA+GDF5 group. Representative images of H&E staining (**D**) and Masson's trichrome staining (**E**) on rat gastrocnemius muscle section at 1 and 3 months post operation. Reprinted with permission from Ref. [134]. Copyright 2019 Wiley.

Table 5. LIPUS + biomaterials for nerve repair.

No	Biomaterials	Constituent	Evidence In Vitro		Evidence In Vivo			Ref.
			Cell	Proliferation	Animal	Nerve Regeneration	Angiogenesis	
1	Composites	iPSCs-NCSCs@PLLA	/	/	Rat	+	+	[133]
2	Autograft	Autograft Nerve	/	/	Rat	+	/	[130]
3	Composites	iPSCs-NCSCs/PFTBA/GDF5A@ADNC	/	/	Rat	+	/	[134]
4	Composites	PCL/F127	/	/	Rat	+	/	[135]
5	Composites	NGF@PCL/F127	/	/	Rat	+	/	[136]
6	Composites	SC@PLGA	/	/	Rat	+	/	[132]
7	Composites	SC@PLGA	Schwann	+	Rat	+	/	[131]

iPSCs-NCSCs@PLLA: induced pluripotent stem cell-derived neural crest stem cells loaded poly(L-lactic acid), iPSCs-NCSCs/PFTBA/GDF5A@ADNC: allogeneic decellularized nerve conduit containing iPSCs-NCSCs, perfluorotributylamine (PFTBA), and growth differentiation factor 5 (GDF5), PCL/F127: Polycaprolactone/Pluronic F127 membrane, NGF@PCL/F127: nerve growth factor loaded PCL/F127 membrane, SC@PLGA: Schwann cells seeded poly(DL-lactic acid-co-glycolic acid) (PLGA). +: Positive, /: not tested.

8. LIPUS Parameters

When compared with other energies, ultrasound has a specific ability to deeply propagate within the human body. Furthermore, it is highly focused, making it an excellent source with high energy for clinical therapy. Based on the parameters and the type of tissue, there are both thermal and non-thermal effects after ultrasonic waves are penetrated into the body [137]. The US frequencies ranged from 1 to 15 MHz in medical application, with 1 MHz frequencies being used for therapeutic applications and 2.5 to 15 MHz frequencies for diagnostic procedures, depending on the depth and tissue type and the mechanics of mechanical wave propagation (Table 6) [38,138].

Table 6. Representative parameters for LIPUS + biomaterials.

Application	Intensity (mW/cm^2)	Frequency (MHz)	Repetition Rate (kHz)	Pulse Burst (μs)	Duty Cycle (%)	Application (min/d)	Constituent	Equipment	Ref.
Bone regeneration	30	1.0	0.1	1000	NG	20	$BaTiO_3$/TC4	Sonicator 740	[34]
Bone regeneration	30	1.5	1.0		20	20	$BaTiO_3$/Ti6Al4V	Ronghai	[28]
Bone regeneration	100	3.0	NG	NG	50	10	cRGD-NBs	2776	[78]
Bone regeneration	300	1.0	NG	NG	NG	20	Collagen I	Agilent	[68]
Spinal fusion	30 ± 30%	1.5 ± 5%	1.0 ± 10%	200 ± 10%	20	20	DFDBA	Exogen	[97]
Spinal fusion	30	1.5	1.0	200	NG	20	AIBG	Exogen	[91]
Cartilage repair	60	1.5	1.0	NG	20	20	PBNPs	Osteotron IV	[119]
Cartilage repair	500	1.0	NG	NG	20	20	L-rapa	Intelect	[116]
Bone-Ligament or -Tendon repair	30 ± 5	1.5	1.0	200	NG	20	Liposomes	NG	[122]
Bone-Ligament or -Tendon repair	30	1.0	NG	NG	NG	20	PET	Osteotron IV	[121]
Nerve repair	250–750	1.0	NG	NG	20	5	Autograft Nerve	Customized device	[130]
Nerve repair	300–500	1.0	0.1	NG	20	45,214	iPSCs-NCSCs/PFTBA/GDF5A@ADNC	US10	[134]

NG: Not given, Ti6Al4V: titanium-6aluminum-4vanadium, cRGD-NBs: cyclic arginine-glycine-aspartic acid-modified nanobubbles, DFDBA: demineralized freeze-dried bone allograft, AIBG: autologous iliac bone graft, PBNPs: Prussian blue nanoparticles, L-rapa: liposome-encapsulated rapamycin, PET: polyethylene terephthalate, iPSCs-NCSCs/PFTBA/GDF5A@ADNC: allogeneic decellularized nerve conduit containing iPSCs-NCSCs.

It is a challenging process to determine optimal LIPUS parameters because of the wide range of biomaterials that have been used. Thus, few studies have been conducted on

LIPUS parameters that could interact with biomaterials on cells or in animal models. One study systematically investigated the acoustic excitation parameters, including intensity, frequency, duty cycle, and excitation duration. The most often used parameters were as follows: 30 mW/cm^2 intensity, 1.5 MHz frequency, and 20% duty cycle, but no differences were observed for excitation durations of 1, 3, and 5 min [120]. After 5 days of LIPUS stimulation, the rBMSC cultured on tissue culture plates could express the highest ALP activity in the group of 30 mW/cm^2 compared to the groups of 2 and 15 mW/cm^2. However, the highest mineralization was observed in the group with 2 mW/cm^2 after 17 days of LIPUS stimulation [139]. Zhou et al. proposed that the LIPUS intensity of 150 mW/cm^2 from the tissue culture plate be used for further study using 3D printed scaffolds, since maximum proliferation of hBMSCs was observed at 150 mW/cm^2 among five intensities (20, 50, 75, 150, and 300 mW/cm^2) [52].

In the porous Ti6Al4V scaffolds, a LIPUS intensity of 30 mW/cm^2 induced improved osteoblast differentiation when compared with 0, 10, and 60 mW/cm^2 [57]. However, no significant differences on MC3T3-E1 cells cultured on Ti6Al4V scaffolds were noted between the 1 and 3.2 MHz frequencies, both in vitro and in vivo [58]. Mouse preosteoblast MC3T3-E1 cells encapsulated in type I collagen hydrogels induced higher COX-2 and PGE2 expression after LIPUS application in the 30 and 150 mW/cm^2 group than in the 0 mW/cm^2 group, and the expression of these two markers was even higher in the 30 mW/cm^2 US intensity group than the 150 mW/cm^2 group. However, COX-2 was an inducer of the expression of PGE2, both of them leading to the aggravation of tissue inflammation [70]. Hsu et al. determined that pulsed ultrasound was more effective in increasing ALP activity and cell proliferation than continuous ultrasound [56].

During the initial inflammatory phase of 2 postoperative weeks, the bone–tendon junction healed more quickly in the LIPUS group than the control group. The prominent relieving effect of LIPUS on local inflammation was validated by decreased mRNA expression of proinflammatory cytokines and increased anti-inflammatory cytokines of patella–patellar tendon complexes [140–142]. Additionally, further study showed that LIPUS initiated during postoperative week 1 had a more noticeable effect on bone–tendon healing compared to immediate postoperative healing and postoperative week 2 [143]. The ultrasonic intensity of 30 mW/cm^2 was insufficient to promote axonal regeneration following nerve injury, as compared to the control and sham groups. It was recommended that the intensity of ultrasound should be adjusted to 200–300 mW/cm^2 for clinical examinations [144]. Domenici et al. identified that a specific range of the exposure energy density (6.3–10.8 J/cm^2) could modulate keratinocytes membrane trafficking with negligible biological damage [145]. After exposure to LIPUS (1 MHz, 65 mW/cm^2) for 1 h, a significant transient deregulation of IL-6 expression and secretion was observed in keratinocytes. High LIPUS intensity could alter membrane permeability and further reduce cell viability [146].

Above all, LIPUS parameters varied with different biomaterials platforms, cell types, initiating exposure time, stimulate duration, and cell culturing techniques. Further systematic studies should be designed in order to confirm the optimal parameters for musculoskeletal tissue engineering. Generally, biomaterial-specific parameters or ranges of effective parameters might be promising solutions.

9. Conclusions and Future Perspective

LIPUS is a safe biophysiotherapy which is effective in the repair of musculoskeletal systems. However, there are several issues that need to be noted in the clinical translation of LIPUS with biomaterials on musculoskeletal tissue repair. Firstly, although the intensity of 30mW/cm^2 is commonly used in most articles [147,148], the optimal LIPUS parameters may differ depending on the biological materials. However, there are few articles that have systematically explored material-specific LIPUS parameters. Secondly, to develop novel LIPUS-responsive biomaterials, a thorough understanding of the mechanisms of LIPUS is required because it could affect the regenerative microenvironments, including various cells, bioactive molecules, and implanted biomaterials. Even though the mechanism of

LIPUS alone on cells has been studied in some papers, there are few studies focusing on the synergistic mechanisms between LIPUS and biomaterials. Thirdly, in addition to bone tissue engineering, researchers should focus on other challenging areas, such as LIPUS on tendon–bone healing, cartilage healing, and skeletal muscle regeneration [149,150].

Combining LIPUS and biomaterials that are approved for bone tissue engineering applications could result in an easier transition to the clinic. However, further exploration is needed in the area of integrating LIPUS with new types of biomaterials. Recently, an emerging trend could be observed of the application of metamaterials in bone repair. Metamaterials could be used to design a patient-specific implant for improving load transfer, or to design a printable tunable stiffness scaffold for bone healing [151,152]. However, there is still a lack of studies on the combination of LIPUS and metamaterials. In order to avoid redundant and meaningless work among peers, it is encouraged to report negative results in the research. LIPUS is recommended for various steps of musculoskeletal tissue engineering, such as the preparation of tissue engineering scaffolds, cell pretreatment, bone remodeling, and other processes. Soft tissues, including muscle, fat, and skin, cover the surface of the musculoskeletal system, but the thickness of the soft tissue varies among the different parts. It is a difficult challenge to convey LIPUS to deeper musculoskeletal tissue [23]. Thus, identifying the optimal LIPUS parameters for specific locations and materials will be crucial to improve the use of LIPUS for musculoskeletal tissue engineering.

Tissue regeneration is a dynamic process involving a bi-directional interaction between cells and the matrix surrounding them. This dynamic reciprocity may be enhanced by exogenous LIPUS stimuli, which could provide more signaling to the microenvironments of bone regeneration. Various changes within cells resulting from LIPUS may explain their alteration in response to local biological signals. In order to achieve new functionalities, specific or extensive strategies could be employed to exploit the physical effects of LIPUS on biomaterials, biological molecules, or cells. Further development also needs to be done on novel biomaterials with adaptable and responsive properties, whether they are stimulated by LIPUS or not.

Author Contributions: Conceptualization, Z.Z. and W.Z.; writing—original draft preparation, W.J.; writing—review and editing, Z.Z. and W.Z.; figures and tables, W.J. and Z.Z. All authors have read and agreed to the published version of the manuscript.

Funding: This article has been supported by National Natural Science Foundation of China (No. 81901892, 82071923).

Institutional Review Board Statement: Not applicable.

Informed Consent Statement: Not applicable.

Data Availability Statement: Not applicable.

Conflicts of Interest: All the authors declare that there are no conflict of interest.

References

1. Dukle, A.; Murugan, D.; Nathanael, A.J.; Rangasamy, L.; Oh, T.-H. Can 3D-Printed Bioactive Glasses Be the Future of Bone Tissue Engineering? *Polymers* **2022**, *14*, 1627. [CrossRef] [PubMed]
2. Bose, S.; Sarkar, N. Natural medicinal compounds in bone tissue engineering. *Trends Biotechnol.* **2020**, *38*, 404–417. [CrossRef] [PubMed]
3. Koons, G.L.; Diba, M.; Mikos, A.G. Materials design for bone-tissue engineering. *Nat. Rev. Mater.* **2020**, *5*, 584–603. [CrossRef]
4. Zhang, K.; Fan, Y.; Dunne, N.; Li, X. Effect of microporosity on scaffolds for bone tissue engineering. *Regen. Biomater.* **2018**, *5*, 115–124. [CrossRef]
5. Xue, X.; Hu, Y.; Wang, S.; Chen, X.; Jiang, Y.; Su, J. Fabrication of physical and chemical crosslinked hydrogels for bone tissue engineering. *Bioact. Mater.* **2021**, *12*, 327–339. [CrossRef]
6. Hao, Z.; Song, Z.; Huang, J.; Huang, K.; Panetta, A.; Gu, Z.; Wu, J. The scaffold microenvironment for stem cell based bone tissue engineering. *Biomater. Sci.* **2017**, *5*, 1382–1392. [CrossRef]
7. Liu, Z.; Li, Y.; Yang, J.; Huang, J.; Luo, C.; Zhang, J.; Yan, W.; Ao, Y. Bone morphogenetic protein 9 enhances osteogenic and angiogenic responses of human amniotic mesenchymal stem cells cocultured with umbilical vein endothelial cells through the PI3K/AKT/m-TOR signaling pathway. *Aging* **2021**, *13*, 24829. [CrossRef]

8. Wang, X.-D.; Li, S.-Y.; Zhang, S.-J.; Gupta, A.; Zhang, C.-P.; Wang, L. The neural system regulates bone homeostasis via mesenchymal stem cells: A translational approach. *Theranostics* **2020**, *10*, 4839. [CrossRef]
9. Sun, S.; Diggins, N.H.; Gunderson, Z.J.; Fehrenbacher, J.C.; White, F.A.; Kacena, M.A. No pain, no gain? The effects of pain-promoting neuropeptides and neurotrophins on fracture healing. *Bone* **2020**, *131*, 115109. [CrossRef]
10. D'Alessandro, D.; Ricci, C.; Milazzo, M.; Strangis, G.; Forli, F.; Buda, G.; Petrini, M.; Berrettini, S.; Uddin, M.J.; Danti, S. Piezoelectric Signals in Vascularized Bone Regeneration. *Biomolecules* **2021**, *11*, 1731. [CrossRef]
11. Massari, L.; Benazzo, F.; Falez, F.; Perugia, D.; Pietrogrande, L.; Setti, S.; Osti, R.; Vaienti, E.; Ruosi, C.; Cadossi, R. Biophysical stimulation of bone and cartilage: State of the art and future perspectives. *Int. Orthop.* **2019**, *43*, 539–551. [CrossRef] [PubMed]
12. Fontana, F.; Iberite, F.; Cafarelli, A.; Aliperta, A.; Baldi, G.; Gabusi, E.; Dolzani, P.; Cristino, S.; Lisignoli, G.; Pratellesi, T. Development and validation of low-intensity pulsed ultrasound systems for highly controlled in vitro cell stimulation. *Ultrasonics* **2021**, *116*, 106495. [CrossRef] [PubMed]
13. Xia, P.; Shi, Y.; Wang, X.; Li, X. Advances in the application of low-intensity pulsed ultrasound to mesenchymal stem cells. *Stem Cell Res. Ther.* **2022**, *13*, 214. [CrossRef]
14. Harrison, A.; Lin, S.; Pounder, N.; Mikuni-Takagaki, Y. Mode & mechanism of low intensity pulsed ultrasound (LIPUS) in fracture repair. *Ultrasonics* **2016**, *71*, 45–52.
15. Jiang, X.; Savchenko, O.; Li, Y.; Qi, S.; Yang, T.; Zhang, W.; Chen, J. A review of low-intensity pulsed ultrasound for therapeutic applications. *IEEE Trans. Biomed. Eng.* **2018**, *66*, 2704–2718. [CrossRef]
16. Padilla, F.; Puts, R.; Vico, L.; Raum, K. Stimulation of bone repair with ultrasound: A review of the possible mechanic effects. *Ultrasonics* **2014**, *54*, 1125–1145. [CrossRef]
17. Li, Y.; Li, W.; Liu, X.; Liu, X.; Zhu, B.; Guo, S.; Wang, C.; Wang, D.; Li, S.; Zhang, Z. Effects of Low-Intensity Pulsed Ultrasound in Tendon Injuries. *J. Ultrasound Med. Off. J. Am. Inst. Ultrasound Med.* **2023**, *42*, 1923–1939. [CrossRef]
18. Pereira, L.F.; Fontes-Pereira, A.J.; de Albuquerque Pereira, W.C. Influence of Low-Intensity Pulsed Ultrasound Parameters on the Bone Mineral Density in Rat Model: A Systematic Review. *Ultrasound Med. Biol.* **2023**, *49*, 1687–1698. [CrossRef]
19. Takase, K.; Fukui, T.; Oe, K.; Sawauchi, K.; Yoshikawa, R.; Yamamoto, Y.; Hayashi, S.; Matsumoto, T.; Kuroda, R.; Niikura, T. Effect of low-intensity pulsed ultrasound on osteogenic differentiation of human induced membrane-derived cells in Masquelet technique. *Injury* **2023**, *54*, 1444–1450. [CrossRef]
20. Cheng, H.; Chiu, W.; Lai, Y.; Truong, T.; Lee, P.; Huang, C. High-frequency noncontact low-intensity pulsed ultrasound modulates Ca-dependent transcription factors contributing to cell migration. *Ultrasonics* **2023**, *127*, 106852. [CrossRef]
21. Fu, S.; Lan, Y.; Wang, G.; Bao, D.; Qin, B.; Zheng, Q.; Liu, H.; Wong, V. External stimulation: A potential therapeutic strategy for tendon-bone healing. *Front. Bioeng. Biotechnol.* **2023**, *11*, 1150290. [CrossRef] [PubMed]
22. Tang, L.; Guo, H.; Wang, K.; Zhou, Y.; Wu, T.; Fan, X.; Guo, J.; Sun, L.; Ta, D. Low-intensity pulsed ultrasound enhances the positive effects of high-intensity treadmill exercise on bone in rats. *J. Bone Miner. Metab.* **2023**, *41*, 592–605. [CrossRef] [PubMed]
23. Palanisamy, P.; Alam, M.; Li, S.; Chow, S.; Zheng, Y. Low-Intensity Pulsed Ultrasound Stimulation for Bone Fractures Healing: A Review. *J. Ultrasound Med. Off. J. Am. Inst. Ultrasound Med.* **2022**, *41*, 547–563. [CrossRef] [PubMed]
24. Zhou, X.; Sun, S.; Chen, Y.; Liu, C.; Li, D.; Cheng, Q.; He, M.; Li, Y.; Xu, K.; Ta, D. Pulsed frequency modulated ultrasound promotes therapeutic effects of osteoporosis induced by ovarian failure in mice. *Ultrasonics* **2023**, *132*, 106973. [CrossRef] [PubMed]
25. Xu, Z.; Li, S.; Wan, L.; Hu, J.; Lu, H.; Zhang, T. Role of low-intensity pulsed ultrasound in regulating macrophage polarization to accelerate tendon-bone interface repair. *J. Orthop. Res. Off. Publ. Orthop. Res. Soc.* **2023**, *41*, 919–929. [CrossRef]
26. Qin, H.; Du, L.; Luo, Z.; He, Z.; Wang, Q.; Chen, S.; Zhu, Y. The therapeutic effects of low-intensity pulsed ultrasound in musculoskeletal soft tissue injuries: Focusing on the molecular mechanism. *Front. Bioeng. Biotechnol.* **2022**, *10*, 1080430. [CrossRef]
27. Raza, H.; Saltaji, H.; Kaur, H.; Flores-Mir, C.; El-Bialy, T. Effect of Low-Intensity Pulsed Ultrasound on Distraction Osteogenesis Treatment Time: A Meta-analysis of Randomized Clinical Trials. *J. Ultrasound Med.* **2016**, *35*, 349–358. [CrossRef]
28. Chen, J.; Li, S.; Jiao, Y.; Li, J.; Li, Y.; Hao, Y.-l.; Zuo, Y. In Vitro Study on the Piezodynamic Therapy with a $BaTiO_3$-Coating Titanium Scaffold under Low-Intensity Pulsed Ultrasound Stimulation. *ACS Appl. Mater. Interfaces* **2021**, *13*, 49542–49555. [CrossRef]
29. Kuang, Y.; Hu, B.; Xia, Y.; Jiang, D.; Huang, H.; Song, J. Low-intensity pulsed ultrasound promotes tissue regeneration in rat dental follicle cells in a porous ceramic scaffold. *Braz. Oral Res.* **2019**, *33*, e0045. [CrossRef]
30. Liang, C.; Liu, X.; Yan, Y.; Sun, R.; Li, J.; Geng, W. Effectiveness and Mechanisms of Low-Intensity Pulsed Ultrasound on Osseointegration of Dental Implants and Biological Functions of Bone Marrow Mesenchymal Stem Cells. *Stem Cells Int.* **2022**, *2022*, 7397335. [CrossRef]
31. Mishima, H.; Sugaya, H.; Yoshioka, T.; Wada, H.; Aoto, K.; Hyodo, K.; Tomaru, Y.; Kumagai, H.; Akaogi, H.; Ochiai, N. The Effect of Combined Therapy, Percutaneous Autologous Concentrated Bone Marrow Grafting and Low-Intensity Pulsed Ultrasound (LIPUS), on the Treatment of Non-Unions. *J. Orthop. Trauma* **2016**, *30*, S2. [CrossRef] [PubMed]
32. Zhu, H.; Cai, X.; Lin, T.; Shi, Z.; Yan, S. Low-intensity pulsed ultrasound enhances bone repair in a rabbit model of steroid-associated osteonecrosis. *Clin. Orthop. Relat. Res.* **2015**, *473*, 1830–1839. [CrossRef] [PubMed]
33. Chen, J.; Li, J.; Hu, F.; Zou, Q.; Mei, Q.; Li, S.; Hao, Y.; Hou, W.; Li, J.; Li, Y. Effect of microarc oxidation-treated Ti6Al4V scaffold following low-intensity pulsed ultrasound stimulation on osteogenic cells in vitro. *ACS Biomater. Sci. Eng.* **2019**, *5*, 572–581. [CrossRef] [PubMed]

34. Cai, K.; Jiao, Y.; Quan, Q.; Hao, Y.; Liu, J.; Wu, L. Improved activity of MC3T3-E1 cells by the exciting piezoelectric $BaTiO_3$/TC4 using low-intensity pulsed ultrasound. *Bioact. Mater.* **2021**, *6*, 4073–4082. [CrossRef]
35. Doinikov, A.A. Acoustic radiation forces: Classical theory and recent advances. *Recent Res. Dev. Acoust.* **2003**, *1*, 39–67.
36. Ninomiya, K.; Noda, K.; Ogino, C.; Kuroda, S.-i.; Shimizu, N. Enhanced OH radical generation by dual-frequency ultrasound with TiO2 nanoparticles: Its application to targeted sonodynamic therapy. *Ultrason. Sonochem.* **2014**, *21*, 289–294. [CrossRef]
37. Fan, C.-H.; Lee, Y.-H.; Ho, Y.-J.; Wang, C.-H.; Kang, S.-T.; Yeh, C.-K. Macrophages as drug delivery carriers for acoustic phase-change droplets. *Ultrasound Med. Biol.* **2018**, *44*, 1468–1481. [CrossRef]
38. O'Brien, W.D., Jr. Ultrasound–biophysics mechanisms. *Prog. Biophys. Mol. Biol.* **2007**, *93*, 212–255. [CrossRef]
39. Wu, J.; Nyborg, W.L. Ultrasound, cavitation bubbles and their interaction with cells. *Adv. Drug Deliv. Rev.* **2008**, *60*, 1103–1116. [CrossRef]
40. Ahmed, S.E.; Martins, A.M.; Husseini, G.A. The use of ultrasound to release chemotherapeutic drugs from micelles and liposomes. *J. Drug Target.* **2015**, *23*, 16–42. [CrossRef]
41. Kimmel, E. Cavitation bioeffects. *Crit. Rev.™ Biomed. Eng.* **2006**, *34*, 105–161. [CrossRef] [PubMed]
42. Jin, Z.; Wen, Y.; Hu, Y.; Chen, W.; Zheng, X.; Guo, W.; Wang, T.; Qian, Z.; Su, B.-L.; He, Q. MRI-guided and ultrasound-triggered release of NO by advanced nanomedicine. *Nanoscale* **2017**, *9*, 3637–3645. [CrossRef] [PubMed]
43. Dong, X.; Li, H.; Lingling, E.; Cao, J.; Guo, B. Bioceramic akermanite enhanced vascularization and osteogenic differentiation of human induced pluripotent stem cells in 3D scaffolds in vitro and vivo. *RSC Adv.* **2019**, *9*, 25462–25470. [CrossRef] [PubMed]
44. Vallet-Regí, M.; Ruiz-Hernández, E. Bioceramics: From bone regeneration to cancer nanomedicine. *Adv. Mater.* **2011**, *23*, 5177–5218. [CrossRef] [PubMed]
45. Jin, F.; Cai, Q.; Wang, W.; Fan, X.; Lu, X.; He, N.; Ding, J. The Effect of Stromal-Derived Factor 1α on Osteoinduction Properties of Porous β-Tricalcium Phosphate Bioceramics. *BioMed Res. Int.* **2021**, *2021*, 8882355. [CrossRef]
46. Brunello, G.; Panda, S.; Schiavon, L.; Sivolella, S.; Biasetto, L.; Del Fabbro, M. The impact of bioceramic scaffolds on bone regeneration in preclinical in vivo studies: A systematic review. *Materials* **2020**, *13*, 1500. [CrossRef]
47. Wang, J.; Wang, C.; Jin, K.; Yang, X.; Gao, L.; Yao, C.; Dai, X.; He, J.; Gao, C.; Ye, J. Simultaneous enhancement of vascularization and contact-active antibacterial activity in diopside-based ceramic orbital implants. *Mater. Sci. Eng. C* **2019**, *105*, 110036. [CrossRef]
48. Walsh, W.R.; Langdown, A.J.; Auld, J.W.; Stephens, P.; Yu, Y.; Vizesi, F.; Bruce, W.J.; Pounder, N. Effect of low intensity pulsed ultrasound on healing of an ulna defect filled with a bone graft substitute. *J. Biomed. Mater. Res. Part B Appl. Biomater. Off. J. Soc. Biomater. Jpn. Soc. Biomater. Aust. Soc. Biomater. Korean Soc. Biomater.* **2008**, *86*, 74–81. [CrossRef]
49. Wang, J.; Wang, J.; Asou, Y.; Paul, F.; Shen, H.; Chen, J.; Sotome, S. Low-intensity pulsed ultrasound prompts tissue-engineered bone formation after implantation surgery. *Chin. Med. J.* **2014**, *127*, 669–674.
50. Iwai, T.; Harada, Y.; Imura, K.; Iwabuchi, S.; Murai, J.; Hiramatsu, K.; Myoui, A.; Yoshikawa, H.; Tsumaki, N. Low-intensity pulsed ultrasound increases bone ingrowth into porous hydroxyapatite ceramic. *J. Bone Miner. Metab.* **2007**, *25*, 392–399. [CrossRef]
51. Nagasaki, R.; Mukudai, Y.; Yoshizawa, Y.; Nagasaki, M.; Shiogama, S.; Suzuki, M.; Kondo, S.; Shintani, S.; Shirota, T. A combination of low-intensity pulsed ultrasound and nanohydroxyapatite concordantly enhances osteogenesis of adipose-derived stem cells from buccal fat pad. *Cell Med.* **2015**, *7*, 123–131. [CrossRef] [PubMed]
52. Zhou, X.; Castro, N.J.; Zhu, W.; Cui, H.; Aliabouzar, M.; Sarkar, K.; Zhang, L.G. Improved human bone marrow mesenchymal stem cell osteogenesis in 3D bioprinted tissue scaffolds with low intensity pulsed ultrasound stimulation. *Sci. Rep.* **2016**, *6*, 32876. [CrossRef] [PubMed]
53. Wu, L.; Lin, L.; Qin, Y.-X. Low-Intensity Pulsed Ultrasound on Osteogenesison Porous Titanium Alloy Scaffolds, and early differentiation in a three-dimensional silicon carbide scaffold using low-intensity pulsed ultrasound. *Tissue Eng. Part A* **2015**, *21*, 53–61. [CrossRef]
54. Higuchi, M.; Moroi, A.; Yoshizawa, K.; Kosaka, A.; Ikawa, H.; Iguchi, R.; Saida, Y.; Hotta, A.; Tsutsui, T.; Ueki, K. Comparison between various densities of pore titanium meshes and e-polytetrafluoroethylene (ePTFE) membrane regarding bone regeneration induced by low intensity pulsed ultrasound (LIPUS) in rabbit nasal bone. *J. Cranio-Maxillofac. Surg.* **2016**, *44*, 1152–1161. [CrossRef]
55. Liu, Q.; Liu, X.; Liu, B.; Hu, K.; Zhou, X.; Ding, Y. The effect of low-intensity pulsed ultrasound on the osseointegration of titanium dental implants. *Br. J. Oral Maxillofac. Surg.* **2012**, *50*, 244–250. [CrossRef] [PubMed]
56. Hsu, S.-K.; Huang, W.-T.; Liu, B.-S.; Li, S.-M.; Chen, H.-T.; Chang, C.-J. Effects of near-field ultrasound stimulation on new bone formation and osseointegration of dental titanium implants in vitro and in vivo. *Ultrasound Med. Biol.* **2011**, *37*, 403–416. [CrossRef]
57. Cao, H.; Feng, L.; Wu, Z.; Hou, W.; Li, S.; Hao, Y.; Wu, L. Effect of low-intensity pulsed ultrasound on the biological behavior of osteoblasts on porous titanium alloy scaffolds: An in vitro and in vivo study. *Mater. Sci. Eng. C* **2017**, *80*, 7–17. [CrossRef]
58. Feng, L.; Liu, X.; Cao, H.; Qin, L.; Hou, W.; Wu, L. A comparison of 1-and 3.2-MHz low-intensity pulsed ultrasound on osteogenesis on porous titanium alloy scaffolds: An in vitro and in vivo study. *J. Ultrasound Med.* **2019**, *38*, 191–202. [CrossRef]
59. Fan, B.; Guo, Z.; Li, X.; Li, S.; Gao, P.; Xiao, X.; Wu, J.; Shen, C.; Jiao, Y.; Hou, W. Electroactive barium titanate coated titanium scaffold improves osteogenesis and osseointegration with low-intensity pulsed ultrasound for large segmental bone defects. *Bioact. Mater.* **2020**, *5*, 1087–1101. [CrossRef]
60. Liu, W.; Yang, D.; Wei, X.; Guo, S.; Wang, N.; Tang, Z.; Lu, Y.; Shen, S.; Shi, L.; Li, X. Fabrication of piezoelectric porous $BaTiO_3$ scaffold to repair large segmental bone defect in sheep. *J. Biomater. Appl.* **2020**, *35*, 544–552. [CrossRef]

61. Wu, H.; Dong, H.; Tang, Z.; Chen, Y.; Liu, Y.; Wang, M.; Wei, X.; Wang, N.; Bao, S.; Yu, D.; et al. Electrical stimulation of piezoelectric $BaTiO_3$ coated Ti6Al4V scaffolds promotes anti-inflammatory polarization of macrophages and bone repair via MAPK/JNK inhibition and OXPHOS activation. *Biomaterials* **2023**, *293*, 121990. [CrossRef] [PubMed]
62. Qin, D.; Wang, N.; You, X.-G.; Zhang, A.-D.; Chen, X.-G.; Liu, Y. Collagen-based biocomposites inspired by bone hierarchical structures for advanced bone regeneration: Ongoing research and perspectives. *Biomater. Sci.* **2022**, *10*, 318–353. [CrossRef] [PubMed]
63. Hu, T.; Lo, A.C. Collagen–Alginate Composite Hydrogel: Application in Tissue Engineering and Biomedical Sciences. *Polymers* **2021**, *13*, 1852. [CrossRef] [PubMed]
64. Benayahu, D.; Benayahu, Y. A Unique Marine-Derived Collagen: Its Characterization towards Biocompatibility Applications for Tissue Regeneration. *Mar. Drugs* **2021**, *19*, 419. [CrossRef]
65. Ishihara, Y.; Ueki, K.; Sotobori, M.; Marukawa, K.; Moroi, A. Bone regeneration by statin and low-intensity pulsed ultrasound (LIPUS) in rabbit nasal bone. *J. Cranio-Maxillofac. Surg.* **2014**, *42*, 185–193. [CrossRef]
66. Angle, S.R.; Sena, K.; Sumner, D.R.; Virkus, W.W.; Virdi, A.S. Combined use of low intensity pulsed ultrasound and rhBMP-2 to enhance bone formation in a rat model of critical-size defect. *J. Orthop. Trauma* **2014**, *28*, 605. [CrossRef]
67. Carina, V.; Costa, V.; Raimondi, L.; Pagani, S.; Sartori, M.; Figallo, E.; Setti, S.; Alessandro, R.; Fini, M.; Giavaresi, G. Effect of low-intensity pulsed ultrasound on osteogenic human mesenchymal stem cells commitment in a new bone scaffold. *J. Appl. Biomater. Funct. Mater.* **2017**, *15*, 215–222. [CrossRef]
68. Assanah, F.; Grassie, K.; Anderson, H.; Xin, X.; Rowe, D.; Khan, Y. Ultrasound-derived mechanical stimulation of cell-laden collagen hydrogels for bone repair. *J. Biomed. Mater. Res. Part A* **2023**, *111*, 1200–1215. [CrossRef]
69. Veronick, J.; Assanah, F.; Nair, L.S.; Vyas, V.; Huey, B.; Khan, Y. The effect of acoustic radiation force on osteoblasts in cell/hydrogel constructs for bone repair. *Exp. Biol. Med.* **2016**, *241*, 1149–1156. [CrossRef]
70. Veronick, J.A.; Assanah, F.; Piscopo, N.; Kutes, Y.; Vyas, V.; Nair, L.S.; Huey, B.D.; Khan, Y. Mechanically loading cell/hydrogel constructs with low-intensity pulsed ultrasound for bone repair. *Tissue Eng. Part A* **2018**, *24*, 254–263. [CrossRef]
71. Wang, Y.; Peng, W.; Liu, X.; Zhu, M.; Sun, T.; Peng, Q.; Zeng, Y.; Feng, B.; Zhi, W.; Weng, J. Study of bilineage differentiation of human-bone-marrow-derived mesenchymal stem cells in oxidized sodium alginate/N-succinyl chitosan hydrogels and synergistic effects of RGD modification and low-intensity pulsed ultrasound. *Acta Biomater.* **2014**, *10*, 2518–2528. [CrossRef] [PubMed]
72. Wang, Y.; Qiu, Y.; Li, J.; Zhao, C.; Song, J. Low-intensity pulsed ultrasound promotes alveolar bone regeneration in a periodontal injury model. *Ultrasonics* **2018**, *90*, 166–172. [CrossRef] [PubMed]
73. Zheng, H.; Lu, L.; Song, J.; Gao, X.; Deng, F.; Wang, Z. Low intensity pulsed ultrasound combined with guided tissue regeneration for promoting the repair of defect at canines periodontal fenestration in Beagle dogs. *Zhonghua Kou Qiang Yi Xue Za Zhi Zhonghua Kouqiang Yixue Zazhi Chin. J. Stomatol.* **2011**, *46*, 431–436.
74. Oh, S.H.; Kim, T.H.; Chun, S.Y.; Park, E.K.; Lee, J.H. Enhanced guided bone regeneration by asymmetrically porous PCL/pluronic F127 membrane and ultrasound stimulation. *J. Biomater. Sci. Polym. Ed.* **2012**, *23*, 1673–1686. [CrossRef]
75. Hasuike, A.; Sato, S.; Makino, N.; Tsunori, K.; Oginuma, T.; Ito, K. The potency of low-intensity pulsed ultrasound in a rat calvarial guided bone-regeneration model. *J. Hard Tissue Biol.* **2011**, *20*, 217–224. [CrossRef]
76. Xinyi, J.; Ji, Y.; Zhaowu, C.; Jinlin, S.; Feng, D.; Zhibiao, W. Low intensity pulsed ultrasound irradiating combined with guided bone regeneration for promoting the repair effect of periodontal bone defect. *West China J. Stomatol.* **2012**, *30*, 487–492.
77. Moonga, S.S.; Qin, Y.-X. MC3T3 infiltration and proliferation in bovine trabecular scaffold regulated by dynamic flow bioreactor and augmented by low-intensity pulsed ultrasound. *J. Orthop. Transl.* **2018**, *14*, 16–22. [CrossRef]
78. Yao, H.; Zhang, L.; Yan, S.; He, Y.; Zhu, H.; Li, Y.; Wang, D.; Yang, K. Low-intensity pulsed ultrasound/nanomechanical force generators enhance osteogenesis of BMSCs through microfilaments and TRPM7. *J. Nanobiotechnol.* **2022**, *20*, 378. [CrossRef]
79. Sirsi, S.; Borden, M. Microbubble compositions, properties and biomedical applications. *Bubble Sci. Eng. Technol.* **2009**, *1*, 3–17. [CrossRef]
80. Osborn, J.; Aliabouzar, M.; Zhou, X.; Rao, R.; Zhang, L.G.; Sarkar, K. Enhanced osteogenic differentiation of human mesenchymal stem cells using microbubbles and low intensity pulsed ultrasound on 3D printed scaffolds. *Adv. Biosyst.* **2019**, *3*, 1800257. [CrossRef]
81. Huang, L.; Cai, Y.; Hu, H.; Guo, P.; Xin, Z. Enhanced regeneration of large cortical bone defects with electrospun nanofibrous membranes and low-intensity pulsed ultrasound. *Exp. Ther. Med.* **2017**, *14*, 525–530. [CrossRef] [PubMed]
82. Jin, L.; Shan, J.; Hao, Y.; Wang, Y.; Liu, L. Enhanced bone regeneration by low-intensity pulsed ultrasound and lipid microbubbles on PLGA/TCP 3D-printed scaffolds. *BMC Biotechnol.* **2023**, *23*, 13. [CrossRef] [PubMed]
83. Tang, L.; Wu, T.; Zhou, Y.; Zhong, Y.; Sun, L.; Guo, J.; Fan, X.; Ta, D. Study on synergistic effects of carboxymethyl cellulose and LIPUS for bone tissue engineering. *Carbohydr. Polym.* **2022**, *286*, 119278. [CrossRef] [PubMed]
84. Elmajee, M.; Munasinghe, C.; Nasser, A.; Nagappa, S.; Mahmood, A. The perceptions of clinicians using low-intensity pulsed ultrasound (LIPUS) for orthopaedic pathology: A national qualitative study. *Injury* **2022**, *53*, 3214–3219. [CrossRef]
85. Wada, K.; Kawano, M.; Hemmi, Y.; Suzuki, R.; Kunoki, K.; Sakagami, H.; Kawazu, H.; Yokose, S. Effect of Low-intensity Pulsed Ultrasound on Healing of Bone Defects in Rat Tibia as Measured by Reconstructed Three-dimensional Analysis of Micro CT Images. *In Vivo* **2022**, *36*, 643–648. [CrossRef]

86. Sawauchi, K.; Fukui, T.; Oe, K.; Kumabe, Y.; Oda, T.; Yoshikawa, R.; Takase, K.; Matsushita, T.; Matsumoto, T.; Hayashi, S. Low-Intensity Pulsed Ultrasound Promotes Osteogenic Differentiation of Reamer-Irrigator-Aspirator Graft-Derived Cells in Vitro. *Ultrasound Med. Biol.* **2022**, *48*, 313–322. [CrossRef]
87. Majeed, H.; Karim, T.; Davenport, J.; Karski, M.; Smith, R.; Clough, T.M. Clinical and patient-reported outcomes following Low Intensity Pulsed Ultrasound (LIPUS, Exogen) for established post-traumatic and post-surgical nonunion in the foot and ankle. *Foot Ankle Surg.* **2020**, *26*, 405–411. [CrossRef]
88. Cook, S.D.; Salkeld, S.L.; Patron, L.P.; Ryaby, J.P.; Whitecloud III, T.S. Low-intensity pulsed ultrasound improves spinal fusion. *Spine J.* **2001**, *1*, 246–254. [CrossRef]
89. Aynaci, O.; Önder, Ç.; Piskin, A.; Özoran, Y. The effect of ultrasound on the healing of muscle-pediculated bone graft in spinal fusion. *Spine* **2002**, *27*, 1531–1535. [CrossRef]
90. Chen, L.-H.; Liao, J.-C.; Chen, W.-J.; Lai, P.-L.; Keorochana, G. Low-intensity pulsed ultrasound enhances healing of laminectomy chip bone grafts on spinal fusion: A model of posterolateral intertransverse fusion in rabbits. *J. Trauma Acute Care Surg.* **2011**, *70*, 863–869. [CrossRef]
91. Liao, J.-C.; Chen, W.-J.; Niu, C.-C.; Chen, L.-H. Effects of Low-Intensity Pulsed Ultrasound on Spinal Pseudarthrosis Created by Nicotine Administration: A Model of Lumbar Posterolateral Pseudarthrosis in Rabbits. *J. Ultrasound Med.* **2015**, *34*, 1043–1050. [CrossRef] [PubMed]
92. Hui, C.F.F.; Chan, C.W.; Yeung, H.Y.; Lee, K.M.; Qin, L.; Li, G.; Leung, K.S.; Hu, Y.Y.; Cheng, J.C.Y. Low-intensity pulsed ultrasound enhances posterior spinal fusion implanted with mesenchymal stem cells-calcium phosphate composite without bone grafting. *Spine* **2011**, *36*, 1010–1016. [CrossRef] [PubMed]
93. Zhuo, X.-L.; Lü, H.-B.; Xu, D.-Q.; Bin, L.; Wang, X.-Y.; Zhang, Y.; Hu, J.-Z. Effects of low-intensity pulsed ultrasound stimulation on porous hydroxyapatite blocks for posterolateral fusion of lumbar spine in rabbits. *Trans. Nonferrous Met. Soc. China* **2010**, *20*, 1921–1927. [CrossRef]
94. Wang, J.; Li, J.-W.; Chen, L. Effect of low–intensity pulsed ultrasound on posterolateral lumbar fusion of rabbit. *Asian Pac. J. Trop. Med.* **2015**, *8*, 68–72. [CrossRef] [PubMed]
95. Xu, X.; Wang, F.; Yang, Y.; Zhou, X.; Cheng, Y.; Wei, X.; Li, M. LIPUS promotes spinal fusion coupling proliferation of type H microvessels in bone. *Sci. Rep.* **2016**, *6*, 20116. [CrossRef] [PubMed]
96. Zhou, X.-Y.; Xu, X.-M.; Wu, S.-Y.; Zhang, Z.-C.; Wang, F.; Yang, Y.-L.; Li, M.; Wei, X.-Z. Low-intensity pulsed ultrasound promotes spinal fusion and enhances migration and proliferation of MG63s through sonic hedgehog signaling pathway. *Bone* **2018**, *110*, 47–57. [CrossRef]
97. Zhang, Z.-C.; Yang, Y.-L.; Li, B.; Hu, X.-C.; Xu, S.; Wang, F.; Li, M.; Zhou, X.-Y.; Wei, X.-Z. Low-intensity pulsed ultrasound promotes spinal fusion by regulating macrophage polarization. *Biomed. Pharmacother.* **2019**, *120*, 109499. [CrossRef]
98. Zaidi, M.; Chambers, T.; Das, R.G.; Morris, H.; MacIntyre, I. A direct action of human calcitonin gene-related peptide on isolated osteoclasts. *J. Endocrinol.* **1987**, *115*, 511–518. [CrossRef]
99. Hukkanen, M.; Konttinen, Y.; Santavirta, S.; Paavolainen, P.; Gu, X.-H.; Terenghi, G.; Polak, J. Rapid proliferation of calcitonin gene-related peptide-immunoreactive nerves during healing of rat tibial fracture suggests neural involvement in bone growth and remodelling. *Neuroscience* **1993**, *54*, 969–979. [CrossRef]
100. Aro, H.; Eerola, E.; Aho, A. Development of nonunions in the rat fibula after removal of periosteal neural mechanoreceptors. *Clin. Orthop. Relat. Res.* **1985**, *199*, 292–299. [CrossRef]
101. Santavirta, S.; Konttinen, Y.T.; Nordström, D.; Mäkelä, A.; Sorsa, T.; Hukkanen, M.; Rokkanen, P. Immunologic studies of nonunited fractures. *Acta Orthop. Scand.* **1992**, *63*, 579–586. [CrossRef] [PubMed]
102. Wang, X.-Y.; Guo, X.; Cheng, J.C.-Y.; Mi, Y.-L.; Lai, P.Y.-M. Involvement of calcitonin gene-related peptide innervation in the promoting effect of low-intensity pulsed ultrasound on spinal fusion without decortication. *Spine* **2010**, *35*, E1539–E1545. [CrossRef] [PubMed]
103. Zhou, X.-Y.; Xu, X.-M.; Wu, S.-Y.; Wang, F.; Zhang, Z.-C.; Yang, Y.-L.; Li, M.; Wei, X.-Z. Low-intensity pulsed ultrasound-induced spinal fusion is coupled with enhanced calcitonin gene-related peptide expression in rat model. *Ultrasound Med. Biol.* **2017**, *43*, 1486–1493. [CrossRef] [PubMed]
104. Saadi, S.B.; Ranjbarzadeh, R.; Amirabadi, A.; Ghoushchi, S.J.; Kazemi, O.; Azadikhah, S.; Bendechache, M. Osteolysis: A Literature Review of Basic Science and Potential Computer-Based Image Processing Detection Methods. *Comput. Intell. Neurosci.* **2021**, *2021*, 4196241. [CrossRef]
105. Yadollahpour, A.; Rashidi, S. Therapeutic applications of low-intensity pulsed ultrasound in osteoporosis. *Asian J. Pharm.* **2017**, *11*, S1–S6.
106. Zamarioli, A.; Butezloff, M.; Ximenez, J.; Volpon, J. Low-intensity pulsed ultrasound partially reversed the deleterious effects of a severe spinal cord injury-induced bone loss and osteoporotic fracture healing in paraplegic rats. *Spinal Cord* **2023**, *61*, 145–153. [CrossRef]
107. Tian, C.; Liu, H.; Zhao, C.; Zhang, C.; Wang, W. A Numerical Study on Mechanical Effects of Low-Intensity Pulsed Ultrasound on Trabecular Bone and Osteoblasts. *J. Biomech. Eng.* **2023**, *145*, 051010. [CrossRef]
108. Zhao, X.; Cai, X.-Z.; Shi, Z.-L.; Zhu, F.-B.; Zhao, G.-S.; Yan, S.-G. Low-intensity pulsed ultrasound (LIPUS) may prevent polyethylene induced periprosthetic osteolysis in vivo. *Ultrasound Med. Biol.* **2012**, *38*, 238–246. [CrossRef]

109. Zhu, H.; Shi, Z.; Cai, X.; Yang, X.; Zhou, C. The combination of PLLA/PLGA/PCL composite scaffolds integrated with BMP-2-loaded microspheres and low-intensity pulsed ultrasound alleviates steroid-induced osteonecrosis of the femoral head. *Exp. Ther. Med.* **2020**, *20*, 126. [CrossRef]
110. Matsumae, G.; Shimizu, T.; Tian, Y.; Takahashi, D.; Ebata, T.; Alhasan, H.; Yokota, S.; Kadoya, K.; Terkawi, M.A.; Iwasaki, N. Targeting thymidine phosphorylase as a potential therapy for bone loss associated with periprosthetic osteolysis. *Bioeng. Transl. Med.* **2021**, *6*, e10232. [CrossRef]
111. Tian, Y.; Terkawi, M.A.; Onodera, T.; Alhasan, H.; Matsumae, G.; Takahashi, D.; Hamasaki, M.; Ebata, T.; Aly, M.K.; Kida, H. Blockade of XCL1/lymphotactin ameliorates severity of periprosthetic osteolysis triggered by polyethylene-particles. *Front. Immunol.* **2020**, *11*, 1720. [CrossRef] [PubMed]
112. Kubo, Y.; Drescher, W.; Fragoulis, A.; Tohidnezhad, M.; Jahr, H.; Gatz, M.; Driessen, A.; Eschweiler, J.; Tingart, M.; Wruck, C.J. Adverse effects of oxidative stress on bone and vasculature in corticosteroid-associated osteonecrosis: Potential role of nuclear factor erythroid 2-related factor 2 in cytoprotection. *Antioxid. Redox Signal.* **2021**, *35*, 357–376. [CrossRef] [PubMed]
113. Liu, L.-H.; Zhang, Q.-Y.; Sun, W.; Li, Z.-R.; Gao, F.-Q. Corticosteroid-induced osteonecrosis of the femoral head: Detection, diagnosis, and treatment in earlier stages. *Chin. Med. J.* **2017**, *130*, 2601–2607. [CrossRef] [PubMed]
114. Chen, T.; Bai, J.; Tian, J.; Huang, P.; Zheng, H.; Wang, J. A single integrated osteochondral in situ composite scaffold with a multi-layered functional structure. *Colloids Surf. B Biointerfaces* **2018**, *167*, 354–363. [CrossRef]
115. Guo, Y.; Ma, Y.; Dong, R.; Liu, S.; Tu, J. Effect of allogeneic chondrocytes-calcium alginate gel composite under intervention of low intensive pulsed ultrasound for repairing rabbit knee articular cartilage defect. *Zhongguo Xiu Fu Chong Jian Wai Ke Za Zhi Zhongguo Xiufu Chongjian Waike Zazhi Chin. J. Reparative Reconstr. Surg.* **2013**, *27*, 928–934.
116. Chen, C.-H.; Kuo, S.M.; Tien, Y.-C.; Shen, P.-C.; Kuo, Y.-W.; Huang, H.H. Steady augmentation of anti-osteoarthritic actions of rapamycin by liposome-encapsulation in collaboration with low-intensity pulsed ultrasound. *Int. J. Nanomed.* **2020**, *15*, 3771. [CrossRef]
117. Vaughan, N.M.; Grainger, J.; Bader, D.L.; Knight, M.M. The potential of pulsed low intensity ultrasound to stimulate chondrocytes matrix synthesis in agarose and monolayer cultures. *Med. Biol. Eng. Comput.* **2010**, *48*, 1215–1222. [CrossRef]
118. Duda, G.N.; Kliche, A.; Kleemann, R.; Hoffmann, J.E.; Sittinger, M.; Haisch, A. Does low-intensity pulsed ultrasound stimulate maturation of tissue-engineered cartilage? *J. Biomed. Mater. Res. Part B Appl. Biomater. Off. J. Soc. Biomater. Jpn. Soc. Biomater. Aust. Soc. Biomater. Korean Soc. Biomater.* **2004**, *68*, 21–28. [CrossRef]
119. Zuo, D.; Tan, B.; Jia, G.; Wu, D.; Yu, L.; Jia, L. A treatment combined prussian blue nanoparticles with low-intensity pulsed ultrasound alleviates cartilage damage in knee osteoarthritis by initiating PI3K/Akt/mTOR pathway. *Am. J. Transl. Res.* **2021**, *13*, 3987.
120. Aliabouzar, M.; Zhang, L.G.; Sarkar, K. Lipid coated microbubbles and low intensity pulsed ultrasound enhance chondrogenesis of human mesenchymal stem cells in 3D printed scaffolds. *Sci. Rep.* **2016**, *6*, 37728. [CrossRef]
121. Liu, X.; Sun, K.; Xu, P.; Yu, Z.; Lei, Z.; Zhou, H.; Li, J.; Li, X.; Zhu, Z.; Wang, H. Effect of Low-Intensity Pulsed Ultrasound on the Graft-Bone Healing of Artificial Ligaments: An In Vitro and In Vivo Study. *Am. J. Sports Med.* **2022**, *50*, 801–813. [CrossRef] [PubMed]
122. Li, S.; Xu, Z.; Wang, Z.; Xiang, J.; Zhang, T.; Lu, H. Acceleration of Bone-Tendon Interface Healing by Low-Intensity Pulsed Ultrasound Is Mediated by Macrophages. *Phys. Ther.* **2021**, *101*, pzab055. [CrossRef] [PubMed]
123. Hu, R.; Yang, Z.; Li, Y.; Zhou, Z. LIPUS Promotes Endothelial Differentiation and Angiogenesis of Periodontal Ligament Stem Cells by Activating Piezo1. *Int. J. Stem Cells* **2022**, *15*, 372–383. [CrossRef]
124. Lin, H.; Wang, Q.; Quan, C.; Ren, Q.; He, W.; Xiao, H. Low-intensity pulsed ultrasound enhances immunomodulation and facilitates osteogenesis of human periodontal ligament stem cells by inhibiting the NF-κB pathway. *Cell Tissue Bank.* **2023**, *24*, 45–58. [CrossRef]
125. Chen, C.; Zhang, T.; Liu, F.; Qu, J.; Chen, Y.; Fan, S.; Chen, H.; Sun, L.; Zhao, C.; Hu, J. Effect of low-intensity pulsed ultrasound after autologous adipose-derived stromal cell transplantation for bone-tendon healing in a rabbit model. *Am. J. Sports Med.* **2019**, *47*, 942–953. [CrossRef] [PubMed]
126. Fontana, F.; Iacoponi, F.; Orlando, F.; Pratellesi, T.; Cafarelli, A.; Ricotti, L. Low-intensity pulsed ultrasound increases neurotrophic factors secretion and suppresses inflammation inmodels of peripheral neuropathies. *J. Neural Eng.* **2023**, *20*, 026033. [CrossRef]
127. Huang, B.; Jiang, Y.; Zhang, L.; Yang, B.; Guo, Y.; Yang, X.; Gong, P. Low-intensity pulsed ultrasound promotes proliferation and myelinating genes expression of Schwann cells through NRG1/ErbB signaling pathway. *Tissue Cell* **2023**, *80*, 101985. [CrossRef]
128. Ye, K.; Li, Z.; Yin, Y.; Zhou, J.; Li, D.; Gan, Y.; Peng, D.; Xiao, M.; Zhao, L.; Dai, Y.; et al. LIPUS-SCs-Exo promotes peripheral nerve regeneration in cavernous nerve crush injury-induced ED rats via PI3K/Akt/FoxO signaling pathway. *CNS Neurosci. Ther.* **2023**. [CrossRef]
129. Song, D.; Chen, X.; Zhou, N.; Yuan, Y.; Geng, S.; Zhang, C.; Zhao, Z.; Wang, X.; Bao, X.; Lan, X.; et al. Low-intensity pulsed ultrasound triggers a beneficial neuromodulation in dementia mice with chronic cerebral hypoperfusion via activation of hippocampal Fndc5/irisin signaling. *J. Transl. Med.* **2023**, *21*, 139. [CrossRef]
130. Jiang, W.; Wang, Y.; Tang, J.; Peng, J.; Wang, Y.; Guo, Q.; Guo, Z.; Li, P.; Xiao, B.; Zhang, J. Low-intensity pulsed ultrasound treatment improved the rate of autograft peripheral nerve regeneration in rat. *Sci. Rep.* **2016**, *6*, 22773. [CrossRef]

131. Chang, C.J.; Hsu, S.h.; Lin, F.t.; Chang, H.; Chang, C.S. Low-intensity-ultrasound–accelerated nerve regeneration using cell-seeded poly (D, L-lactic acid-co-glycolic acid) conduits: An in vivo and in vitro study. *J. Biomed. Mater. Res. Part B Appl. Biomater. Off. J. Soc. Biomater. Jpn. Soc. Biomater. Aust. Soc. Biomater. Korean Soc. Biomater.* **2005**, *75*, 99–107. [CrossRef] [PubMed]
132. Chang, C.-J.; Hsu, S.-H. The effects of low-intensity ultrasound on peripheral nerve regeneration in poly (DL-lactic acid-co-glycolic acid) conduits seeded with Schwann cells. *Ultrasound Med. Biol.* **2004**, *30*, 1079–1084. [CrossRef] [PubMed]
133. Lv, Y.; Nan, P.; Chen, G.; Sha, Y.; Xia, B.; Yang, L. In vivo repair of rat transected sciatic nerve by low-intensity pulsed ultrasound and induced pluripotent stem cells-derived neural crest stem cells. *Biotechnol. Lett.* **2015**, *37*, 2497–2506. [CrossRef]
134. Xia, B.; Chen, G.; Zou, Y.; Yang, L.; Pan, J.; Lv, Y. Low-intensity pulsed ultrasound combination with induced pluripotent stem cells-derived neural crest stem cells and growth differentiation factor 5 promotes sciatic nerve regeneration and functional recovery. *J. Tissue Eng. Regen. Med.* **2019**, *13*, 625–636. [CrossRef]
135. Park, S.C.; Oh, S.H.; Seo, T.B.; Namgung, U.; Kim, J.M.; Lee, J.H. Ultrasound-stimulated peripheral nerve regeneration within asymmetrically porous PLGA/Pluronic F127 nerve guide conduit. *J. Biomed. Mater. Res. Part B Appl. Biomater.* **2010**, *94*, 359–366. [CrossRef] [PubMed]
136. Kim, J.R.; Oh, S.H.; Kwon, G.B.; Namgung, U.; Song, K.S.; Jeon, B.H.; Lee, J.H. Acceleration of peripheral nerve regeneration through asymmetrically porous nerve guide conduit applied with biological/physical stimulation. *Tissue Eng. Part A* **2013**, *19*, 2674–2685. [CrossRef]
137. Paris, J.L.; Manzano, M.; Cabañas, M.V.; Vallet-Regí, M. Mesoporous silica nanoparticles engineered for ultrasound-induced uptake by cancer cells. *Nanoscale* **2018**, *10*, 6402–6408. [CrossRef]
138. Fateh, S.T.; Moradi, L.; Kohan, E.; Hamblin, M.R.; Dezfuli, A.S. Comprehensive review on ultrasound-responsive theranostic nanomaterials: Mechanisms, structures and medical applications. *Beilstein J. Nanotechnol.* **2021**, *12*, 808–862. [CrossRef]
139. Angle, S.; Sena, K.; Sumner, D.; Virdi, A. Osteogenic differentiation of rat bone marrow stromal cells by various intensities of low-intensity pulsed ultrasound. *Ultrasonics* **2011**, *51*, 281–288. [CrossRef]
140. Lu, H.; Liu, F.; Chen, H.; Chen, C.; Qu, J.; Xu, D.; Zhang, T.; Zhou, J.; Hu, J. The effect of low-intensity pulsed ultrasound on bone-tendon junction healing: Initiating after inflammation stage. *J. Orthop. Res.* **2016**, *34*, 1697–1706. [CrossRef]
141. Li, X.; Zhong, Y.; Zhou, W.; Song, Y.; Li, W.; Jin, Q.; Gao, T.; Zhang, L.; Xie, M. Low-intensity pulsed ultrasound (LIPUS) enhances the anti-inflammatory effects of bone marrow mesenchymal stem cells (BMSCs)-derived extracellular vesicles. *Cell. Mol. Biol. Lett.* **2023**, *28*, 9. [CrossRef]
142. Li, X.; Zhong, Y.; Zhang, L.; Xie, M. Recent advances in the molecular mechanisms of low-intensity pulsed ultrasound against inflammation. *J. Mol. Med.* **2023**, *101*, 361–374. [CrossRef]
143. Lu, H.; Chen, C.; Qu, J.; Chen, H.; Chen, Y.; Zheng, C.; Wang, Z.; Xu, D.; Zhou, J.; Zhang, T. Initiation timing of low-intensity pulsed ultrasound stimulation for tendon-bone healing in a rabbit model. *Am. J. Sports Med.* **2016**, *44*, 2706–2715. [CrossRef] [PubMed]
144. Daeschler, S.C.; Harhaus, L.; Bergmeister, K.D.; Boecker, A.; Hoener, B.; Kneser, U.; Schoenle, P. Clinically available low intensity ultrasound devices do not promote axonal regeneration after peripheral nerve surgery—A preclinical investigation of an FDA-approved device. *Front. Neurol.* **2018**, *9*, 1057. [CrossRef] [PubMed]
145. Domenici, F.; Brasili, F.; Giantulli, S.; Cerroni, B.; Bedini, A.; Giliberti, C.; Palomba, R.; Silvestri, I.; Morrone, S.; Paradossi, G. Differential effects on membrane permeability and viability of human keratinocyte cells undergoing very low intensity megasonic fields. *Sci. Rep.* **2017**, *7*, 16536. [CrossRef] [PubMed]
146. Giantulli, S.; Tortorella, E.; Brasili, F.; Scarpa, S.; Cerroni, B.; Paradossi, G.; Bedini, A.; Morrone, S.; Silvestri, I.; Domenici, F. Effect of 1-MHz ultrasound on the proinflammatory interleukin-6 secretion in human keratinocytes. *Sci. Rep.* **2021**, *11*, 19033. [CrossRef]
147. Lou, S.; Lv, H.; Li, Z.; Tang, P.; Wang, Y. Effect of low-intensity pulsed ultrasound on distraction osteogenesis: A systematic review and meta-analysis of randomized controlled trials. *J. Orthop. Surg. Res.* **2018**, *13*, 205. [CrossRef]
148. McCarthy, C.; Camci-Unal, G. Low Intensity Pulsed Ultrasound for Bone Tissue Engineering. *Micromachines* **2021**, *12*, 1488. [CrossRef]
149. Chen, Y.; Yang, H.; Wang, Z.; Zhu, R.; Cheng, L.; Cheng, Q. Low-intensity pulsed ultrasound promotes mesenchymal stem cell transplantation-based articular cartilage regeneration via inhibiting the TNF signaling pathway. *Stem Cell Res. Ther.* **2023**, *14*, 93. [CrossRef]
150. Qin, H.; Luo, Z.; Sun, Y.; He, Z.; Qi, B.; Chen, Y.; Wang, J.; Li, C.; Lin, W.; Han, Z.; et al. Low-intensity pulsed ultrasound promotes skeletal muscle regeneration via modulating the inflammatory immune microenvironment. *Int. J. Biol. Sci.* **2023**, *19*, 1123–1145. [CrossRef]
151. Hashemi, M.S.; Mccrary, A.; Kraus, K.H.; Sheidaei, A. A novel design of printable tunable stiffness metamaterial for bone healing. *J. Mech. Behav. Biomed. Mater.* **2021**, *116*, 104345. [CrossRef] [PubMed]
152. Singh, A.P.; Rana, M.; Pal, B.; Datta, P.; Majumder, S.; Roychowdhury, A. Patient-specific femoral implant design using metamaterials for improving load transfer at proximal-lateral region of the femur. *Med. Eng. Phys.* **2023**, *113*, 103959. [CrossRef] [PubMed]

Journal of
Functional
Biomaterials

MDPI

Article

A Synergic Strategy: Adipose-Derived Stem Cell Spheroids Seeded on 3D-Printed PLA/CHA Scaffolds Implanted in a Bone Critical-Size Defect Model

Gabriela S. Kronemberger [1,2,3], Thiago Nunes Palhares [4], Alexandre Malta Rossi [4], Brunno R. F. Verçosa [1], Suelen C. Sartoretto [5], Rodrigo Resende [5], Marcelo J. Uzeda [5], Adriana T. N. N. Alves [5], Gutemberg G. Alves [5], Mônica D. Calasans-Maia [5], José Mauro Granjeiro [2,3,5,*] and Leandra Santos Baptista [1,2,3,*]

1 Nucleus of Multidisciplinary Research in Biology (Numpex-Bio), Federal University of Rio de Janeiro (UFRJ) Xerém, Duque de Caxias 25245-390, RJ, Brazil; soareskg@tcd.ie (G.S.K.); brunno.vercoza@gmail.com (B.R.F.V.)

2 Laboratory of Eukariotic Cells, National Institute of Metrology, Quality and Technology (Inmetro), Duque de Caxias 25250-020, RJ, Brazil

3 Post-Graduation Program of Translational Biomedicine (Biotrans), Unigranrio, Campus I, Duque de Caxias 25071-202, RJ, Brazil

4 Brazilian Center for Physics Research, Xavier Sigaud 150, Urca 22290-180, RJ, Brazil; thiagonup@gmail.com (T.N.P.); rossi@cbpf.br (A.M.R.)

5 Laboratory of Clinical Research in Odontology, Fluminense Federal University (UFF), Niterói 24020-140, RJ, Brazil; susartoretto@hotmail.com (S.C.S.); resende.r@hotmail.com (R.R.); mjuzeda@gmail.com (M.J.U.); aterezinhanovellino@gmail.com (A.T.N.N.A.); gutopepe@yahoo.com.br (G.G.A.); monicacalasansmaia@gmail.com (M.D.C.-M.)

* Correspondence: jmgranjeiro@inmetro.gov.br (J.M.G.); leandrabaptista@xerem.ufrj.br (L.S.B.)

Abstract: Bone critical-size defects and non-union fractures have no intrinsic capacity for self-healing. In this context, the emergence of bone engineering has allowed the development of functional alternatives. The aim of this study was to evaluate the capacity of ASC spheroids in bone regeneration using a synergic strategy with 3D-printed scaffolds made from poly (lactic acid) (PLA) and nanostructured hydroxyapatite doped with carbonate ions (CHA) in a rat model of cranial critical-size defect. In summary, a set of results suggests that ASC spheroidal constructs promoted bone regeneration. In vitro results showed that ASC spheroids were able to spread and interact with the 3D-printed scaffold, synthesizing crucial growth factors and cytokines for bone regeneration, such as VEGF. Histological results after 3 and 6 months of implantation showed the formation of new bone tissue in the PLA/CHA scaffolds that were seeded with ASC spheroids. In conclusion, the presence of ASC spheroids in the PLA/CHA 3D-printed scaffolds seems to successfully promote bone formation, which can be crucial for a significant clinical improvement in critical bone defect regeneration.

Keywords: bone; osteogenesis; critical-size defects; tissue engineering; spheroids; synergic strategy; scaffolds; 3D printing; bone regeneration

1. Introduction

Bone tissue has an excellent capacity for self-healing from non-critical-size defects [1]. However, most cases of bone critical-size defects and non-union fractures caused by trauma, tumor resection, infection, or congenital malformations will lead to a clinical intervention [2]. The gold standard treatment for these defects is autologous bone graft [3]. However, this treatment has some drawbacks, such as donor site morbidity, which is considered the biggest challenge. Allogenic bone grafts have garnered attention in the clinic [4]. However, they are devitalized prior to implantation, and the process starts with the loss of osteoinductive potential together with immunogenicity issues.

"*Scaffold-free*" techniques are being largely explored in tissue engineering. In this approach, cells interact directly with one another through a process called "*self-assembly*", which leads to the formation of spheroids [5]. The main advantages related to the use of spheroids are their ability to provide superior cellular heterogeneity, nutrient and oxygen gradients, extracellular matrix deposition, and tissue-like gene expression profiles [6]. Spheroids produced from bone marrow mesenchymal stem cells (bmMSCs) and adipose-derived stem/stromal cells (ASCs) have been already applied as successful developmental models to recapitulate osteogenesis in vitro [7–9]. They are capable of fusing with other spheroids, which allows them to be used as building blocks for bone tissue engineering [10,11]. Although bmMSCs and ASC spheroids have great potential to form bones, few regenerative pre-clinical studies have been performed to achieve this goal throughout the years [7,12,13].

Suenaga and collaborators (2015) [7] fabricated osteogenically induced bmMSC spheroids, combining them with β-tricalcium phosphate (β-TCP) granules for their implantation into rat calvaria critical-size bone defect. The results showed a higher proportion of new bone formation in the group administered osteogenically induced bmMSC spheroids. Lee and collaborators (2020) [12] delivered ASC spheroids, formed by 3D-printed microchamber system, into rat calvaria critical-size bone defects. The microchamber was coated with platelet-derived growth factors (PDGF) and morphogenetic protein 2 (BMP-2). The main results showed that this group was able to form a larger area of new bone in these defects, compared with the control groups. Findeisen and collaborators (2021) [13] produced bmMSC spheroids and implanted them into rat femoral defects and compared them with a cell suspension. The results showed that the spheroid group produced a significantly higher density of bone mineral.

As discussed, few studies have been published with ASC spheroids that regenerate bone tissues through the years. Gurumurthy and collaborators (2017) [14] showed that ASC spheroids had a higher osteogenic potential when compared to traditional monolayer techniques. The ASC spheroids also showed a high viability after the culture period. Ahmad and collaborators (2018) [15] developed composite ASC spheroids combined with polymeric fibers. Their main results showed that the constructs formed a mineralized functional tissue. Yamada and collaborators (2022) [16] implanted ASC spheroids in a rat calvaria defect. They showed a significant bone regeneration in the group containing the spheroids when compared to the control group.

One innovative approach that is being explored in tissue engineering consists of a synergic strategy that combines spheroids and scaffolds to achieve better biomechanical properties, higher cell density, and extracellular matrix production, resulting in an improvement in functionality of the final construct [17]. In this context, nanostructured hydroxyapatite scaffolds are being highly explored for bone tissue engineering applications due to their high osteoinductive potential [18,19]. These scaffolds can also be 3D printed using an extrusion-based technique to develop more versatile and custom-sized scaffolds that will fit better into a patient's bone defect, thereby improving the bone regeneration process [20,21].

So far, to the best of our knowledge, only two studies have applied this synergic approach to regenerate bone tissue in vivo [7,12]. Therefore, the aim of this study was to evaluate the capacity of ASC spheroids seeded on the surface of poly (lactic acid) (PLA) and nanostructured hydroxyapatite doped with carbonate ions (CHA) on 3D-printed scaffolds to regenerate bone critical-size defects in rats.

2. Material and Methods

2.1. Monolayer 2D Culture of Human Adipose Stem/Stromal Cells (ASCs)

ASCs were isolated from healthy human donors (aged 18–55 years old) and cryopreserved as previously described [22]. The isolation procedure of the human ASCs was performed according to the Research Ethics Committee of Clementino Fraga Filho University Hospital, Federal University of Rio de Janeiro, Brazil (25818719.4.0000.5257). Next, the ASCs were seeded into 75 cm^2 flasks and maintained in culture as previously described [10,11]. For their growth, the ASC cell suspension at passage one was thawed and seeded into 175 cm^2 flasks and maintained in chemically defined TheraPEAK™ MSCGM-CD™, Mesenchymal Stem Cell Medium (Lonza, São Paulo, Brazil), at 37 °C in a humid atmosphere with 5% carbon dioxide (CO_2). After passage three, the cells were used for spheroid fabrication. For all the experiments performed in this study, we have used one donor and three independent experiments for each assay.

2.2. Spheroid 3D Culture

The ASC spheroids were fabricated in a micro-molded non-adhesive hydrogel system. The micro-molded non-adhesive hydrogel wells were fabricated using a solution of 2% (*w/v*) Ultrapure Agarose (Invitrogen, São Paulo, Brazil) in a 0.9% NaCl solution, formed from silicone molds containing 81 resections (MicroTissues® 3D Petri Dish®, Sigma Aldrich, St. Louis, MO, USA). Next, a total cell suspension of 2×10^6 ASCs at passage three was gently seeded into the micro-molded non-adhesive hydrogels. After 40 min of the seeding, the culture media was added in each well. The culture media was made using low glucose Dulbecco's modified eagle medium (DMEM) supplemented with Insulin-Transferrin-Selenium (ITS) (1×; Sigma), 1.25 μg/mL human albumin (Farma Biagini SPA, Rio de Janeiro, Brazil), 50 μg/mL ascorbic acid (Sigma), 100 μg/mL penicillin, and 100 μg/mL streptomycin (Sigma). The ASC spheroids were then maintained for 2 days in the micro-molded non-adhesive hydrogels at 37 °C in a humid atmosphere with 5% carbon dioxide (CO_2) and 21% oxygen (O_2).

2.3. Spheroid Diameter Measurements

After 24 h of ASC spheroid formation and up to 1 week after the culture, a total of 15 images were acquired using a phase contrast microscope (Primo Vert, Zeiss, Pittsburgh, Pennsylvania, USA) equipped with a digital camera. The width and length measurements of each spheroid were determined using the AxioVision Rel. 4.6 software (Zeiss, Jena, Germany). Next, the diameter ratio of each spheroid was obtained by dividing width by length to obtain the spheroid sphericity. These analyses were performed using one cell donor and three independent experiments using fifteen spheroids from each sample.

2.4. 3D Printing of PLA/CHA Scaffolds

The 3D printing process for the fabrication of the PLA/CHA scaffolds has been described previously in detail [23]. Briefly, the 3D printing filaments of the Filament Deposition Modeling (FDM) printing technique were produced in a twin-screw extruder printer (Haake Rheomex OS Prw16, Thermo Scientific, Waltham, MA, USA). For the preparation of the composite, the PLA pellets were manually mixed according to the desired proportion (% by mass) of 10% CHA. Next, these pellets were added to the extruder feed hopper, where they were mixed using a melting temperature profile of 150/160/170/180/185/190 °C for the 3D printing process. At the extruder outlet, a circular matrix was used in order to achieve a filament shape for the 3D printing process.

2.5. Seeding of ASC Spheroids on the Surface of 3D-Printed PLA/CHA Scaffolds

Initially, the 3D-printed PLA/CHA scaffolds were placed in wells of a 48-well plate. Next, a total of 162 ASC spheroids, which corresponds to a total density of 4×10^6 cells, were gradually dispersed at once on the surface of 3D-printed PLA/CHA scaffolds in a total volume of 200 μL. The manual dispersion of spheroids was performed in all areas of the 3D-printed scaffolds. Next, the 3D-printed scaffolds containing the seeded ASC spheroids were maintained in a humid atmosphere at 37 °C, at 5% CO_2 and 21% CO_2 for 24 h, in a final volume of 100 μL of culture media of the spheroids. The low volume of media was used in order to achieve a better adhesion of the ASC spheroids on the surfaces of the 3D-printed PLA/CHA scaffolds. After 24 h, 500 μL of the culture media for spheroids was slowly added in each well, and the constructs (ASC spheroids + PLA/CHA scaffolds) were maintained at 37 °C, at 5% CO_2 and 21% CO_2. Finally, the constructs were maintained for a total period of 7 days in culture for in vitro analyses and for the surgical procedures (Figure 1A).

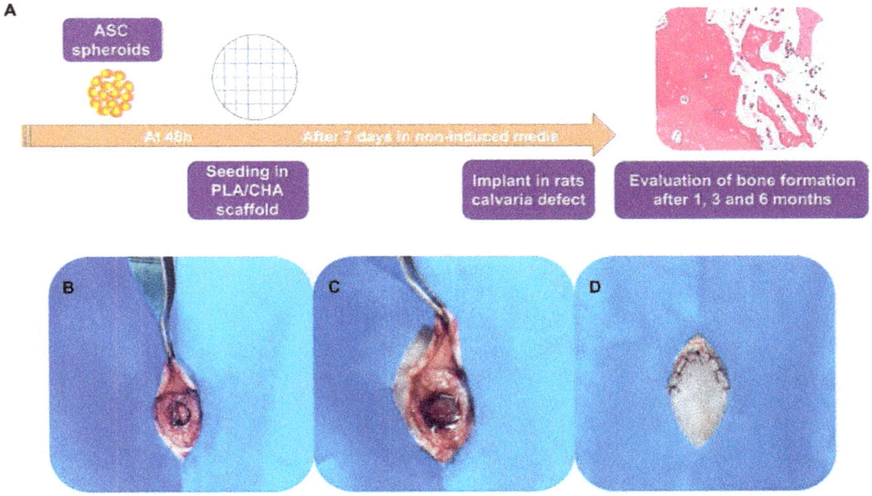

Figure 1. Graphical abstract and surgical procedure of constructs implanted in 8 mm calvaria defects in rats. (**A**) ASC spheroids were fabricated using non-adherent micromolded agarose hydrogel technique. After 48 h, spheroids were seeded on the surface of PLA/CHA scaffolds. The constructs (PLA/CHA scaffolds + ASC spheroids) were maintained for 1 week in culture media and then implanted in calvaria defects of rats. The evaluation of new bone formation was conducted by histological analysis after 1, 3, and 6 months of the implantation period. (**B**) Surgical bone defect exposure. (**C**) ASC spheroidal constructs implanted in situ. (**D**) Defect sutured. ASC: adipose-derived stem/stromal cell; PLA/CHA: poly (lactic acid) and nanostructured hydroxyapatite doped with carbonate ions.

2.6. Scanning Electron Transmission (SEM) Analysis

In order to evaluate the interaction of the ASC spheroids on the surface of the 3D-printed PLA/CHA scaffolds, SEM analyses were performed. Initially, the constructs were washed three times with 0.01 M PBS for 5 min and then fixed in a solution of 2.5% glutaraldehyde (Sigma) in 0.1 M sodium cacodylate buffer (Sigma) for 2 h in the dark. Next, the samples were washed twice with sodium cacodylate buffer for 5 min and post-fixed in osmium tetroxide (Sigma) 1% diluted in 0.1 M sodium cacodylate and kept for 40 min in the dark. Sequentially, the samples were washed 3 times for 5 min with 0.1 M sodium cacodylate buffer. After that period, samples were dehydrated by serial washes in ethanol solutions for 10 min each, in concentrations of 30%, 50%, 70%, 90%, and finally 100% ethanol (three times). All steps were performed at room temperature. After dehydration, the samples were slowly dried in a critical point device (Leica/CPDO30, São Paulo, Brazil) and afterwards covered with 10 nm gold layer by the sputtering equipment (DENTON VACUUM, Leica, São Paulo, Brazil).

In addition, the energy-dispersive X-ray spectroscopy (EDS) analyses were performed for the PLA/CHA scaffolds to identify the presence of calcium (Ca) and phosphorus in the samples. For EDS analyses, the detector used was the Oxford X-MaxN 20 mm^2.

2.7. Analysis of Multiple Secreted Proteins

At week 1 of ASC spheroidal constructs culture, the medium was changed to fresh medium, and the samples were maintained in culture for 24 h. In addition, a scaffold without spheroids seeded was maintained in culture for the same period as a control condition. After that period, the total volume of supernatant was collected in Eppendorfs and immediately frozen and stored at −80 °C. In order to quantify the proteins secreted in the supernatant, a Luminex xMAP technology based on a magnetic bead panel was used for recognition of the following mediators: human MIP-1β, IFN-y, interleukin-1ra (IL-1ra), IL-5, GM-CSF, TNFα, RANTES, IL-2, IL-1β, Eotaxin, bFGF, VEGF, PDGF-BB, IP-10, IL-13, IL-4, IL-5, MCP-1, IL-8, MIP-1a, IL-10, G-CSF, IL-15, IL-7, IL-12p70, IL-17ra, and IL-9 (27-plex panel, Bio Rad Laboratories Inc., Hercules, CA, USA). The quantification was performed using the Bio-Plex Magpix apparatus (Bio Rad Laboratories Inc., Hercules, CA, USA), following the manufacturer's instructions. Next, the concentration of each secreted product was quantified using the xPONENT v3.1 software (LuminexCorp®, Austin, TX, USA), and the results were expressed as picograms per milliliter (ρg/mL). The control group (scaffold without seeded spheroids) values were used to normalize the results of the ASC spheroids seeded in the PLA/CHA group. The assay was performed from one cell donor and three independent experiments using five replicates from each sample.

2.8. In Vivo Study

This study was approved by the Ethics Committee on the Use of Animals at the Universidade Federal Fluminense (CEUA-UFF) (approval number: 1474030618). In this study, 72 female Wistar rats between two and four months old, weighing between 180 and 300 g, were provided by the Laboratory Animal Nucleus (NAL, Niteroi, Rj, Brazil). All surgical procedures on the animals were performed at the Animal Experimentation Laboratory (LEA, Niteroi, RJ, Brazil). The animals were kept during the entire period of this study in mini-isolators ($n = 3$) and fed with pelleted feed and water at will. The animals were randomly distributed using envelopes for their allocation to each experimental group ($n = 6$), totaling 24 animals for each experimental period (1, 3, and 6 months).

For the surgeries, the animals were deprived of solid food six hours before the procedures and then were administered general anesthesia. Anesthetic induction and maintenance were performed with the administration of 100 mg/kg of Ketamine (Francotar®—Virbac—Jurubatura, SP, Brazil), 10 mg/kg of Xylazine (Sedazine®—Fort Dodge, Iowa, USA), and 5 mg/kg of Midazolam (Eurofarma), which were administered intramuscularly. After observing the absence of pain reflexes, trichotomy was performed in the animal's skullcap region using a sharp razor blade (Enox platinum®, Enox, São Paulo, SP, Brazil), and the region was cleaned with 2% chlorhexidine soap (Riohex 2%®, Rio de Janeiro, Rj, Brazil). Then, the animals were placed in the ventral decubitus position, and a semilunar incision of 2 cm length was made with a surgical blade, number 15C (Solidor®, Tampa, Florida, USA), followed by a periosteal mucus detachment of the region using Molt #9 (Molt Quinelato®, Tampa, Florida, USA) and the bone exposure of the entire skull of the animal. The bone defect was performed with a trephine bur with an 8 mm internal diameter (Harte group®, Doylestown, PA, USA) (Figure 1B–D).

Subsequently, the division of the implants was carried out, and the animals were allocated into three experimental groups for this study: (1) PLA/CHA; (2) PLA/CHA seeded with ASC spheroids; and (3) Clot. After the end of the surgical procedures, the tissues were repositioned by suturing with a 5.0 Nylon thread (Mononylon®, Cidade Monções, SP, Brazil). Subsequently, the animals received Meloxicam (Maxicam®, Croydon, United Kingdom) injection 15 mg/1.5 mL, 1 mg/kg every 24 h, and Tramadol Hydrochloride (Tramal®, Milwaukee, WI, USA), 10 mg/kg, subcutaneously every 8 h for three days. The animals were euthanized at 1, 3, and 6 months post-surgical procedures by applying a lethal dose of general anesthetic (Thiopental, 150 mg/kg).

2.9. Histological Processing

For the ASC spheroid samples, at the end of 7 days of culture, the spheroids were initially harvested and washed twice using 0.01 M PBS (Sigma). Next, samples were fixed in 4% paraformaldehyde (Sigma) solution for 1 h. Sequentially, ASC spheroids were washed twice with 0.01 M PBS solution and dehydrated with serial washes in ethanol (Sigma) for 20 min each, in concentrations of 70%, 90%, and 100%. Later, samples were washed twice in xylene (Sigma) for 20 min and embedded in wax. The wax blocks were cut with a microtome (Slee Medical, Cut 5062, Nieder-Olm, Germany) and stained with Hematoxylin (Sigma) and Eosin (Sigma) solutions.

The calvaria bone block samples containing the constructs were collected with a long carbide spherical drill #6 (Low Spherical #6 FG®, São Paulo, SP, Brazil), coupled with a micromotor (Marathon 3 Champion Micromotor – Talma, São Paulo, SP, Brazil). Next, the samples from each experimental group were dissected to remove the soft tissue and fixed for 48 h in 4% formalin solution (pH 7.4) and subjected to the standard histological procedure. Briefly, samples were washed for 1 h in running water, demineralized in decalcifier solution (Allkimia®—Campinas, São Paulo, SP, Brazil) for 48 h, again washed in running water, dehydrated in increasing concentrations of ethanol (70, 80, 90, and 100%, Rialcool®, Rioquímica, Brazil) for 1 h each, cleared in two xylene baths (Xylene®, Ehningen, Germany) of 1 h each, and finally embedded in paraffin. Furthermore, these samples were cut with a microtome (Jung-Leica RM 2054) into slices of 5 µm thickness and stained with Hematoxylin (Sigma) and Eosin (Sigma) and Masson's Trichrome (Sigma) for the descriptive evaluation of newly formed bone and connective tissue. For the descriptive histological analysis, a Light Field Light Microscope (OLYMPUS®, Tokyo, Japan) was used. Images were obtained with an optical microscope-coupled camera (OLYMPUS® SC100, Tokio, Japan), associated with the CELLSENS®1.9 DigitalImage software (Tokio, Japan).

2.10. Statistical Analysis

Non-parametric Mann–Whitney t-test was utilized for the comparison between 0 h and week 1 ASC spheroid diameter values. For the in vitro analysis of multiple secreted proteins, the comparisons of growth factors, interleukins, and chemokines secreted by the ASC spheroids at week 1 of culture were performed with a *one-way* ANOVA test. The results in the graphs are expressed as mean ± standard error. Differences were considered statistically significant when $p < 0.05$. Statistical analyses were performed using GraphPad Prism 6.0 software (GraphPad Inc., La Jolla, CA, USA).

3. Results

3.1. ASC Spheroids Are Homogeneous in Size

After 7 days of culture, it was observed that ASC spheroids had an average diameter of 400 µm (Figure 2A,B). The Hematoxylin and Eosin staining showed a rounded cell morphology at the center of the spheroid and a fibroblastic morphology at its periphery after 7 days of culture (Figure 2C).

The PLA/CHA scaffolds have homogeneous filaments and porous sizes (Figure 2D). The EDS analysis showed the presence of Ca and P in the surface of the scaffold (Figure 2E), confirming the efficiency of CHA incorporation.

3.2. ASCs Are Released from Spheroids Spread on Most Areas of the PLA/CHA Scaffolds

Initially, the interaction of ASC spheroids with PLA/CHA scaffolds in vitro was evaluated using scanning electron microscope (SEM) analysis (Figure 3). After being seeded on the surface of the PLA/CHA scaffold, the ASC spheroids showed a high capacity of adhesion in all areas of the PLA/CHA scaffold (Figure 3A,B). In addition, ASCs derived from spheroids were able to interact with the surface of the PLA/CHA scaffolds (Figure 3C–H). These migrating cells mostly showed a fibroblastic morphology after adhering to the surface of the PLA/CHA scaffold (Figure 3F–H).

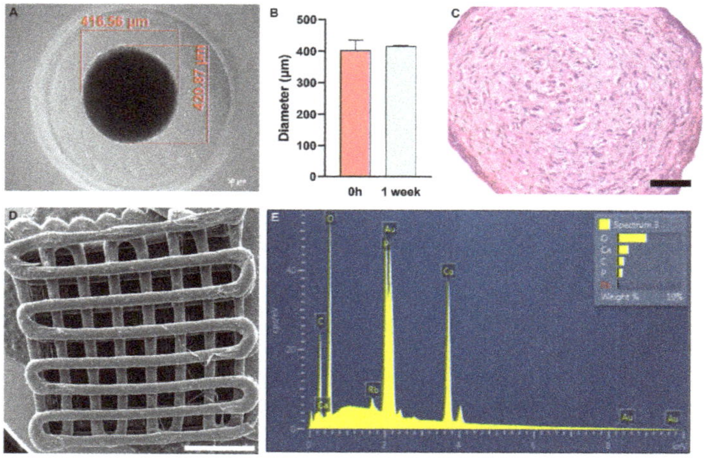

Figure 2. ASC spheroids are homogeneous in size, and 3D-printed PLA/CHA scaffolds contain calcium (Ca) and phosphorus (P) in their surface. (**A**) Representative image of one ASC spheroid inside one agarose resection. The red lines represent the measurements of width and length. (**B**) Diameter ratio of ASC spheroids after 0 h and 1 week of culture. (**C**) Hematoxylin and Eosin staining of one ASC spheroid after 1 week of culture. Notably, a rounded cell morphology at the center and a fibroblastic morphology at the periphery of spheroids are observed. (**D**) Scanning electron microscopy (SEM) image of one 3D-printed PLA/CHA scaffold. (**E**) Energy-dispersive X-ray spectroscopy (EDS) analysis of the PLA/CHA scaffolds showing the presence of Ca and P element signals originating from their surface. ASC: adipose-derived stem/stromal cell; and PLA/CHA: poly (lactic acid) and nanostructured hydroxyapatite doped with carbonate ions. Scale bars: (**C**): 50 µm; and (**D**): 2 mm.

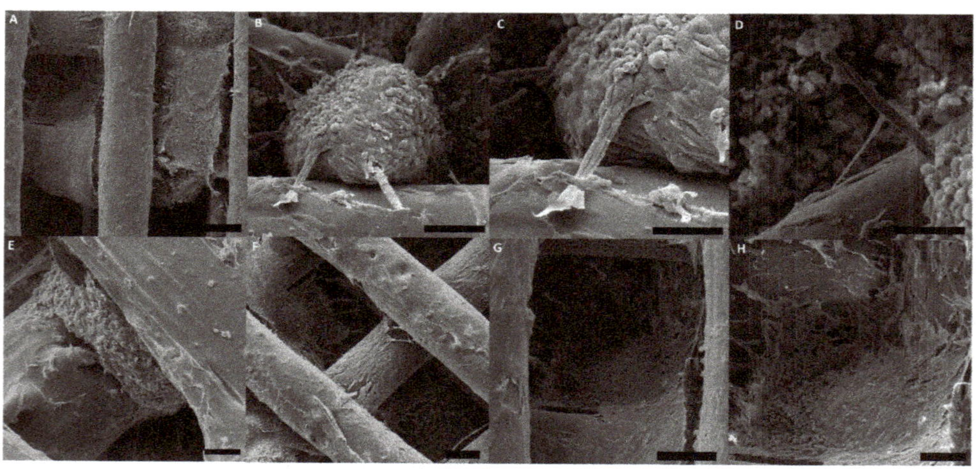

Figure 3. ASCs derived from spheroids were able to spread on the surface of the PLA/CHA scaffolds. (**A–H**) ASC spheroids seeded on the surface of PLA/CHA scaffold at week 1. (**A**) ASC spheroids fused

and spread on different areas of the PLA/CHA scaffold. (**B–D**) ASCs derived from spheroids were able to migrate and adhere to the surface of the PLA/CHA scaffold. (**E**) ASC spheroids fused to interact with the filaments of the PLA/CHA scaffold. (**F**) Cells from the ASC spheroids were able to interact and adhere inside the pores of the PLA/CHA scaffold. (**G,H**) The cells showed mostly a fibroblastic morphology when they made contact with the PLA/CHA scaffold. Scale bars: (**A**)—500 µm; (**B,C**)—50 µm; (**D**)—20 µm; (**E,F**)—200 µm; (**G**)—100 µm; and (**H**)—50 µm. ASC: adipose-derived stem/stromal cell; PLA/CHA: poly (lactic acid) and nanostructured hydroxyapatite doped with carbonate ions.

3.3. ASC Spheroidal Constructs Show a Low Secretion of Pro-Inflammatory Mediators and a High Secretion of VEGF In Vitro

The levels of the pro-inflammatory cytokines, i.e., GM-CSF (Figure 4A), IFNy (Figure 4B), IL-15 (Figure 4B), and IL-12p70 (Figure 4B), were more than 10 times reduced in PLA/CHA scaffolds seeded with ASC spheroids at week 1 of in vitro culture when compared to the levels of VEGF (Figure 4A) and IL-8 (Figure 4B), which are well known to promote angiogenesis, and IL-6 (Figure 4B) that plays a crucial role during osteogenesis.

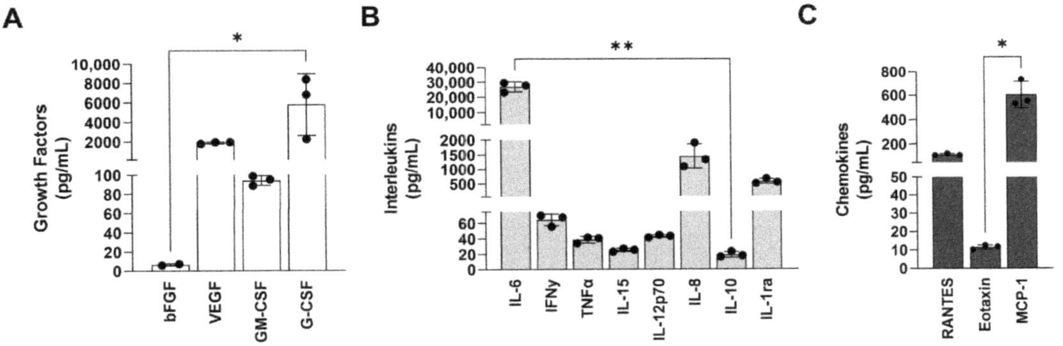

Figure 4. ASC spheroid constructs showed a high secretion of VEGF at week 1 of culture. Quantification of soluble mediators secreted by ASC spheroids seeded on PLA/CHA scaffold at week 1 of culture. (**A**) Growth factors. (**B**) Interleukins. (**C**) Chemokines. The data are expressed as mean ± SD. The asterisks indicate *p*-values obtained by non-paired *one-way* ANOVA followed by Kruskal–Wallis multiple comparisons (* $p < 0.05$; ** $p < 0.001$). ASC: adipose-derived stem/stromal cells; PLA/CHA: poly (lactic acid) and nanostructured hydroxyapatite doped with carbonate ions; IL: interleukin; IL-6: interleukin-6; IL-8: interleukin-8; IL-10: interleukin-10; IL-12p70: interleukin-12; IL-15: interleukin-15; IFN-y: interferon-γ; MCP-1: monocyte chemoattractant protein-1; VEGF: vascular endothelial growth factor; GM-CSF: granulocyte-macrophage colony-stimulating factor; and G-CSF: granulocyte colony-stimulating factor.

3.4. ASC Spheroidal Constructs Were Able to Produce New Bone Tissue at the Center of Critical-Size Defects

The biological response and new bone formation between the implanted groups were evaluated at months 1, 3, and 6 using histological analysis (Figure 5). Initially, we observed that none of the scaffolds lost their structural integrity, and no significant degradation was found after the in vivo assay. In addition, no inflammatory response was observed throughout the analyzed period by any of the implanted groups.

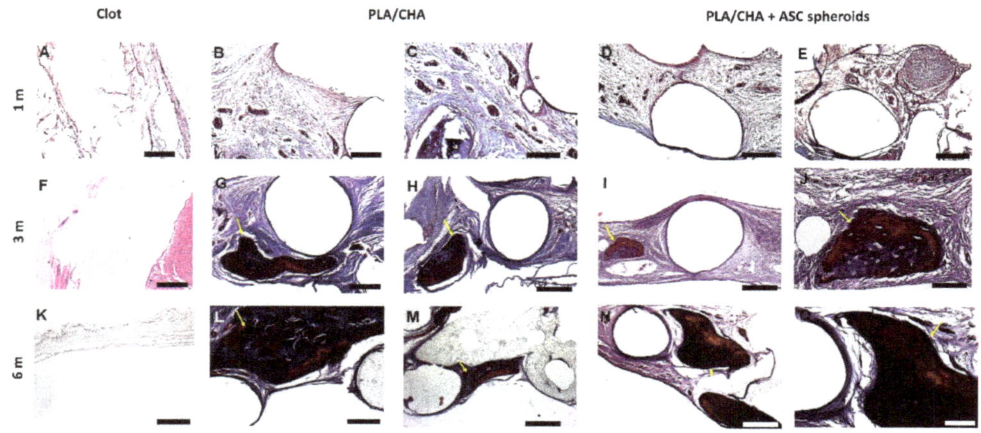

Figure 5. Histological analysis of calvaria defects after 1, 3, and 6 months of implantation. (**A**,**F**,**G**) Masson's Trichrome staining of Clot group after 1, 3, and 6 months. (**B**,**C**,**G**,**H**,**L**,**M**) Masson's Trichrome staining of PLA/CHA group after 1, 3, and 6 months, respectively. (**D**,**E**,**I**,**J**,**N**,**O**) Masson's Trichrome staining of with ASC spheroids construct after 1, 3 and 6 months, respectively. The new bone tissue formed at month 3 in PLA/CHA ((**G**,**H**), arrow) and PLA/CHA seeded with ASC spheroidal constructs groups ((**I**,**J**), arrow). The new bone tissue formed at month 6 in PLA/CHA ((**L**,**M**), arrow) and PLA/CHA seeded with ASC spheroidal constructs groups ((**N**,**O**), arrow). Results are representative of 6 rats/groups. Scale bars: (**A**)—200 μm; (**B**)—200 μm; (**C**)—100 μm; (**D**)—200 μm; (**E**)—200 μm; (**F**)—200 μm; (**G**)—200 μm; (**H**)—200 μm; (**I**)—200 μm; (**J**)—50 μm; (**K**)—200 μm; (**L**)—200 μm; (**M**)—200 μm; (**N**)—200 μm; and (**O**)—100 μm. ASC: adipose-derived stem/stromal cell; PLA/CHA: poly (lactic acid) and nanostructured hydroxyapatite doped with carbonate ions.

At month 1, in the PLA/CHA (Figure 5B,C) and in the PLA/CHA seeded with ASC spheroids groups (Figure 5D,E), no bone tissue was formed. In addition, in both groups, we observed connective tissue in close contact with areas of virtual biomaterial (Figure 5B–E). At month 3, in the PLA/CHA (Figure 5G,H) and in the PLA/CHA seeded with ASC spheroids groups (Figure 5I,J), it was possible to observe areas of new bone tissue in close contact with areas of virtual biomaterial, compared with month 1. At month 6 (Figure 5L–O), both groups presented a mature bone tissue formation. In the Clot groups at 1, 3, and 6 months, it was not possible to observe the formation of bone tissue (Figure 5A,F,K).

4. Discussion

In this study, ASC spheroids seeded on the surface of 3D-printed PLA/CHA scaffolds showed a high potential to promote bone regeneration in a rat calvaria critical-size defect model. Our ASC spheroidal constructs showed an active secretion of cytokines and growth factors when cultured in vitro, together with new bone tissue formed in the central area of the calvarial defect. To the best of our knowledge, this is the first study based on the synergic strategy (scaffold-free combined with scaffold-based) to promote bone regeneration using ASC spheroids combined with a 3D-printed PLA/CHA scaffold.

Initially, ASC spheroids were able to adhere and spread onto the PLA/CHA scaffold. In the past, several studies explored the seeding of MSCs on the surface of polymeric scaffolds for bone tissue engineering applications [12,24,25]. However, the level of cell density that they achieved was lower compared to the use of spheroids. Previous studies also showed the interaction between ASC and MSC spheroids with hydrogels and polymeric scaffolds [13,26–29]. However, these spheroids remained as single spheroids throughout the period of culture. In the study developed by Shanbhag and collaborators (2021) [29], it is possible to observe that, although the bmMSC spheroids show a high cell viability in the 3D-printed polymeric scaffolds, the spheroids showed a low adhesion on their surface. The pore size of polymeric scaffolds can influence the behavior and differentiation potential of stem/stromal cells [30–32]. Here, we used 3D printing technique to fabricate the PLA/CHA scaffolds to achieve a higher reproducibility in terms of the pores and overall scaffolds sizes. 3D printing allows the deposition of cells, growth factors, and biomaterials layer by layer in high resolution [33,34]. The ASC spheroids had an average size of 400 μm, and the pore sizes of the PLA/CHA scaffolds had an average of 500 μm. In addition, the scaffolds had a diameter of 8 mm to match the size of the rat critical-size calvaria defect model [35].

The secretory capacity of soluble mediators with the ASC spheroidal constructs was evaluated, and collectively, the levels of the pro-inflammatory cytokines, i.e., GM-CSF [36], IFNy [37], IL-15 [38], and IL-12p70 [39], showed a low secretion with ASC spheroidal constructs at 1 week of culture. In addition, the secretory levels of IL-8 and VEGF, described as key soluble mediators that stimulate angiogenesis during bone development [40–42], were high for ASC spheroidal constructs at 1 week of culture. In addition, it was observed that a high secretion of IL-6 played a relevant role during osteogenesis [43]. Few studies have explored the inflammatory profile of ASC or bmMSC spheroids for tissue engineering applications. Souza and collaborators (2019) [44] showed a low amount of pro-inflammatory cytokines produced with SAOS-2 spheroids after 3 days of culture. Similarly, Shanbhag and collaborators (2020) [29] also observed a low amount of pro-inflammatory cytokines with bmMSC spheroids after 2 days of culture.

Other studies showed that constructs based on hydrogels or printed polymers, aiming at bone tissue engineering strategies, can successfully release a high quantity of VEGF and anti-inflammatory cytokines in vitro [45–47]. However, the number of studies that explored the functional response of spheroids seeded in scaffolds is still scarce. Maia-Pinto and collaborators (2021) [28] produced osteoblast spheroids that could be seeded on PLA scaffolds. When the secretory activity was evaluated, it was possible to observe the secretion of VEGF, PDGF, and FGF. However, the levels of VEGF were 10 times lesser when compared to those observed in this study.

Following the implantation, the constructs were analyzed after 1, 3, and 6 months for inflammatory response and connective tissue and new bone tissue formation at the defect area. So far, few studies have been published that investigate bone regeneration using spheroids [7,12,29,48] and 3D-printed scaffolds seeded with spheroids [12,29].

Previously, Lee and collaborators (2020) [12] explored the regenerative capacity of calvaria defects with 3D-printed micro-chambers containing ASC spheroids. The authors also observed that the group containing the ASC spheroids resulted in a larger regenerated bone area. Shanbhag and collaborators (2021) [29] explored the regeneration capacity of calvaria defects of rats with MSC spheroids encapsulated in 3D-printed polymeric scaffolds. As the main result, the authors observed bone formation in the groups containing the spheroids. Yanagi and collaborators (2021) [48] showed that dedifferentiated fat cell spheroids were also able to form new bone tissue in calvaria defects of rats. Similarly, Yamada and collaborators (2022) [16] explored the capacity of ASC spheroids only to regenerate calvaria defects. The authors observed a significantly higher regeneration capacity when compared to that of the control group. However, these authors did not explore a synergic strategy for bone regeneration. In addition, the bone formation was evaluated up to 3 months after implantation, while in this study, the follow-up was performed for up to 6 months to better investigate mature bone formation.

In addition, previous studies showed the potential of PLA scaffolds to promote bone regeneration in critical-size models of rats [49–51]. Accordingly, we also observed bone formation in the PLA/CHA only group at 3 and 6 months, showing that the scaffold alone is indeed osteoinductive and can be successfully used in bone tissue engineering. Based on previous studies, the PLA/CHA's osteoinductive capacity can be attributed to the presence of CHA [52]. More importantly, the use of ASC spheroids in a synergic strategy revealed similar results, including the absence of inflammatory response, suggesting a regenerative potential of this strategy for bone critical-size defects. However, in a future study, biofabrication approaches such as 3D bioprinting can be applied to achieve a higher homogeneity for the ASC spheroid distribution on the surface of the PLA/CHA scaffold.

In summary, a set of results suggests that ASC spheroidal constructs hold the potential to promote bone regeneration in critical-size calvarial defects due to the following reasons [1]: a high adhesion and interaction with most areas of the 3D-printed PLA/CHA scaffold in vitro; [2] a high secretion of VEGF, IL-6, and IL-8, which suggests a strong osteogenic and angiogenic potential; [3] a low secretion of proinflammatory soluble mediator factors as GM-CSF, IFNy, IL-15, and IL-12p70; and [4] a higher quantity of new bone tissue together with a lower quantity of connective tissue formed after 6 months. To optimize results in the presence of ASC spheroids, we hypothesize that two critical factors need to be improved: [1] the seeding of ASC spheroids using 3D bioprinting to increase the homogeneous distribution of spheroids on scaffolds; [2] and the use of polymers that show a higher biodegradability capacity compared with PLA.

5. Conclusions

Therefore, the synergic strategy explored in this study, combining ASC spheroids and the 3D-printed PLA/CHA scaffolds, allowed a high cell density, due to the successful adhesion of the spheroids on the surface of the scaffold, together with a high secretion of VEGF, IL-6, and IL-8. Both the PLA/CHA and ASC spheroidal constructs were able to form new bone tissue in vivo in a critical-size defect rat model. Future studies using 3D bioprinting approaches are necessary to optimize the seeding of the ASC spheroids on the surface of PLA/CHA scaffolds, leading to a more homogeneous organization of the constructs and better in vivo regenerative responses. In addition, the use of PCL scaffolds seeded with ASC spheroids can optimize the bone formation in vivo as they have higher biodegradability properties.

Author Contributions: Conceptualization: G.S.K., A.M.R., M.D.C.-M., J.M.G. and L.S.B.; Methodology: G.S.K., T.N.P., B.R.F.V., S.C.S., R.R., M.J.U., A.T.N.N.A. and G.G.A.; Formal analysis: J.M.G. and L.S.B.; Investigation: G.S.K.; Resources: M.D.C.-M., J.M.G. and L.S.B.; Writing—original draft: G.S.K. and L.S.B.; Writing—review & editing: G.S.K., J.M.G. and L.S.B.; Visualization: B.R.F.V. and A.T.N.N.A.; Supervision: L.S.B.; Project administration: J.M.G. and L.S.B.; Funding acquisition: A.M.R., M.D.C.-M., J.M.G. and L.S.B. All authors have read and agreed to the published version of the manuscript.

Funding: This study was supported by the National Council for Scientific and Technological Development (CNPq) (Finance Code: 307460/2019-3), Carlos Chagas Filho Foundation for Research Support of the State of Rio de Janeiro (Faperj) (Finance Code: E26/202.682/2018), and the Office of Naval Research (ONR) (Finance Code: N62909-21-1-2091). The authors also would like to thank the Coordenação de Aperfeiçoamento de Pessoal de Nível Superior, Brazil (CAPES) (Finance Code: 88882.366181/2019-01), and the National Centre of Science and Technology on Regenerative Medicine (CNPq/Faperj) (INCT Regenera) (http://www.inctregenera.org.br/, accessed on 6 October 2023).

Data Availability Statement: Data is contained within the article.

Acknowledgments: We thank the National Institute of Metrology, Quality and Technology (INMETRO, RJ, Brazil), the Nucleus of Multidisciplinary Research in Biology (Numpex-Bio, Federal University of Rio de Janeiro, RJ, Brazil), and LEA/LPCO/UFF for the use of their facilities.

Conflicts of Interest: The authors declare no conflict of interest.

References

1. Pereira, H.; Cengiz, I.; Silva, F.; Reis, R.; Oliveira, J. Scaffolds and coatings for bone regeneration. *J. Mater. Sci. Mater. Med.* **2020**, *31*, 27. [CrossRef] [PubMed]
2. Majidinia, M.; Sadeghpour, A.; Yousefi, B. The roles of signaling pathways in bone repair and regeneration. *J. Cell. Physiol.* **2018**, *233*, 2937–2948. [CrossRef]
3. Baldwin, P.; Li, D.; Auston, D.; Mir, D.; Yoon, R.; Koval, K. Autograft, allograft, and bone graft substitutes: Clinical evidence and indications for use in the setting of orthopaedic trauma surgery. *J. Orthop. Trauma* **2019**, *33*, 203–213. [CrossRef] [PubMed]

4. Martin, W.B.; Sicard, R.; Namin, S.; Ganey, T. Methods of cryoprotectant preservation: Allogeneic cellular bone grafts and potential effects. *BioMed Res. Int.* **2019**, *2019*, 5025398. [CrossRef] [PubMed]

5. Achilli, T.M.; Meyer, J.; Morgan, J.R. Advances in the formation, use and understanding of multi-cellular spheroids. *Expert Opin. Biol. Ther.* **2012**, *12*, 1347–1360. [CrossRef] [PubMed]

6. Desai, K.; Salve, P.; Sapkal, N.; Dave, J.; Tomar, J. Scaffold-free spheroids derived from stem cells for tissue-engineering applications. *Crit. Rev. Biomed. Eng.* **2018**, *46*, 469–493. [CrossRef] [PubMed]

7. Suenaga, H.; Furukawa, K.; Suzuki, Y.; Takato, T.; Ushida, T. Bone regeneration in calvarial defects in a rat model by implantation of human bone marrow-derived mesenchymal stromal cell spheroids. *J. Mater. Sci. Mater. Med.* **2015**, *26*, 254. [CrossRef]

8. Shen, F.H.; Werner, B.C.; Liang, H.; Shang, H.; Yang, N.; Li, X.; Shimer, A.L.; Balian, G.; Katz, A.J. Implications of adipose-derived stromal cells in a 3D culture system for osteogenic differentiation: An in vitro and in vivo investigation. *Spine J.* **2013**, *13*, 32–43. [CrossRef]

9. Lee, J.; Lee, S.; Ahmad, T.; Madhurakkat Perikamana, S.K.; Lee, J.; Kim, E.M.; Shin, H. Human adipose-derived stem cell spheroids incorporating platelet-derived growth factor (PDGF) and bio-minerals for vascularized bone tissue engineering. *Biomaterials* **2020**, *255*, 120192. [CrossRef]

10. Mironov, V.; Visconti, R.P.; Kasyanov, V.; Forgacs, G.; Drake, C.J.; Markwald, R.R. Organ printing: Tissue spheroids as building blocks. *Biomaterials* **2009**, *30*, 2164–2174. [CrossRef]

11. Lee, J.; Lee, S.; Huh, S.J.; Kang, B.J.; Shin, H. Directed regeneration of osteochondral tissue by hierarchical assembly of spatially organized composite spheroids. *Adv. Sci.* **2022**, *9*, e2103525. [CrossRef] [PubMed]

12. Lee, J.; Seok, J.M.; Huh, S.J.; Byun, H.; Lee, S.; Park, S.A.; Shin, H. 3D printed micro-chambers carrying stem cell spheroids and pro-proliferative growth factors for bone tissue regeneration. *Biofabrication* **2020**, *13*, 015011. [CrossRef]

13. Findeisen, L.; Bolte, J.; Vater, C.; Petzold, C.; Quade, M.; Müller, L. Cell spheroids are as effective as single cells suspensions in the treatment of critical-sized bone defects. *BMC Musculoskelet. Disord.* **2021**, *22*, 401. [CrossRef] [PubMed]

14. Gurumurthy, B.; Bierdeman, P.C.; Janorkar, A.V. Spheroid model for functional osteogenic evaluation of human adipose derived stem cells. *J. Biomed. Mater. Res. Part A* **2017**, *105*, 1230–1236. [CrossRef]

15. Ahmad, T.; Shin, H.J.; Lee, J.; Shin, Y.M.; Perikamana, S.K.M.; Park, S.Y.; Jung, H.S.; Shin, H. Fabrication of in vitro 3D mineralized tissue by fusion of composite spheroids incorporating biomineral-coated nanofibers and human adipose-derived stem cells. *Acta Biomater.* **2018**, *74*, 464–477. [CrossRef] [PubMed]

16. Yamada, Y.; Okano, T.; Orita, K.; Makino, T.; Shima, F.; Nakamura, H. 3D-cultured small size adipose-derived stem cell spheroids promote bone regeneration in the critical-sized bone defect rat model. *Biochem. Biophys. Res. Commun.* **2022**, *603*, 57–62. [CrossRef]

17. Ovsianikov, A.; Khademhosseini, A.; Mironov, V. The synergy of scaffold-based and scaffold-free tissue engineering strategies. *Trends Biotechnol.* **2018**, *36*, 348–357. [CrossRef]

18. Geng, Z.; Yuan, Q.; Zhuo, X.; Li, Z.; Cui, Z.; Zhu, S.; Liang, Y.; Liu, Y.; Bao, H.; Li, X.; et al. Synthesis, Characterization, and Biological Evaluation of Nanostructured Hydroxyapatite with Different Dimensions. *Nanomaterials* **2017**, *7*, 38. [CrossRef]

19. Mosaad, K.E.; Shoueir, K.R.; Saied, A.H.; Dewidar, M.M. New Prospects in Nano Phased Co-substituted Hydroxyapatite Enrolled in Polymeric Nanofiber Mats for Bone Tissue Engineering Applications. *Ann. Biomed. Eng.* **2021**, *49*, 2006–2029. [CrossRef]

20. Avanzi, I.R.; Parisi, J.R.; Souza, A.; Cruz, M.A.; Martignago, C.C.S.; Ribeiro, D.A.; Braga, A.R.C.; Renno, A.C. 3D-printed hydroxyapatite scaffolds for bone tissue engineering: A systematic review in experimental animal studies. *J. Biomed. Mater. Res. Part B Appl. Biomater.* **2023**, *111*, 203–219. [CrossRef]

21. Soleymani, S.; Naghib, S.M. 3D and 4D printing hydroxyapatite-based scaffolds for bone tissue engineering and regeneration. *Heliyon* **2023**, *9*, e19363. [CrossRef]

22. Baptista, L.S.; Amaral, R.; Carias, R.; Aniceto, M.; Claudio-da-Silva, C.; Borojevic, R. An alternative method for the isolation of mesenchymal stromal cells derived from lipoaspirate samples. *Cytotherapy* **2009**, *11*, 706–715. [CrossRef]

23. Palhares, T.N.; Menezes, L.M.; Kronemberger, G.S.; Borchio, P.; Baptista, L.S.; Pereira, L.; Silva, E. Production and characterization of poly (lactic acid)/nanostructured carboapatite for 3D printing of bioactive scaffolds for bone tissue engineering. *3D Print. Addit. Manuf.* **2021**, *8*, 227–237. [CrossRef] [PubMed]

24. Johnson, Z.M.; Yuan, Y.; Li, X.; Jashashvili, T.; Jamieson, M.; Urata, M.; Chen, Y.; Chai, Y. Mesenchymal stem cells and three-dimensional-osteoconductive scaffold regenerate calvarial bone in critical size defects in swine. *Stem Cells Transl. Med.* **2021**, *10*, 1170–1183. [CrossRef] [PubMed]

25. Lu, G.; Xu, Y.; Liu, Q.; Chen, M.; Sun, H.; Wang, P.; Li, X.; Wang, Y.; Li, X.; Hui, X.; et al. An instantly fixable and self-adaptive scaffold for skull regeneration by autologous stem cell recruitment and angiogenesis. *Nat. Commun.* **2022**, *13*, 2499. [CrossRef] [PubMed]

26. Laschke, M.; Schank, T.; Scheuer, C.; Kleer, S.; Shadmanov, T.; Eglin, D.; Alini, M.; Menger, M. In vitro osteogenic differentiation of adipose-derived mesenchymal stem cell spheroids impairs their in vivo vascularization capacity inside implanted porous polyurethane scaffolds. *Acta Biomater.* **2014**, *10*, 4226–4235. [CrossRef]

27. Ho, S.; Hung, B.; Heyrani, N.; Lee, M.; Leach, J. Hypoxic preconditioning of mesenchymal stem cells with subsequent spheroid formation accelerates repair of segmental bone defects. *Stem Cells* **2018**, *36*, 1393–1403. [CrossRef] [PubMed]

28. Maia-Pinto, M.; Brochado, A.C.; Teixeira, B.N.; Sartoretto, S.; Uzeda, M.; Alves, A.; Alves, G.G.; Calasans-Maia, M.D.; Thiré, R. Biomimetic mineralization on 3D printed pla scaffolds: On the response of human primary osteoblasts spheroids and in vivo implantation. *Polymers* **2020**, *13*, 74. [CrossRef]

29. Shanbhag, S.; Suliman, S.; Mohamed-Ahmed, S.; Kampleitner, C.; Hassan, M.N.; Heimel, P.; Dobsak, T.; Tangl, S.; Bolstad, A.I. Mustafa, K. Bone regeneration in rat calvarial defects using dissociated or spheroid mesenchymal stromal cells in scaffold hydrogel constructs. *Stem Cell Res. Ther.* **2021**, *12*, 575. [CrossRef]
30. Matsiko, A.; Gleeson, J.P.; O'Brien, F.J. Scaffold mean pore size influences mesenchymal stem cell chondrogenic differentiation and matrix deposition. *Tissue Eng. Part A* **2015**, *21*, 486–497. [CrossRef]
31. Chen, X.; Fan, H.; Deng, X.; Wu, L.; Yi, T.; Gu, L.; Zhou, C.; Fan, Y.; Zhang, X. Scaffold Structural Microenvironmental Cues to Guide Tissue Regeneration in Bone Tissue Applications. *Nanomaterials* **2018**, *8*, 960. [CrossRef] [PubMed]
32. Qi, P.; Ning, Z.; Zhang, X. Synergistic effects of 3D chitosan-based hybrid scaffolds and mesenchymal stem cells in orthopaedic tissue engineering. *IET Nanobiotechnol.* **2023**, *17*, 41–48. [CrossRef] [PubMed]
33. Rahman, Z.; Barakh Ali, S.F.; Ozkan, T.; Charoo, N.A.; Reddy, I.K.; Khan, M.A. Additive Manufacturing with 3D Printing Progress from Bench to Bedside. *AAPS J.* **2018**, *20*, 101. [CrossRef] [PubMed]
34. Vaz, V.M.; Kumar, L. 3D Printing as a Promising Tool in Personalized Medicine. *AAPS PharmSciTech* **2021**, *22*, 49. [CrossRef]
35. Vajgel, A.; Mardas, N.; Farias, B.C.; Petrie, A.; Cimões, R.; Donos, N. A systematic review on the critical size defect model. *Clin Oral Implant. Res.* **2014**, *25*, 879–893. [CrossRef]
36. Bhattacharya, P.; Budnick, I.; Singh, M.; Thiruppathi, M.; Alharshawi, K.; Elshabrawy, H.; Holterman, M.J.; Prabhakar, B.S. Dual Role of GM-CSF as a Pro-Inflammatory and a Regulatory Cytokine: Implications for Immune Therapy. *J. Interferon Cytokine Res.* **2015**, *5*, 585–599. [CrossRef] [PubMed]
37. Kopitar-Jerala, N. The role of interferons in inflammation and inflammasome activation. *Front. Immunol.* **2017**, *8*, 873. [CrossRef]
38. Warner, S.; Nair, A.; Marpadga, R.; Chubinskaya, S.; Doherty, M.; Valdes, A.; Scanzello, C. IL-15 and IL15RA in osteoarthritis: Association with symptoms and protease production, but not structural severity. *Front. Immunol.* **2020**, *11*, 1385. [CrossRef]
39. Gee, K.; Guzzo, C.; Mat, N.F.; Ma, W.; Kumar, A. The IL-12 family of cytokines in infection, inflammation and autoimmune disorders. *Inflamm. Allergy-Drug Targets* **2009**, *8*, 40–52. [CrossRef]
40. Hankenson, K.; Dishowitz, M.; Gray, C.; Schenker, M. Angiogenesis in bone regeneration. *Injury* **2011**, *42*, 556–561. [CrossRef]
41. Hou, Y.; Ryu, C.H.; Jun, J.; Kim, S.M.; Jeong, C.M.; Jeun, S.-S. IL-8 enhances the angiogenic potential of human bone marrow mesenchymal stem cells by increasing vascular endothelial growth factor. *Cell Biol. Int.* **2014**, *38*, 1050–1059. [CrossRef] [PubMed]
42. Hu, K.; Olsen, B. Osteoblast-derived VEGF regulates osteoblast differentiation and bone formation during bone repair. *J. Clin. Investig.* **2015**, *126*, 509–526. [CrossRef]
43. Huh, J.E.; Lee, S.Y. IL-6 is produced by adipose-derived stromal cells and promotes osteogenesis. *Biochim. Biophys. Acta* **2013**, *1833*, 2608–2616. [CrossRef] [PubMed]
44. Souza, W.; Piperni, S.G.; Laviola, P.; Rossi, A.L.; Rossi, M.I.D.; Archanjo, B.S.; Leite, P.E.; Fernandes, M.H.; Rocha, L.A.; Granjeiro, J.M.; et al. The two faces of titanium dioxide nanoparticles bio-camouflage in 3D bone spheroids. *Sci. Rep.* **2019**, *9*, 9309. [CrossRef]
45. Bao, X.; Zhu, L.; Huang, X.; Tang, D.; He, D.; Shi, J.; Xu, G. 3D biomimetic artificial bone scaffolds with dual-cytokines spatiotemporal delivery for large weight-bearing bone defect repair. *Sci. Rep.* **2017**, *7*, 7814. [CrossRef] [PubMed]
46. Zehnder, T.; Boccaccini, A.R.; Detsch, R. Biofabrication of a co-culture system in an osteoid-like hydrogel matrix. *Biofabrication* **2017**, *9*, 025016. [CrossRef]
47. Bian, J.; Cai, F.; Chen, H.; Tang, Z.; Xi, K.; Tang, J.; Wu, L.; Xu, Y.; Deng, L.; Gu, Y.; et al. Modulation of local overactive inflammation via injectable hydrogel microspheres. *Nano Lett.* **2021**, *21*, 2690–2698. [CrossRef]
48. Yanagi, T.; Kajiya, H.; Fujisaki, S.; Maeshiba, M.; Yanagi, A.; Yamamoto-M, N.; Kakura, K.; Kido, H.; Ohno, J. Three-dimensional spheroids of dedifferentiated fat cells enhance bone regeneration. *Regen. Ther.* **2021**, *18*, 472–479. [CrossRef]
49. Zhang, H.; Mao, X.; Du, Z.; Jiang, W.; Han, X.; Zhao, D.; Han, D.; Li, Q. Three dimensional printed macroporous polylactic acid/hydroxyapatite composite scaffolds for promoting bone formation in a critical-size rat calvarial defect model. *Sci. Technol. Adv. Mater.* **2016**, *17*, 136–148. [CrossRef]
50. Khoobi, M.M.; Naddaf, H.; Hoveizi, E.; Mohammadi, T. Silymarin effect on experimental bone defect repair in rat following implantation of the electrospun PLA/carbon nanotubes scaffold associated with Wharton's jelly mesenchymal stem cells. *J. Biomed. Mater. Res. Part A* **2020**, *108*, 1944–1954. [CrossRef]
51. Bahraminasab, M.; Talebi, A.; Doostmohammadi, N.; Arab, S.; Ghanbari, A.; Zarbakhsh, S. The healing of bone defects by cell-free and stem cell-seeded 3D-printed PLA tissue-engineered scaffolds. *J. Orthop. Surg. Res.* **2022**, *17*, 320. [CrossRef] [PubMed]
52. Lemos, V.P.A.; Rossi, A.M.; Granjeiro, J.M.; Calasans-Maia, M.D.; Sartoretto, S.C.; Sesterheim, P.; Boeckel, D.G.; Grivicich, I.; Camassola, M. Osteogenic potential of nanostructured carbonated hydroxyapatite microspheres associated with mesenchymal stem cell in vitro and in vivo. *Cytotherapy* **2021**, *23*, 29. [CrossRef]

Journal of
Functional Biomaterials

MDPI

Article

Analysis of Bone Mineral Density and Bone Quality of Cortical Bone in the Human Hyoid Body and Histological Observation of the Entheses

Masaaki Kasahara [1,*], Tomoko Someya [1], Kei Kitamura [2], Genji Watanabe [3], Satoru Matsunaga [3], Shinichi Abe [3] and Masayuki Hattori [1]

1 Department of Dental Materials Science, Tokyo Dental College, Tokyo 101-006, Japan; someyatomoko@tdc.ac.jp (T.S.); hattori@tdc.ac.jp (M.H.)
2 Department of Histology and Developmental Biology, Tokyo Dental College, Tokyo 101-006, Japan; kitamurakei@tdc.ac.jp
3 Department of Anatomy, Tokyo Dental College, Tokyo 101-006, Japan; watanabegenji@tdc.ac.jp (G.W.); matsuna@tdc.ac.jp (S.M.); abesh@tdc.ac.jp (S.A.)
* Correspondence: kasaharamasaaki@tdc.ac.jp; Tel.: +81-3-6380-9266

Abstract: The hyoid is the only bone in the human body that is completely independent, not forming a joint with any other bone; its position is maintained by the suprahyoid and infrahyoid muscles, as well as several ligaments. The purpose of this study was to ascertain the effect of the functional pressure arising from these muscles and ligaments on the hyoid body structure from its bone mineral density, bone quality, and histological observations. The area between the mesial-most part of each lesser horn and the center of the hyoid body was divided equally into four measurement regions. We conducted histological investigations at each measurement region and observed the entheses. To analyze bone mass and bone quality, we also measured bone mineral density (BMD) and analyzed biological apatite (BAp) crystallite orientation in the same regions. Histological observations identified periosteal insertions and fibrocartilaginous entheses. There was no significant difference in BMD between any of the measurement regions, but the preferential orientation of BAp crystallites was stronger in the infrahyoid muscles and ligaments, where fibrocartilaginous entheses are found, than in other places. This suggests that the functional pressure at these sites might exert a major effect not only on the morphological characteristics of the entheses but also on bone quality.

Keywords: hyoid bone; entheses; bone quality; apatite crystallite orientation; bone mineral density

1. Introduction

The human hyoid bone is a horseshoe-shaped bone that is involved in major functional movements in the maxillofacial region, including respiration, mastication, and swallowing. It is also the only bone in the human body that is known to be completely independent, not forming a joint with any other bone. Ten different muscles (making a total of 20 including those on both sides) and ligaments contribute to maintaining its position [1]. The hyoid is, thus, a unique bone that is connected via these muscles and ligaments to the mandible and temporal bones, scapulae, tongue, manubrium of the sternum, and thyroid cartilage. The muscles that attach to the hyoid are broadly divided into the suprahyoid and infrahyoid muscles, the extrinsic muscles of the tongue, and the pharyngeal constrictors. Of these, the muscles with attachments to the body of the hyoid are the stylohyoid muscle, mylohyoid muscle (MH), and geniohyoid muscle (GH) among the suprahyoid muscles, and the sternohyoid muscle (SH), thyrohyoid muscle (TH) and omohyoid muscle among the infrahyoid muscles. Pearson et al. reported that the GH and MH muscles are involved in the movement of the hyoid during swallowing [2]. Gehrking et al. reported that the infrahyoid muscles depress the hyoid during swallowing and vocalization [3]. The functional pressure of the muscles from these physiological movements can be expected to have an effect on the morphology and structure of the hyoid. Capasso et al. reported that the morphology of the homo erectus hyoid differs from that of contemporary humans in having tiny greater horns and being without impressions from the attachment of the major suprahyoid muscles [4]. They suggested that this hyoid shape is similar to those of non-human and pre-human genera and that they would have had a reduced capacity for elevating the hyoid bone compared with that of contemporary humans. Shimizu et al. suggested that the force exerted by the muscles that attach to the hyoid might affect the symphysis between the hyoid body (HB) and the greater horns in older individuals [5]. These reports indicate that the hyoid bone, which is constantly subjected to functional pressure from the suprahyoid and infrahyoid muscles, has optimized its structure and morphology to acquire the bone strength needed to cope with such a mechanical environment. However, the different types of muscle functional pressure and where and how they exert their effects remain unknown.

Recently, attention has been focused on bone quality, in the evaluation of bone. Bone quality is defined as a factor other than bone mineral density that affects bone strength as a promising index for ascertaining the relationship between the strength of bone and its structural characteristics [6]. Bone quality includes structural and material properties, with structural properties primarily including macroscopic bone structure and microscopic cortical bone porosity, and material properties including microdamage, degree of calcification, bone matrix, and bone metabolic turnover. Biological apatite (BAp) crystallite/collagen orientation in the bone matrix of material properties responds particularly strongly to local stress, and evaluating this orientation can reportedly enable the identification of regions under mechanical stress and its direction [7–11]. Nakano et al. used microbeam X-ray diffraction system (XRD) analysis to conduct a quantitative evaluation of BAp crystallite orientation in the trunk and limb bones of laboratory animals and found that mechanical load on the bone was strongly associated with BAp crystallite orientation [8]. A similar tendency has also been reported in human bones [9,10]. In a previous study, the authors showed that the structural characteristics of the tendon–bone entheses of the masticatory muscles in the mandibular coronoid process are associated with BAp crystallite orientation and mechanical strength and suggested that the structure and morphology of the coronoid process might be optimized to provide bone strength sufficient to withstand muscle functional pressure [12]. It should, therefore, be possible to predict the effects of functional pressure from muscles and ligaments on the local structure of the hyoid bone with a high degree of accuracy by conducting a qualitative analysis of the hyoid. Our purpose in this study was to conduct histological investigations of the entheses at the attachment sites of muscles, tendons, and ligaments to the HB, and to carry out bone mineral density and bone quality evaluations of the cortical bone at these attachment sites in order to identify the structural characteristics of the HB under the load environment imposed by the functional pressure of muscles.

2. Materials and Methods

2.1. Samples

Prior to these experiments, the hyoid bones of eight adult Japanese cadavers (mean age 82.4 ± 12.1 years; 5 men and 3 women) in the collection of the Department of Anatomy of Tokyo Dental College were harvested as specimens. To prevent spoilage, all cadavers used in the study were perfusion-fixed in 10% neutral formalin from the femoral artery and dehydrated in 70% ethanol. Only hyoid bones from patients with no history of metabolic bone disease or morphological abnormalities were used. In addition, in the women's cadavers, those with no history of postmenopausal osteoporosis or other noted bone diseases were used. This study was approved by the Ethics Committee of Tokyo Dental College (No. 1001).

2.2. Preparation of Samples

The hyoid bones were divided into specimens for histological observation and specimens for bone mineral density (BMD) and bone quality evaluation. Before each of these evaluations, three reference axes were designated, with the X-axis defined as the long axis of the HB, the Y-axis as the direction perpendicular to the long axis of the HB, and the Z-axis as the anterior–posterior direction (Figure 1). The hyoid bones for use in histological observations were immersion-fixed in 4% paraformaldehyde phosphate buffer solution and decalcified in 10% ethylenediaminetetraacetic acid (EDTA) for 4 weeks. The decalcified specimens were embedded in paraffin by the usual method. Thin slices of the embedded specimens with a thickness of 5 μm were prepared as sagittal sections parallel to the Y-axis. Hematoxylin-eosin (H-E) staining was conducted to enable morphological histological observations, and toluidine blue staining was performed to identify the fibrocartilaginous layer (n = 2). All the hyoid bones for use in bone quality evaluation were scanned by microscopic computed tomography (micro-CT) (μCT-50, Scanco Medical AG, Wangen-Brüttisellen, Switzerland), their internal structure was observed, and their bone mineral density (BMD) was measured. They were then embedded in autopolymerizing resin (SCANDIPLEX, SCANDIA, Hagen, Germany), after which they were sliced with a saw microtome (SP1600, Leica, Wetzlar, Germany) with a 300 μm wide blade in the sagittal direction, parallel to the Y axis. The roughness of the cut surface was removed with a polisher (ECOMET3, BUEHLER, Uzwil, Switzerland), and they were polished with waterproof abrasive paper (#400 → #800 → #1200). Specimens for use in microbeam X-ray diffraction system (XRD) were prepared with a thickness of 200 μm.

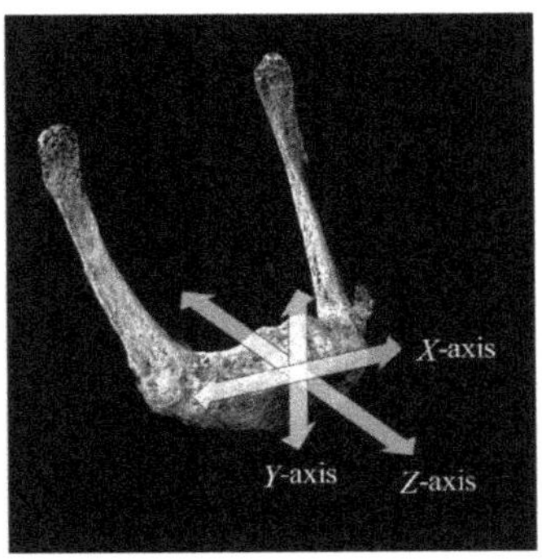

Figure 1. Setting of the coordinate axes.

Three axes were designated in relation to the hyoid bone, with the X-axis defined as the long axis of the hyoid body (HB), the Y-axis as the direction perpendicular to the long axis of the HB, and the Z-axis as the anterior–posterior direction.

2.3. Setting of Measurement Sites

The region of interest was the body of the hyoid bone. The measurement regions comprised the area between the mesial-most parts of each lesser horn and the center of the HB divided into four equal parts. These quarters were designated as α, β, γ, and δ, respectively (Figure 2A), starting from the mesial-most part of the lesser horn and going toward the center of the HB. In each quarter, the BMD and BAp crystallite orientation were measured at four designated points around the circumference of the cortical bone of the HB. The uppermost part of the center of the cortical bone of the HB was designated as point a (superior HB) and the bottommost part as point c (inferior HB), after which a perpendicular line bisecting the line joining points a and c was drawn, with the ends being designated as point b (anterior HB) and point d (posterior HB) (Figure 2B).

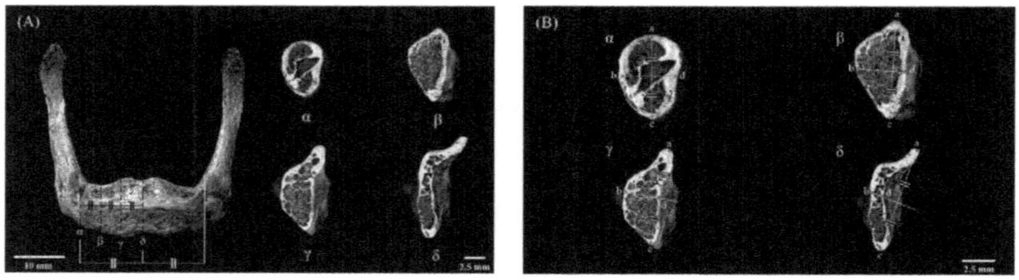

Figure 2. Designation of the region of interest in the human hyoid bone. (**A**) The region of interest was the body of the hyoid between the mesial-most parts of the lesser horns bilaterally. The measurement regions comprised the area of the hyoid body (HB) between the mesial-most part of the lesser horn and the center of the HB divided into four equal parts (α, β, γ, and δ). (**B**) Four points around the circumference of the cortical bone of the HB were used for analysis. Point a; uppermost part of the center of the cortical bone of the HB (superior HB), Point b; anterior-most part of the center of the cortical bone of the HB (anterior HB), Point c; bottommost part of the center of the cortical bone of the HB (inferior HB), Point d; posterior -most part of the center of the cortical bone of the HB (posterior HB) The equals sign (=) indicates lines of equal length.

2.4. Scanning Electron Microscopy

The cortical bone structure was observed by scanning the cut surfaces of the specimens with a scanning electron microscope (SEM) (SU6600, Hitachi, Tokyo, Japan). The specimen for surface observation was taken from the remaining HB after the sections for evaluation of BAp crystallite orientation were taken. The specimens were polished with waterproof abrasive paper up to #1200 and were treated with Au-Pd sputtering SEM observation. The setting conditions were an acceleration voltage of 25.0 kV and a working distance of 10.9 mm. The measurement points were the sites of interstitial lamellae containing osteons, The measurement points were the sites of interstitial lamellae containing osteons, as BAp crystallite does not exist in the Havers canal [8].

2.5. Measurement of Bone Mineral Density

All the specimens were scanned by micro-CT (μCT50, Scanco Medical AG, Wangen-Brüttisellen, Switzerland) with the following scanning conditions: tube voltage 90 kV, tube current 200 μA, matrix size 1024 × 1024, and voxel size 48.4 μm. To observe the internal structure, the images thus obtained were used for three-dimensional (3D) reconstruction using 3D structural analysis software ver. 10.01 (TRI/3D-BON, Ratoc system Engineering Corporation, Tokyo, Japan). For quantitative analysis of BMD (mgHA/cm^3) of the HB cortical bone at each measurement point, a set of BMD phantoms with X-ray absorption coefficients equivalent to BMDs of 800, 700, 600, 500, 400, 300, and 200 mgHA/cm^3 (Ratoc System Engineering Corporation, Tokyo, Japan) were scanned with the same condition. The CT gray value of the samples was converted to BMD using the scan data of the BMD phantoms by 3D analysis software ver. 6.6 (Scanco Medical AG, Wangen-Brüttisellen, Switzerland).

2.6. Biological Apatite (BAp) Crystallite Orientation

Qualitative analysis of BAp crystallite orientation was performed using an optical-system curved imaging plate (IP) XRD unit (D/MAX RAPIDII-CMF, Rigaku Corporation, Tokyo, Japan). Measurements were made by two methods, one using a transmission-based optical system and the other a reflection-based optical system, with Cu-Kα radiation used as the radiation source in both cases. The device was set to a tube voltage of 40 kV and a tube current of 30 mA. The incident beam was set to a microcircle with a of diameter 100 μm. The diffraction X-ray beam was detected with a curved IP. The measurements for each specimen were used to calculate the X-ray diffraction intensity ratios in the directions of the *X*-, *Y*-, and *Z*-axis for quantitative analysis. The measurement conditions were based on Miyabe et al.'s method for the transmission-based optical system and Nakano et al.'s method for the reflection-based optical system [8,9]. Diffraction intensity along the *X*-axis was measured with the reflection-based optical system, and those along the *Y*- and *Z*-axis were measured with the transmission-based optical system. The incident X-ray beam took account of the Bragg angles of the (002) and (310) peaks. The surface of the specimen was irradiated with the X-ray beam taking into account the Bragg angles θ = 12.95– and 19.9º of the (310) peak. The measurement time at each measurement point was 120 s. The XRD intensity ratio of the two diffraction peaks of the (002) and (310) planes was calculated from the acquired data using two-dimensional data processing software ver. 2.1.6 (2DP, Rigaku Corporation, Tokyo, Japan). This has previously been reported to be a suitable index for evaluating apatite orientation [8,13].

Measurements were conducted thrice at each measurement point, and the mean value was calculated. The XRD intensity ratio of non-oriented apatite powder was 1.53 with the transmission-based system and 1.06 with the reflection-based system.

2.7. Statistical Analysis

Two-way analysis of variance (ANOVA) of the mean values for each measurement point was conducted with the measurement region and measurement point as factors, and Tukey's multiple comparison test was performed. Statistical analysis was carried out using statistical software ver. 4.05 (BellCurve for Excel, Social Survey Research Information, Tokyo, Japan), with $p < 0.05$ regarded as statistically significant.

3. Results

3.1. Observation of Cortical Bone Surface

Figure 3 shows an SEM image of a cross-section of the cortical bone of the HB. Healthy osteons and surrounding interstitial lamellae were evident. This confirmed that the interstitial lamellae could be adequately targeted by an incident beam with a 100 μm diameter.

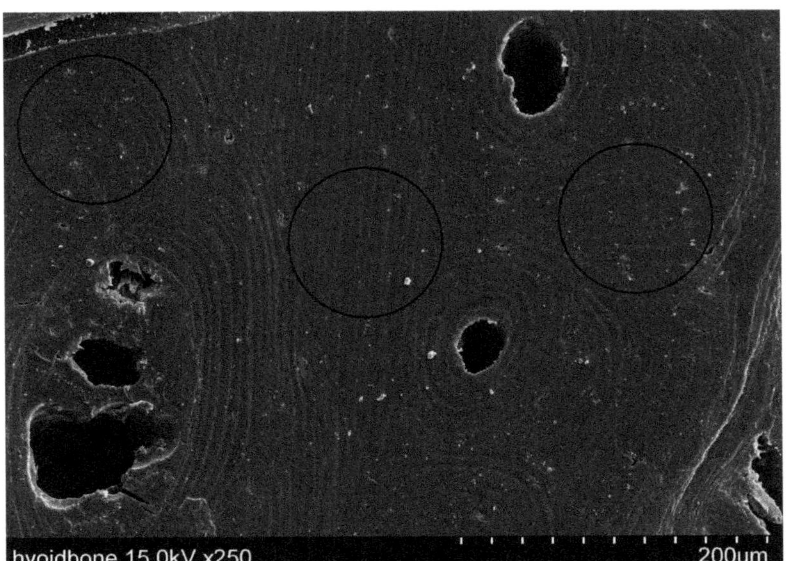

Figure 3. Evaluation of scanning electron microscope (SEM) images.

This SEM image shows the cortical bone of the hyoid body (HB). The black circles represent 100 μm, which is the size of the collimator beam. Osteons and surrounding interstitial lamellae can be seen.

3.2. Histological Observation

We observed muscles and ligaments attached to the HB in the region of interest (Figure 4). In the α region, a septum consisting of cartilaginous tissue was evident between the HB and the greater horn of the hyoid (GHH) (Figure 4A arrowhead). The GH was attached to the anterior and superior parts of the HB, the SH to the inferior part, and the TH to the posterior part (Figure 4A). In the β region, cartilaginous tissue was evident in the inferior and posterior parts of the HB (Figure 4B arrowhead). GH was attached to the anterior part of the HB, SH to the inferior part, and the hyoepiglottic ligament (HEL) and thyrohyoid ligament (THL) to the posterior part (Figure 4B). In the γ region, cartilaginous tissue was evident in the inferior and posterior parts of the HB (Figure 4C arrowhead). The muscle and ligament attachments to the bone were the same as those in the β region (Figure 4C). In the δ region, cartilaginous tissue was evident in the superior and inferior parts of the HB (Figure 4D arrowhead). SH was attached to the inferior part of the HB. TH could not be seen as a muscle attachment to the bone. The attachments of the HEL and THL were localized to the superior and posterior parts where cartilaginous tissue is present (Figure 4D).

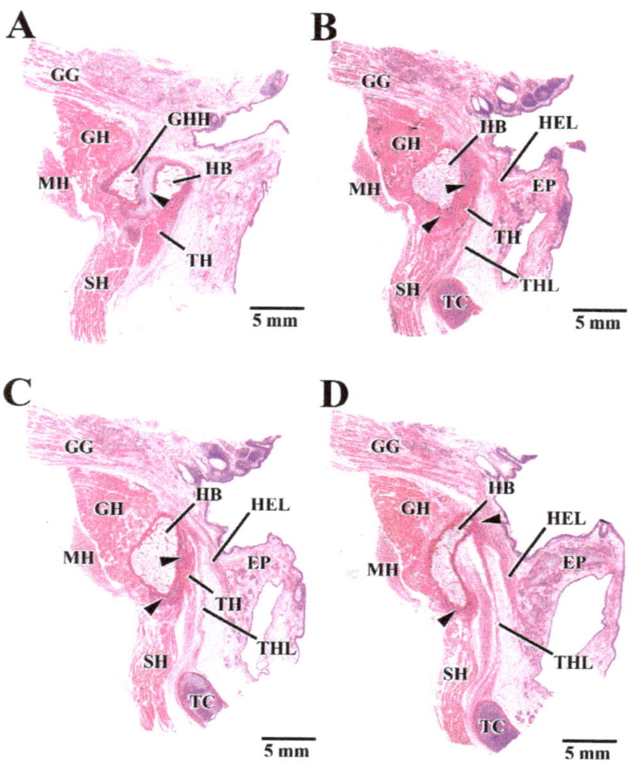

Figure 4. Hematoxylin-eosin-stained sagittal sections from each measurement region (α, β, γ, δ).

A comprehensive evaluation of Figure 4 shows that GH was always attached to the anterior part of the HB, and there was no cartilage intervening between GH and HB. Also, SH was attached to the inferior part of the HB, and cartilage was always present at the SH attachment. TH was attached to the posterior part of the HB. TH attachment contained cartilaginous tissue on the outside of the HB, although the attachment tended to disappear further medially. HEL and THL were attached to the superior part and the superior-posterior part of the HB in the medial panel. A thin layer of cartilage was often present at the attachments of these ligaments. The MH was continuous with the SH and did not attach to the HB. The fact that neither the digastric nor the stylohyoid, the other two suprahyoid muscles, was seen at any of the measurement regions indicated that they were both attached outside the region of interest.

Figure 4A–D show regions α, β, γ, and δ in the hyoid body (HB) region of interest, respectively. Panel A is the most lateral side and panel D is the most median side. In all regions of interest (α, β, γ, δ), the hyoid bone also had muscles or ligaments attached to it. HB; hyoid body, GHH; greater horn of the hyoid, GH; geniohyoid muscle, SH; geniohyoid muscle, TH; thyrohyoid muscle, MH; mylohyoid muscle, GG; genioglossus muscle, HEL; hyoepiglottic ligament, THL; thyrohyoid ligament, TC; thyroid cartilage, EP; epiglottic.

We observed multiple enthesis patterns within the region of interest (Figure 5). The first pattern was the direct attachment of the muscle fibers to the periosteum. The muscle fibers were not aggregated but were inserted individually into the periosteum (Figure 5A). In the enlarged image of enthesis, there were no metachromatic acid mucopolysaccharides (cartilage substrate: purple) evident within the periosteum (Figure 5B), and the cells within the periosteum were spindle-shaped (Figure 5B small window). The second pattern was the attachment by a tendon via the periosteum. Multiple (fewer than ten) muscle fibers have aggregated to form a tendon, and this tendon was inserted into the periosteum (Figure 5C). In the enlarged image of enthesis, there were no metachromatic acid mucopolysaccharides (cartilage substrate: purple) evident within the periosteum (Figure 5C), and the cells within the periosteum are spindle-shaped (Figure 5D small window). The third pattern was the attachments in which a tendon was transformed into fibrocartilage connecting with the bone. Multiple (several tens of) muscle fibers have aggregated to form a tendon, and the tendon was transformed into fibrocartilage connecting with the bone (Figure 5E). In the enlarged image of enthesis, there were metachromatic acid mucopolysaccharides (cartilage substrate: purple) evident within the fibrocartilage layer (purple) (Figure 5F), and large chondrocytes were present within the lacunae of the cartilage (Figure 5F small window). The fourth pattern was the attachments in which a ligament was transformed into fibrocartilage connecting with the bone. The ligament fibers were inserted obliquely into the fibrocartilage layer, and the individual fibers continue into the bone as if entwining with it (Figure 5G). In the enlarged image of the enthesis, there were metachromatic acid mucopolysaccharides (cartilage substrate: purple) evident within the fibrocartilage layer (purple), and the metachromatic reaction became stronger closer to the bone (Figure 5H). Large chondrocytes were also present within the lacunae of the cartilage (Figure 5H small window).

Direct attachment of the muscle fibers to the periosteum (Figure 5A,B) was localized to the GH–HB attachment (anterior hyoid body). Attachment of a tendon to the periosteum (Figure 5C,D) was seen in every part of the HB (anterior, posterior, superior, and inferior). Entheses of this type were seen supplementing the other types at every location. Attachments in which a tendon (Figure 5E,F) was transformed into fibrocartilage connecting with the bone were observed at the SH–HB (inferior) and TH–HB (posterior) attachments. Attachments in which a ligament was transformed into fibrocartilage connecting with the bone (Figure 5G,H) were observed at the HEL and THL–HB (superior-posterior) attachments. The ligament typically entered at a slant, forming a thin fibrocartilage layer.

Panel A, C, E, and G are H-E-stained images; Panel B, D, F, and H are magnified images of Panel A, C, E, and G, stained with toluidine blue. Direct attachment of the muscle fibers to the periosteum (Panel A and B), attachment by a tendon via the periosteum (Panel C and D), attachments in which a tendon was transformed into fibrocartilage connecting with the bone (Panel E and F), and attachments in which a ligament was transformed into fibrocartilage connecting with the bone (Panel G and H) are shown.

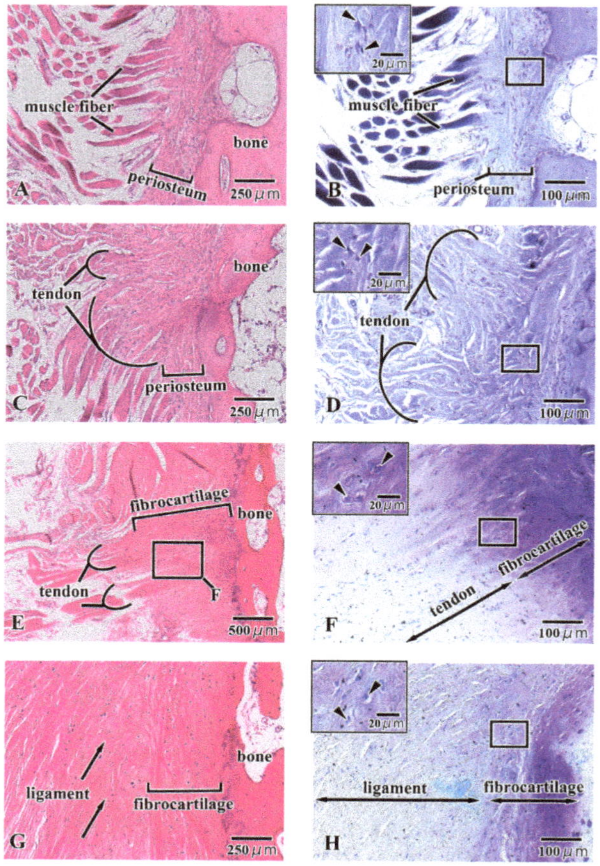

Figure 5. Hematoxylin-eosin-stained and toluidine blue-stained sections showing the morphology of the entheses around the hyoid body (HB).

3.3. Bone Mineral Density

Figure 6 shows the results of the quantitative evaluation of BMD at each of the measurement points (a, b, c, d) in each measurement region (α, β, γ, δ). The results of two-way ANOVA revealed no significant difference between any of the measurement regions nor between any of the measurement points.

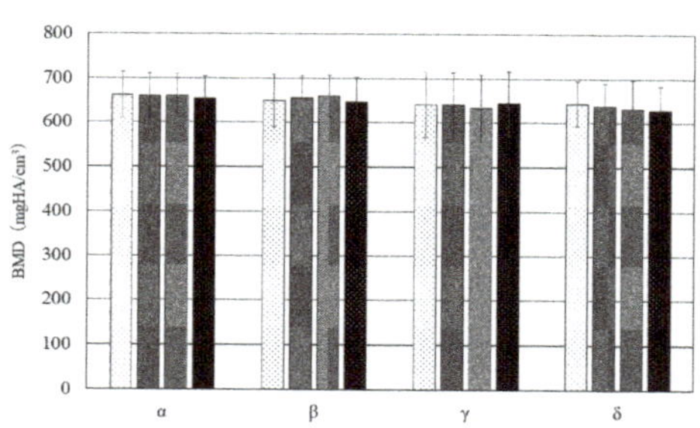

Figure 6. Evaluation of BMD at each measurement point in each measurement region of the hyoid.

3.4. Bap Crystallite Orientation

Figure 7 shows the diffraction intensity ratios along the X-, Y-, and Z-axis, which indicate the preferential orientation of BAp crystallites at each measurement point in each measurement region. The results of two-way ANOVA revealed that a high preferential orientation was observed in the X-axis direction at point b in the anterior part of the hyoid body in the β, γ, and δ regions. Compared with the other measurement regions, a high preferential orientation was seen in the δ region at point a in the superior part of the hyoid body and point c in the inferior part of the hyoid body (Figure 7A). In the Y-axis direction, a tendency for a high preferential orientation was seen at point d in the posterior part of the hyoid body at all measurement regions (Figure 7B). The highest preferential orientation was evident at point a in the β region. In the Z-axis direction, there was a tendency for a high preferential orientation at points c and d at all the measurement regions (Figure 7C) ($p < 0.05$).

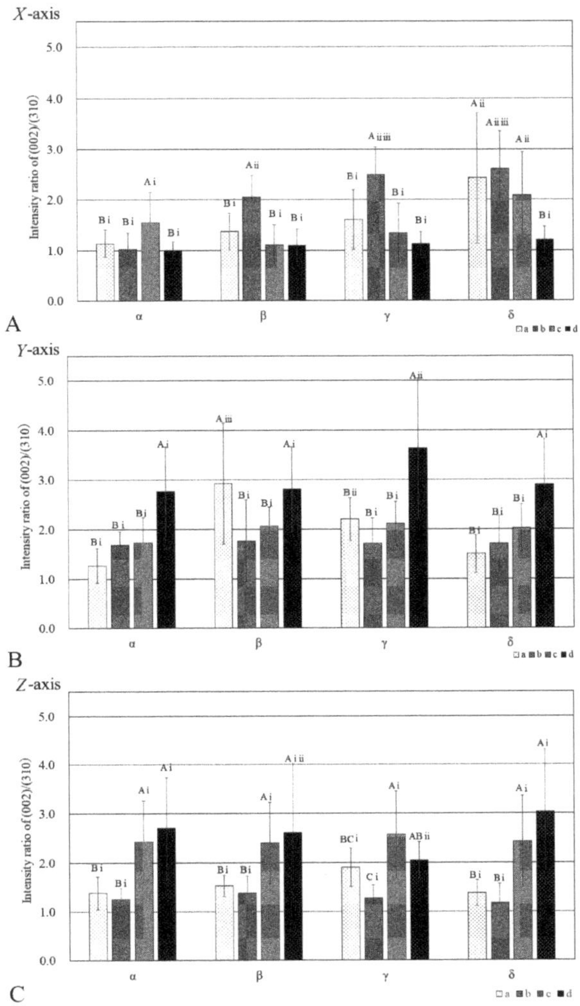

Figure 7. Diffraction intensity ratios along the X-, Y-, and Z-axis, indicating the preferential orientation of BAp at each measurement point in each measurement region of the hyoid.

The same letter of the alphabet indicates the absence of a significant difference between measurement points, and the same lower-case Roman numeral indicates the absence of a significant difference between measurement regions.

4. Discussion

The hyoid bone is located between the mandible and thyroid cartilage and is attached by muscles and ligaments to the base of the skull, the mandible, the tongue, and the larynx. As a result, it is known to be involved in vocalization, respiration, mastication, and swallowing. The hyoid is, therefore, believed to be subjected to forces in a variety of different directions as a result of these functional movements. In particular, the entheses, which represent the attachments where muscles, tendons, and ligaments are inserted into bone, transmit the load imposed by movement to the bone and are thus believed to be the sites of stress concentration [14]. To alleviate this stress concentration and enable efficient load transfer, the tissue layers comprising the entheses vary substantially in composition, structure, and mechanical properties and exhibit complex hierarchical structures [15]. Entheses are generally divided into three types: periosteal insertions, fibrocartilaginous entheses, and fibrous entheses with no intervening fibrocartilaginous layers [15–19].

4.1. Local Structure of the Hyoid Body from Histological Observations

The histological observations made in this study revealed periosteal insertions and fibrocartilaginous entheses in the HB. Two types of periosteal insertions in the HB were observed: one with insertion of muscle fibers alone directly into the periosteum and insertions in which multiple muscle fibers fused to form a tendon that was inserted into the periosteum being seen. Direct attachment of the muscle fibers to the periosteum was localized to the attachment from the geniohyoid. Fibrocartilaginous entheses were seen with the SH and TH muscles. This suggests that the entheses of the suprahyoid and infrahyoid muscles might be different. In the trunk and limbs, fibrocartilaginous entheses are seen at the epiphyses of the long bones and at sites where the angle of insertion of the tendon is large, such as the rotator cuff and Achilles tendon, at which the chondrocytes reportedly act as a "stretching brake" that prevents the concentration of stress at the interface with bone [15,20]. The infrahyoid muscles are mainly responsible for pulling the hyoid downwards. They are also reportedly involved in flexing the neck [21]. It has further been suggested that the SH might play an important role in the dilation of the upper airway [22]. Leelamanit et al. have suggested that the TH might play an important role in laryngeal elevation [23]. This muscle is also said to transmit force from the suprahyoid muscles to the cricothyroid complex to open the upper esophageal sphincter [24]. This suggests that, of the infrahyoid muscles, the entheses of the SH and TH muscles in the HB might be sites where the stress generated by these muscle functions is concentrated. Fibrocartilaginous entheses were also observed in the hyoepiglottic and TH ligaments. The hyoepiglottic ligament, which connects the hyoid to the epiglottis, and the TH ligament, which connects the hyoid to the thyroid cartilage, reportedly play important roles in the movement of the epiglottis during the swallowing reflex [25–27]. The hyoepiglottic ligament, in particular, is extended during epiglottic movements and is responsible for the movements of bending the epiglottis backward so that it becomes horizontal, and it pulls the tip of the epiglottis forward together with the hyoid, meaning that it plays a major role during the act of swallowing. This suggests that the entheses of these ligaments might thus be sites where the stress generated by swallowing reflex movements is concentrated. However, Vandaele et al. reported that attachment of the hyoepiglottic ligament to the hyoid covers a wide area and that it can be broadly divided into the median hyoepiglottic ligament, which is attached to the hyoid body, and the lateral hyoepiglottic ligament, which is attached to the greater horn [28]. They also reported that the median and lateral hyoepiglottic ligaments have different functions during epiglottic movements and that similar investigations of the greater horn of the hyoid are, therefore, required.

4.2. BMD of the Hyoid Body

To our knowledge, although one previous study of the BMD of the hyoid bone has compared measurements in the HB and the greater horn [29], there has been no comparative investigation of localized BMD values in the HB. Our results indicated that there was no significant difference in BMD between any of the measurement regions or measurement points. We previously evaluated the BMD of cortical bone directly below the entheses of the human coronoid process of the mandible and found that there was no significant difference between the BMD of the cortical bone below fibrocartilaginous entheses and fibrous entheses [13]. Previous studies in laboratory animals have also shown that BMD is unaffected by the presence or absence of functional pressure from the muscles [30,31]. This suggests that functional pressure from muscles and ligaments also has a negligible effect on BMD in the human hyoid.

4.3. BAp Crystallite Orientation in the Hyoid Body

In our previous study, we reported that BAp crystallite orientation varied depending on the direction of the muscle-tendon attachment [12]. Our results in this study also confirmed that the BAp crystallite preferential orientation varied depending on the measurement region and measurement point. In particular, preferential orientations along the Y-axis and Z-axis were evident at points c and d (inferior and posterior hyoid body), corresponding to the attachments of the SH and TH muscles and the hyoepiglottic and thyrohyoid ligaments, and these orientations resembled the directions of the courses of the muscles, tendons, and ligaments. The X-axis in this study was defined as the long axis of the hyoid bone, and the preferential orientation tended to be more pronounced closer to the center of the hyoid body. At the measurement points, a preferential orientation in the X-axis direction was observed in the superior and anterior hyoid body, to which the geniohyoid muscle is attached, as compared to other measurement points. Previous reports have found that long bones such as the femur and ulna reportedly exhibit a uniaxial preferential orientation in the direction of the long axis of the bone [7,8]. Furuya et al. measured BAp crystallite orientation in the cortical bone of human dentulous mandibles and reported that mandibular cortical bone in the vicinity of teeth subject to occlusal pressure exhibited a preferential orientation in the direction of this occlusal pressure, whereas a preferential orientation toward the long axis of the bone was evident at the base of the mandible [10]. In their discussion, they noted that the base of the mandible is horseshoe-shaped and has a structure similar to that of a long bone and that it exhibits a similar preferential orientation to that of long bones in areas not subject to occlusal pressure. Like the mandible, the HB also is horseshoe-shaped and has a structure similar to that of long bones, and as the geniohyoid attachment site has a direct insertion of muscle fibers into the periosteum, the concentration of stress at this site might be comparatively lower than it is at sites with fibrocartilaginous connections, meaning that a uniaxial preferential orientation in the direction of the long axis of the bone was observed.

The above results suggest that the microstructural characteristics of the tendon–bone and ligament–bone attachment sites in the HB might be locally optimized for fibrocartilaginous connections, in order to bear the functional pressure imposed by the infrahyoid muscles and ligaments, that sites with fibrocartilaginous connections have high bone quality in a different direction than the long axis of the bone.

The limitations of this study include the relatively high mean age of the eight Japanese adult cadavers used (82.4 $\pm$ 12.1 years), the fact that the number of cadavers was not standardized among the sexes, with five males and three females, and the small sample size. Therefore, we believe it is necessary to increase the number of samples, classify hyoid bones by age and sex, and conduct similar searches for comparison. According to Shimizu et al., the morphology of the hyoid differs depending on age and sex [5]. They also reported that the degree of fusion between the HB and the greater horn increased with age, and the form of the entheses of the hyoid bone and bone quality of the HB might also differ with age and sex. There is also reportedly great variation in the size and shape of the hyoid bone [1], and comparisons of the local structural characteristics of different morphologies might therefore be required.

5. Conclusions

Periosteal insertions and fibrocartilaginous entheses are evident at the muscle, tendon, and ligament attachment sites in the human HB. A stronger preferential orientation of BAp crystallites was evident at the sites of infrahyoid muscles and ligaments forming fibrocartilaginous connections than at other sites, and the direction of this orientation was generally consistent with the courses of these tendons and ligaments. This suggests that the functional pressure they impose might have a major effect not only on the morphological characteristics of the entheses but also on bone quality.

Author Contributions: Conceptualization, M.K. and K.K.; methodology, M.K., S.M., K.K. and S.A.; data curation, M.K. and T.S.; resources, S.A., S.M. and G.W.; investigation, M.K., K.K. and G.W.; formal analysis, T.S.; writing—original draft preparation, M.K.; writing—review and editing, S.A. and M.H.; visualization, M.K. and K.K.; project administration, M.H. All authors have read and agreed to the published version of the manuscript.

Funding: This work was supported by JSPS KAKENHI Grant Number 20K18616.

Data Availability Statement: All the data used and analyzed for the current study are contained in the article.

Conflicts of Interest: The authors declare no conflicts of interest.

References

1. Ronald, C.A.; Nathan, J.P. The hyoid bone: An overview. *Cranio* **2020**, *38*, 6–14.
2. Pearson, W.G., Jr.; Langmore, S.E.; Zumwalt, A.C. Evaluating the Structural Properties of Suprahyoid Muscles and their Potential for Moving the Hyoid. *Dysphagia* **2011**, *26*, 345–351. [CrossRef] [PubMed]
3. Gehrking, E.; Klostermann, W.; Wessel, K.; Remmert, S. Electromyography of the infrahyoid muscles—Part 1: Normal findings *Laryngorhinootologie* **2001**, *80*, 662–665. [CrossRef]
4. Capasso, L.; Michetti, E.; D'Anastasio, R. A homo erectus hyoid bone: Possible implications for origin of the human capability of speech. *Coll. Antropol.* **2008**, *32*, 1007–1011. [PubMed]
5. Shimizu, Y.; Kanetaka, H.; Kim, Y.H.; Okayama, K.; Kano, M.; Kikuchi, M. Age-related morphological changes in the human hyoid bone. *Cells Tissues Organs* **2005**, *180*, 185–192. [CrossRef]
6. NIH Consensus. Development Panel on Osteoporosis Prevention Diagnosis, and Therapy Osteoporosis prevention, diagnosis, and therapy. *J. Am. Med. Assoc.* **2001**, *285*, 785–795. [CrossRef]
7. Sasaki, N.; Matsushima, N.; Ikawa, T.; Yamamura, H.; Fukuda, A. Orientation of bone mineral and its role in the anisotropic mechanical properties of bone-transverse anisotropy. *J. Biomech.* **1989**, *22*, 157–164. [CrossRef]
8. Nakano, T.; Kaibara, K.; Tabata, Y.; Nagata, N.; Enomoto, S.; Marukawa, E.; Umakoshi, Y. Unique alignment and texture of biological apatite crystallites in typical calcified tissues analyzed by microbeam X-ray diffractometer system. *Bone* **2002**, *31*, 479–487. [CrossRef] [PubMed]
9. Miyabe, N.T.; Ishimoto, T.; Takano, N.; Adachi, T.; Iwaki, H.; Kobayashi, A.; Umakoshi, Y. Two-dimensional quantitative analysis of preferential alignment of BAp c-axis for isolated human trabecular bone using microbeam X-ray diffractometer with a transmission optical system. *Mater. Trans.* **2007**, *48*, 343–347. [CrossRef]
10. Furuya, H.; Matsunaga, S.; Tamatsu, Y.; Nakano, T.; Yoshinari, M.; Abe, S.; Ide, Y. Analysis of biological apatite crystal orientation in anterior cortical bone of human mandible using microbeam X-ray diffractometry. *Mater. Trans.* **2012**, *53*, 980–984. [CrossRef]
11. Ishimoto, T.; Nakano, T.; Umakoshi, Y.; Yamamoto, M.; Tabata, Y. Degree of biological apatite c-axis orientation rather than bone mineral density controls mechanical function in bone regenerated using recombinant bone morphogenetic protein-2. *J. Bone Miner. Res.* **2013**, *28*, 1170–1179. [CrossRef]
12. Kasahara, M.; Matsunaga, S.; Someya, T.; Kitamura, K.; Odaka, K.; Ishimoto, T.; Nakano, T.; Abe, S.; Hattori, M. Micro- and nano-bone analyses of the human mandible coronoid process and tendon-bone entheses. *J. Biomed. Mater. Res. B Appl. Biomater.* **2020**, *108*, 2799–2806. [CrossRef]
13. Nakano, T.; Kaibara, K.; Ishimoto, T.; Tabata, Y.; Umakoshi, Y. Biological apatite (BAp) crystallographic orientation and texture as a new index for assessing the microstructure and function of bone regenerated by tissue engineering. *Bone* **2012**, *51*, 741–747. [CrossRef]
14. Shaw, H.M.; Benjamin, M. Structure-function relationships of entheses in relation to mechanical load and exercise. *Scand. J. Med. Sci. Sports* **2007**, *17*, 303–315. [CrossRef]

15. Apostolakos, J.; Durant, T.J.; Dwyer, C.R.; Russell, R.P.; Weinreb, J.H.; Alaee, F.; Beitzel, K.; McCarthy, M.B.; Cote, M.P.; Mazzocca, A.D. The enthesis: A review of the tendon-to-bone insertion. *Muscles Ligaments Tendons J.* **2014**, *4*, 33–42. [CrossRef]
16. Lu, H.H.; Thomopoulos, S. Functional attachment of soft tissues to bone: Development, healing, and tissue engineering. *Annu. Rev. Biomed. Eng.* **2013**, *15*, 201–226. [CrossRef] [PubMed]
17. Doschak, M.R.; Zernicke, R.F. Structure, function and adaptation of bone-tendon and bone-ligament complexes. *J. Musculoskelet. Neuronal Interact.* **2005**, *5*, 35–40.
18. Benjamin, M.; Ralphs, J.R. The skeletal attachment of tendons—Tendon "entheses". *Comp. Biochem. Physiol. A Mol. Integr. Physiol.* **2002**, *133*, 931–945. [CrossRef] [PubMed]
19. Benjamin, M.; Ralphs, J.R. Fibrocartilage in tendons and ligaments—An adaptation to compressive load. *J. Anat.* **1998**, *193*, 481–494. [CrossRef] [PubMed]
20. Hems, T.; Tillmann, B. Tendon entheses of the human masticatory muscles. *Anat. Embryol.* **2000**, *202*, 201–208. [CrossRef]
21. Chancey, V.C.; Nightingale, R.W.; Van Ee, C.A.; Knaub, K.E.; Myers, B.S. Improved estimation of human neck tensile tolerance: Reducing the range of reported tolerance using anthropometrically correct muscles and optimized physiologic initial conditions. *Stapp Car Crash J.* **2003**, *47*, 135–153. [PubMed]
22. Roberts, J.L.; Reed, W.R.; Thach, B.T. Pharyngeal airway-stabilizing function of sternohyoid and sternothyroid muscles in the rabbit. *J. Appl. Physiol. Respir. Environ. Exerc. Physiol.* **1984**, *57*, 1790–1795. [CrossRef] [PubMed]
23. Leelamanit, V.; Limsakul, C.; Geater, A. Synchronized electrical stimulation in treating pharyngeal dysphagia. *Laryngoscope* **2002**, *112*, 2204–2210. [CrossRef] [PubMed]
24. Matsuo, K.; Palmer, J.B. Anatomy and physiology of feeding and swallowing: Normal and abnormal. *Phys. Med. Rehabil. Clin. N. Am.* **2008**, *19*, 691–707. [CrossRef] [PubMed]
25. Fink, B.R.; Martin, R.W.; Rohrmann, C.A. Biomechanics of the human epiglottis. *Acta Otolaryngol.* **1979**, *87*, 554–559. [CrossRef] [PubMed]
26. Reidenbach, M.M. The periepiglottic space: Tomographic relations and histological organisation. *J. Anat.* **1996**, *188*, 173–182.
27. Kitamura, K.; Watanabe, T.; Yamamoto, M.; Ishikawa, N.; Kasahara, N.; Abe, S.; Yamamoto, H. A Newly Discovered Tendon Between the Genioglossus Muscle and Epiglottic Cartilage Identified by Histological Observation of the Pre-Epiglottic Space. *Dysphagia* **2023**, *38*, 315–329. [CrossRef]
28. Vandaele, D.J.; Perlman, A.L.; Cassell, M.D. Intrinsic fibre architecture and attachments of the human epiglottis and their contributions to the mechanism of deglutition. *J. Anat.* **1995**, *186*, 1–15.
29. Wang, X.; Wang, C.; Zhang, S.; Wang, W.; Li, X.; Gao, S.; Li, K.; Chen, J.; Wang, H.; Chen, L.; et al. Microstructure of the hyoid bone based on micro-computed tomography findings. *Medicine* **2020**, *99*, e22246. [CrossRef]
30. Picard, S.; Lapointe, N.; Brown, J.; Guertin, P. Histomorphometric and densitometric changes in the femora of spinal cord transected mice. *Anat. Rec.* **2008**, *291*, 303–307. [CrossRef]
31. Tsai, C.; Huang, R.; Lee, C.; Hsiao, W.; Yang, L. Morphologic and bony structural changes in the mandible after a unilateral injection of botulinum Neurotoxin in adult rats. *J. Oral. Maxillofac. Surg.* **2010**, *68*, 1081–1087. [CrossRef] [PubMed]

Journal of
Functional
Biomaterials

Article

In Vitro Biocompatibility Assessment of Bioengineered PLA-Hydrogel Core–Shell Scaffolds with Mesenchymal Stromal Cells for Bone Regeneration

Federica Re [1,2,3,*], Luciana Sartore [3,4], Chiara Pasini [3,4], Matteo Ferroni [3,5,6], Elisa Borsani [3,7,8], Stefano Pandini [3,4], Andrea Bianchetti [3,9], Camillo Almici [3,9], Lorena Giugno [7], Roberto Bresciani [10,11], Silvia Mutti [1,2,3], Federica Trenta [1,2,3], Simona Bernardi [1,2,3,12], Mirko Farina [1] and Domenico Russo [1,3]

1 Unit of Blood Diseases and Cell Therapies, Department of Clinical and Experimental Sciences, University of Brescia, "ASST-Spedali Civili" Hospital of Brescia, 25123 Brescia, Italy; silvia.mutti@unibs.it (S.M.); federica.trenta@unibs.it (F.T.); simona.bernardi@unibs.it (S.B.); mirko.farina@unibs.it (M.F.); domenico.russo@unibs.it (D.R.)
2 Centro di Ricerca Emato-Oncologica AIL (CREA), ASST Spedali Civili, 25123 Brescia, Italy
3 University Center of Research "STem cells, bioENgineering and regenerative MEDicine"—STENMED, University of Brescia, 25123 Brescia, Italy; luciana.sartore@unibs.it (L.S.); chiara.pasini1@unibs.it (C.P.); matteo.ferroni@unibs.it (M.F.); elisa.borsani@unibs.it (E.B.); stefano.pandini@unibs.it (S.P.); andrea.bianchetti@asst-spedalicivili.it (A.B.); camillo.almici@asst-spedalicivili.it (C.A.)
4 Materials Science and Technology Laboratory, Department of Mechanical and Industrial Engineering, University of Brescia, 25123 Brescia, Italy
5 Department of Civil, Environmental, Architectural Engineering and Mathematics (DICATAM), University of Brescia, Via Valotti 9, 25123 Brescia, Italy
6 National Research Council (CNR)—Institute for Microelectronics and Microsystems, Via Gobetti 101, 40129 Bologna, Italy
7 Division of Anatomy and Physiopathology, Department of Clinical and Experimental Sciences, University of Brescia, 25123 Brescia, Italy; lorena.giugno@unibs.it
8 Interdepartmental University Center of Research "Adaption and Regeneration of Tissues and Organs (ARTO)", University of Brescia, 25123 Brescia, Italy
9 Laboratory for Stem Cells Manipulation and Cryopreservation, Department of Transfusion Medicine, ASST Spedali Civili di Brescia, 25123 Brescia, Italy
10 Department of Molecular and Translational Medicine, University of Brescia, 25123 Brescia, Italy; roberto.bresciani@unibs.it
11 Highly Specialized Laboratory, ASST Spedali Civili di Brescia, 25123 Brescia, Italy
12 National Center for Gene Therapy and Drugs based on RNA Technology—CN3, 35122 Padua, Italy
* Correspondence: federica.re@unibs.it

Abstract: Human mesenchymal stromal cells (hMSCs), whether used alone or together with three-dimensional scaffolds, are the best-studied postnatal stem cells in regenerative medicine. In this study, innovative composite scaffolds consisting of a core–shell architecture were seeded with bone-marrow-derived hMSCs (BM-hMSCs) and tested for their biocompatibility and remarkable capacity to promote and support bone regeneration and mineralization. The scaffolds were prepared by grafting three different amounts of gelatin–chitosan (CH) hydrogel into a 3D-printed polylactic acid (PLA) core (PLA-CH), and the mechanical and degradation properties were analyzed. The BM-hMSCs were cultured in the scaffolds with the presence of growth medium (GM) or osteogenic medium (OM) with differentiation stimuli in combination with fetal bovine serum (FBS) or human platelet lysate (hPL). The primary objective was to determine the viability, proliferation, morphology, and spreading capacity of BM-hMSCs within the scaffolds, thereby confirming their biocompatibility. Secondly, the BM-hMSCs were shown to differentiate into osteoblasts and to facilitate scaffold mineralization. This was evinced by a positive Von Kossa result, the modulation of differentiation markers (osteocalcin and osteopontin), an expression of a marker of extracellular matrix remodeling (bone morphogenetic protein-2), and collagen I. The results of the energy-dispersive X-ray analysis (EDS) clearly demonstrate the presence of calcium and phosphorus in the samples that were incubated in OM, in the presence of FBS and hPL, but not in GM. The chemical distribution maps of calcium and phosphorus indicate that these elements are co-localized in the same areas of the sections, demonstrating the formation of hydroxyapatite. In conclusion, our findings show that the combination of BM-hMSCs and PLA-CH, regardless of the amount of hydrogel content, in the presence of differentiation stimuli, can provide a construct with enhanced osteogenicity for clinically relevant bone regeneration.

Keywords: scaffold design; PLA; human mesenchymal stromal cells; gelatin–chitosan hydrogel; human platelet lysate; bone regeneration; tissue engineering; 3D printing

1. Introduction

Bone defects often result from trauma, revision arthroplasty, or tumor resection. Current bone reconstruction procedures include autologous, allogeneic, and xenogeneic bone grafts [1]. Due to the challenges associated with the current bone grafts, regenerative medicine has been integrated into clinical practice [2]. Regenerative medicine offers a revolutionary approach to tissue and organ repair, harnessing the power of human mesenchymal stromal cells (hMSCs) and three-dimensional (3D) scaffolds [3,4].

hMSCs are currently the most promising cell population for clinical applications in bone disease. They are being extensively studied for their potential in tissue engineering and regenerative medicine [5]. hMSCs are present in a multitude of tissues, including bone marrow, adipose tissue, and synovium [6]. hMSCs are multipotent adult stem cells with the remarkable ability to self-renew and differentiate into a variety of cell types within the mesodermal and other embryonic lineages, including osteocytes [7], neurons, muscle cells, hepatocytes, and epithelial cells [8,9]. However, the regenerative potential of hMSCs is intricately tied to their tissue of origin [10]. The potential efficacy of cell therapy based on hMSCs alone or in combination with scaffolds has been demonstrated in several clinical trials, although their efficacy remains limited [4]. Most of the current clinical trials on bone regeneration therapy have used bone-marrow-derived hMSCs (BM-hMSCs) [4,11–13]. Moreover, BM-hMSCs exhibit superior abilities for bone and cartilage formation using standard differentiation methods [14]. hMSCs stand out as the perfect cell source for various regenerative medicine applications with their unique properties: ease of isolation, ability to expand, differentiation capacity, self-renewal capacity, immunological properties, antimicrobial ability, and ability to migrate to injured sites [7,15–18]. In addition to these properties, autocrine or paracrine functions that generate growth factors have been proposed as the main mechanism that contributes to tissue repair [19,20].

Human platelet lysate (hPL) serves as an excellent substitute for fetal bovine serum (FBS) due to its rich content of growth factors, making it ideal for expanding hMSCs. By eliminating the risks of potential immune reactions to animal antigens, hPL allows for large-scale cell expansion for clinical use while satisfying all ISCT criteria [21]. hPL has proven to be highly effective in applications such as regenerative medicine, tissue engineering, cell culture, and cell therapy. However, its production remains controversial due to the numerous variables involved in the manufacturing process [22]. hPL stimulates proliferation, cell growth, and differentiation towards the osteogenic lineage of hMSCs from different sources [22].

Numerous biomaterials have undergone preclinical testing in conjunction with hMSCs for the purpose of bone regeneration [23]. The overall positive results of the bone regenerative medicine approaches have been confirmed by several approved clinical trials [24]. In most cases, frameworks made of ceramic cement were used, which merely fill bone defects [25]. However, bone graft substitutes should simultaneously fulfill several requirements, including biocompatibility, biodegradability, porosity, osteogenicity, osteoconductivity, and osteoinductivity to support the regeneration of bone tissue at the defect site by degrading in place and allowing newly formed bone to take its position [24]. For this reason, hybrid combinations of several classes of materials with different properties have received considerable attention in the inducement of osteogenic and chondrogenic differentiation [22,26–29].

Several composite bio-scaffolds, based on both synthetic and natural polymers, are potential candidates for bone tissue regeneration [30,31]. Synthetic polymers (e.g., polyethylene glycol, polyglycolic acid, poly-L-lactic acid (PLA), poly(ε-caprolactone) (PCL), and polyurethanes) allow for easier control of mechanical and degradation behaviors but are not bioactive, while bioactive ceramics (e.g., hydroxyapatite, corals, sulfate, tricalcium phosphate, bioactive glass, and calcium silicate) and natural polymers (e.g., chitosan, fibrin, hyaluronic acid, and collagen) mimic the extracellular matrix (ECM) of the bone but generally are mechanically weaker [2,30,32]. Current development is especially moving towards the use of hydrogels that mimic ECM excellently thanks to adequate porosity, surface morphology, and bioactive properties [33,34]. Natural biopolymers, such as Chitosan being added to the network of hybrid hydrogels, have an active role in promoting cell growth and osteo-differentiation [30–35]. Chitosan enhances osteogenesis in MSCs and upregulates osteopontin and collagen I [36]. Our recent studies showed higher stress relaxation in gelatin and CH-based scaffolds, which allow better cell adhesion, spreading, viability, and osteochondral differentiation [22,35,37]. However, given the complex functions of ECM and the fact that they are mineralized tissues, the mechanical properties of hydrogels are generally unsuitable for load-bearing applications [38]. Looking for enhanced mechanical support and higher control of mechanical properties and degradation behavior, stiffer polyesters such as PLA and PCL, which are already commercially available in medical grades, have been explored [39,40]. Based on these premises, the combination of synthetic polymers, such as PLA or PCL, with natural polymers, such as hydrogels, could lead to improvements of both the mechanical properties and bioactivity of the scaffolds; furthermore, the type and proportion of the components could be tailored for the specific purposes of tissue regeneration [41].

In a previous study, the authors developed a solvent-free fabrication process to create composite scaffolds. These scaffolds consist of a porous PLA-PCL core and a bioactive gelatin–CH hydrogel shell. It has been demonstrated that these scaffolds support the proliferation and osteogenic differentiation of BM-hMSCs [42]. However, more detailed knowledge and precise control of their structure–property correlations are required for the optimization of core–shell scaffolds [43]. Based on these results, the authors have recently utilized advanced additive manufacturing techniques to develop bioresorbable scaffolds with a core–shell architecture. These scaffolds are constructed using 3D-printed PLA cores with diverse lattice structures and are coated with gelatin–CH hydrogel. This innovative approach aims to fulfill the varied requirements for applications in bone tissue engineering [44–46].

In this study, the novel PLA-CH core–shell scaffolds were synthetized and investigated in more detail for bone tissue regeneration. We analyzed three different types of core–shell scaffolds with different core lattice structures, corresponding to different amounts of incorporated gelatin–CH hydrogel. The physical–mechanical and degradation properties of the scaffolds were assessed. In order to assess their biocompatibility and ability to replicate the natural bone microenvironment, the scaffolds were seeded with BM-hMSCs in the presence of FBS or hPL with or without differentiation stimuli. We then thoroughly evaluated the cell viability, proliferation, osteogenic differentiation, and scaffold mineralization.

2. Materials and Methods

2.1. Materials

For the scaffold core, poly(L-lactic acid) (PLA) filament was employed (Raise3D PLA Premium by Raise3D Technologies, Inc., Irvine, CA, USA). For the scaffold shell, the following materials were procured as the initial supply: poly(ethylene glycol) diglycidyl ether (PEGDGE) (molecular weight 526 Da) and chitosan (molecular weight 50,000 ÷ 190,000 Da, degree of deacetylation 75–85%) by Sigma-Aldrich Co (Milan, Italy); ethylene diamine (EDA) and acetic acid by Fluka (Milan, Italy). Pharmaceutical-grade type A gelatin (280 bloom, viscosity 4.30 mPs) was kindly supplied by ©Italgel S.p.A. (Cuneo, Italy). Dulbecco's modified Eagle's medium (DMEM), penicillin–streptomycin, l-glutamine, and sodium pyruvate were provided by Sigma-Aldrich Co (Milan, Italy). Amphotericin B and minimum essential medium (MEM) non-essential amino acid solution were purchased by Gibco, ThermoFisher Scientific (Milan, Italy).

2.2. Scaffold Fabrication

Lattice core structures with parallelepipedal unit cells were realized with the same strut thickness (t = 0.6 mm) but different hole height (h) and hole width (w), as indicated in Figure 1a. They were designed by using the software Solidworks (Dassault Systèmes, Vélizy-Villacoublay, France) and named after the amount of hydrogel they could contain (L: low; M: medium; H: high, corresponding to PLA-CH(L), PLA-CH(M), PLA-CH(H)). Cubic specimens with a side of about 10 mm, bar specimens with a cross-section of around 45–60 mm^2, and a length of about 100 mm were 3D-printed by fused deposition modeling (FDM) of PLA. The slicing software Ideamaker and the 3D-printer Raise 3D Pro2 were both supplied by Raise3D Technologies, Inc. (Irvine, CA, USA). The following setup was employed: bed temperature of 60 °C; nozzle temperature of 205 °C; nozzle diameter of 0.2 mm; layer thickness of 0.1 mm (along the build direction z illustrated in Figure 1a).

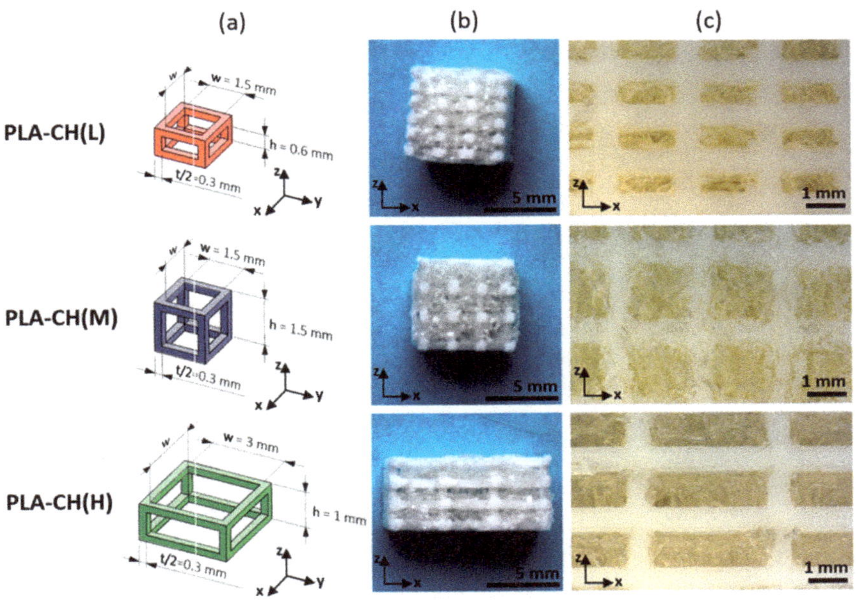

Figure 1. Geometry of lattice unit cells (**a**), dry specimens employed in in vitro experiments (**b**), and magnified details of wet specimens (**c**), relative to scaffolds hosting low (L), medium (M), or high (H) hydrogel content.

The hydrogel shell was prepared by dissolving 6 g of gelatin in 65 mL of distilled water and successively adding 1.4 g of PEGDGE, 32.5 g of chitosan solution (2 wt% in acetic acid 1%), and 70 mg of EDA, resulting in a nominal composition of the final dry hydrogel equal to 74.3 wt% gelatin, 17.6 wt% PEGDGE, and 8.1 wt% chitosan. The process was carried out at 45 °C under mild magnetic stirring until the dissolution of the reactants and the initiation of crosslinking reactions between the epoxy groups of PEGDGE and the amino groups of gelatin and chitosan. The core structures were incorporated into the hydrogel forming solution at a temperature of 45 °C, and any entrapped air bubbles were removed through the application of three cycles of vacuum/air. After waiting for proper hydrogel crosslinking, the specimens were frozen and freeze-dried in a lyophilizer (HyperCOOL HC3055, LabTech Srl, Sorisole, Italy). The dry scaffolds were subjected to the removal of the excess hydrogel, post-curing in oven (2 h at 45 °C under vacuum), washing with distilled water, and a further freeze-drying treatment. In order to conduct the in vitro experiments, the bar specimens were cut along their length into slices of approximately 5 mm in thickness (Figure 1b), which were sealed in polypropylene bags under vacuum and sterilized by gamma irradiation (25 kGy of Cobalt 60 gamma rays, concordant to UNI EN ISO 1113 standard).

2.3. Characterization of Scaffolds

Physical–mechanical characterization and degradation experiments were carried out on cubic specimens. The void volume fraction (V_v) of the core was determined through the equation below:

$$V_v[\%] = \left(1 - \frac{V_c}{V}\right) \times 100 \quad (1)$$

where $V_c = m_c/\rho_c$ is the volume occupied by the lattice struts, m_c is the core mass (evaluated by weight), ρ_c is the PLA declared density [1], and V is the core total volume.

The hydrogel content by weight (Hy) and the water uptake (W) of the core–shell scaffolds were calculated as:

$$Hy[\%] = \frac{m_{cs,dry} - m_c}{m_{cs,dry}} \times 100 \quad (2)$$

$$W[\%] = \frac{m_{cs,wet} - m_{cs,dry}}{m_{cs,dry}} \times 100 \quad (3)$$

where $m_{cs,dry}$ and $m_{cs,wet}$ correspond to the mass of the composite specimens in dry conditions and after 24-h immersion in water, respectively. Core and core–shell specimens were subjected to mechanical tests at room temperature. The core–shell specimens have been previously immersed in distilled water at 37 °C for 24 h and for 7 weeks. A compressive load was applied perpendicularly to the z direction through an electromechanical dynamometer (Instron 3366 with 10 kN load cell, Illinois Tool Works Inc., Norwood, MA, USA), moving the crosshead at 2 mm/min. Compressive stiffness was evaluated as the apparent modulus (E_{app}), derived from the initial slope of the stress–strain curves. The term apparent highlights that stress and strain are not those locally experienced by the material, but macromechanical parameters describing the response of the whole structure as an equivalent homogeneous material. Furthermore, the mass loss of the specimens was quantified following their immersion in distilled water at 37 °C for 28 days and subsequent drying under vacuum in an oven at 40 °C. The data were expressed as a percentage with respect to the initial mass of the dry scaffold, as well as with respect to the initial mass of the dry hydrogel shell only:

$$\text{mass loss/total mass}[\%] = \frac{m_{cs,dry} - m_{deg}}{m_{cs,dry}} \times 100 \quad (4)$$

$$\text{mass loss/hydrogel mass}[\%] = \frac{m_{cs,dry} - m_{deg}}{m_{cs,dry} - m_c} \times 100 \quad (5)$$

where m_{deg} is the mass of the scaffolds after hydrolytic degradation and drying.

2.4. Bone-Marrow-Derived Human Mesenchymal Stromal Cell (BM-hMSC) Culture

For the study, commercial human mesenchymal stromal cells derived from bone-marrow (BM-hMSCs) (PromoCell, Heidelberg, Germany) were expanded as previously reported [22] using a growth medium (GM) consisting of Dulbecco's modified Eagle's medium (DMEM), with glucose and 2% L-glutamine/penicillin–streptomycin/amphotericin B solution, MEM non-essential amino acids solution 1X, and 1 mM sodium pyruvate. GM was obtained by adding 10% of fetal bovine serum (GM FBS) or 5% of human platelet lysate (GM hPL). The cells were cultured in an incubator at 37 °C/5% CO_2. hPL for the expansion of hMSCs was prepared according to standardized clinical procedures in closed systems as previously described [22]. Details of the technical procedure can be found in F. Re et al. [22].

2.5. BM-hMSC Seeding on Scaffolds

Once the BM-hMSCs reached 80% confluence, they were detached from the flask with trypsin ethylenediaminetetraacetic acid and neutralized with GM FBS. They were then centrifuged at 1100 rpm for 5 min and resuspended in GM before being counted. All experiments were performed with cells at passages 3 and 4. The dry scaffolds were used for cell seeding and placed in 24-well non-adherent plates (Corning, Sigma-Aldrich, St. Louis, MO, USA). The required cell number was concentrated in a volume of 50 µL. A small droplet (7×10^5/50 µL viable BM-hMSCs) was applied to the top of the dry scaffold and left for 2 h in an incubator for complete absorption of the droplet. We have applied a static seeding method, namely "dry", to increase the efficiency of cell penetration and distribution within the scaffold. Then, 1ml of GM FBS or GM hPL was added to each well. The medium for culturing the cells was changed three times per week. We examined each cell-seeded scaffold for cell viability and cell proliferation on day 28. On the day of the viability and proliferation analysis (day 28), the 24-well culture plates were replaced with new, sterile plates to remove any debris from the wells.

2.6. BM-hMSC Osteogenic Differentiation on Scaffolds

A cell suspension of 7×10^5 viable BM-hMSCs was seeded onto the scaffold for osteogenic differentiation under static conditions. After 48 h, one milliliter of GM FBS or GM hPL was replaced with osteogenic medium (OM). For details, please referred to F. Re et al. [22]. Samples were also maintained in GM FBS or GM hPL without additional osteogenic factors for 4 weeks as controls. After 28 days of culture, the scaffolds were fixed with formaldehyde 4% for 1 h at 4 °C and dehydrated for subsequent analysis.

2.7. BM-hMSC Cell Viability and Cell Proliferation Assay

To assess the viability of BM-hMSCs in the scaffolds, a live/dead mammalian cell kit (ThermoFisher, Waltham, MA, USA) was applied following the manufacturer's instructions. NucBlue® Live Reagent (2 drops/mL) was added to the cultures for nuclei staining. Cells were analyzed using a Zeiss Observer Z1 fluorescence microscope. Several images of three different replicates of each sample were taken. Cell proliferation was determined using a Cell Counting Kit-8 (CCK-8, Sigma-Aldrich, St. Louis, MO, USA) on day 28 of cell culture according to the manufacturer's instructions. The cell-cultured samples (three replicates) were incubated with the CCK-8 reagent at 37 °C for 2 h and 30 min. The absorbance of the supernatant that was transferred to a new cell culture plate was then measured at 450 nm using an Infinite 200 PRO plate reader (Tecan, Männedorf, Switzerland). The absorbance at 450 nm directly correlates with the number of viable cells in each sample.

2.8. Histological and Immunohistochemical Analysis

An automatic processor (Donatello series 2, Diapath S.p.A., Bergamo, Italy) was employed for the embedding of scaffolds in paraffin. A semi-automatic microtome, the Galileo semi-series 2 (Diapath S.p.A.), was employed to cut serial paraffin sections (7 µm thick) of each sample. The alternate sections were deparaffinized and rehydrated in accordance with standard procedures. Sections were then stained with a hematoxylin–eosin stain (automatic stainer Giotto; Diapath S.p.A.) for general morphological analysis. Moreover, the sections were evaluated histochemically, with von Kossa staining (Bio-Optica, Milan, Italy) for calcium deposits (brown/black dots) on PLA-CH sections, and immunohistochemically with a focus on the PLA-CH(M) sections for osteopontin (OSP), an early marker of osteoinduction, osteocalcin (OSC), a late marker of osteoinduction, bone morphogenetic protein-2 (BMP2), a marker of extracellular matrix remodeling, and collagen

I. For immunohistochemistry, three alternate sections were incubated in primary antiserum anti-OSP (rabbit polyclonal antibody, 1:200, Abcam, Cambridge, UK), anti-OSC (mouse monoclonal antibody, 1:200, Santa Cruz Biotechnology, Dallas, TX, USA), anti-BMP-2 (goat polyclonal antibody, 1:200, Santa Cruz Biotechnology), and anti-collagen I (goat polyclonal antibody, 1:100, Santa Cruz Biotechnology). The sections were incubated in the primary antiserum, then in the appropriated biotinylated secondary antibodies, and, finally, in the avidin–biotin peroxidase complex (Vector Laboratories, Burlingame, CA, USA). The visualization of the immunopositivity reaction was achieved using a hydrogen peroxide and diaminobenzidine (chromogen) mixture, resulting in a brown coloration (Sigma, St. Louis, MO, USA). To ensure the accuracy of the results, an immuno-histochemical control was conducted by omitting the primary antibody and incubating the sections with non-immune rabbit serum and isotype-matched irrelevant rat IgGs, which served as a negative control. For all treatments, the staining for each antibody was performed simultaneously. Sections were counterstained with Carazzi's hematoxylin (blue/violet color) to better visualize the positive reaction. The sections were dehydrated and mounted with DPX. The micrographs were performed using an optical microscope (Olympus BX50 Microscope, Hamburg, Germany) equipped with an image analyzer (Image-Pro Premier 9.3; 2018, Media Cybernetics, Rockville, MD, USA). For the analysis, the histochemical reaction was evaluated quantitatively, while the immunohistochemical reaction was evaluated qualitatively (positive vs. negative), both at a final 400× magnification, and digitally fixed images were analyzed blind. The analysis for each staining was performed on three sections in the center of each sample, spacing 50 mm each other, and considering three areas. The percentage of positive area within the scaffold meshes was evaluated for calcium deposits according to F. Re et al. (2021) [2]. As a control for histological staining, the scaffolds without cells were also analyzed.

2.9. Morphological and Microstructural Analysis

Scanning electron microscopy (SEM) was used to detect both the adhesion on all scaffold typologies and the differentiation toward the osteogenic lineage of the cells focusing on PLA-CH(M) sections. The samples were dehydrated through immersion in increasing alcohol solutions without any additional preparation. The scanning electron microscope (SEM) (ZEISS EVO LS-10, Oberkochen, Germany) was operated in environmental mode, with a pressure of 0.1–0.2 Torr at the specimen. The electron beam was accelerated to a range of 10–20 kV, and backscattered electron imaging was employed. In combination with SEM imaging, the elemental identification and localization provided by energy dispersive X-ray spectrometry (EDS) was used to pursue the identification of biomineral deposits. A Bruker SSD spectrometer was used for this purpose (Bruker, Billerica, MA, USA, Quantax 200 with a detector active area of 30 mm^2). As a control for SEM analysis, the scaffolds without cells were also analyzed.

2.10. Statistical Analysis

The statistical analyses were conducted using two one-way analysis of variance (ANOVA) with Tukey's multiple comparisons test. The numerical results are presented as mean ± standard deviation (SD). The graphical results were produced using GraphPad Prism (version 10). Three replicates of each sample were employed. The level of statistical significance was set at $p < 0.05$.

3. Results

3.1. Physical–Mechanical and Degradation Properties of the Scaffolds

The three types of core–shell scaffolds investigated are illustrated in Figure 1, showing their lattice unit cells (a) and their appearance in dry (b) and wet (c) conditions. In all the cases, the hydrogel is proved to be highly porous and to penetrate throughout the PLA lattice holes until the complete occupation of the core void volume. The lattices differ for the values of the width and height of their holes, corresponding to the different fractions of void volume and, therefore, to the different weight fractions of hydrogel in the final scaffolds, as reported in Table 1 for cubic specimens. The table also displays the overall water uptake of the specimens, being directly proportional to the hydrogel content since PLA water absorption can be neglected. The shell of the scaffolds is particularly important for the bioactive properties previously observed in the gelatin–chitosan hydrogels [37], of which it shares the same interconnected porous structure (~80% porosity) and water absorption (800% ca.).

Table 1. Geometry and composition of cubic scaffolds hosting low (L), medium (M), or high (H) hydrogel content: width (w) and height (h) of lattice holes; void volume fraction in the core; hydrogel content; water uptake after 24 h in water.

Label	w × h [mm]	Core Void Volume Fraction [%]	Hydrogel Content [%]	Water Uptake (24 h) [%]
PLA-CH(L)	1.5 × 0.6	68.8 ± 0.3	11.6 ± 0.4	103 ± 7
PLA-CH(M)	1.5 × 1.5	77.0 ± 0.3	17.6 ± 0.4	137 ± 4
PLA-CH(H)	3.0 × 1.0	83.4 ± 0.2	22.0 ± 0.9	201 ± 9

Moreover, the three types of scaffolds present different mechanical and degradation properties (Table 2), which can be modulated by acting on the lattice design. In fact, as the core void volume fraction and the hydrogel content increase, the stiffness (apparent modulus) decreases, covering a range of values comparable with those of bone tissue [3]; the same results were found for both core and core–shell specimens, meaning that the hydrogel incorporation does not affect them. Furthermore, the PLA lattices were found capable of providing long-term mechanical support in the first phases of bone tissue regeneration, since the scaffolds maintained their stiffness after immersion in water at 37 °C for prolonged times (7 weeks). Conversely, the scaffold mass loss increases with the hydrogel content, as the shell earlier leaves space for the growth of new bone tissue, undergoing gradual hydrolysis in the aqueous environment (e.g., after 28 days in water at 37 °C, about 20 wt% of the hydrogel is degraded).

Table 2. Mechanical and degradation properties of cubic scaffolds hosting low (L), medium (M), or high (H) hydrogel content: apparent modulus (E_{app}) of core specimens, core–shell scaffolds in wet conditions (24 h immersion), and core–shell scaffolds maintained in water at 37 °C for 7 weeks; mass loss of core–shell scaffolds maintained in water at 37 °C for 28 days.

Label	E_{app}, Core [MPa]	E_{app}, Core-Shell [MPa]	E_{app}, Core-Shell 7 Weeks [MPa]	Mass Loss, Core-Shell 28 d [%]
PLA-CH(L)	381 ± 14	378 ± 17	367 ± 39	11.4
PLA-CH(M)	223 ± 34	230 ± 6	231 ± 6	6.8
PLA-CH(H)	204 ± 1.3	174 ± 7	179 ± 6	5.4

3.2. Assessment of In Vitro Viability, Proliferation, and Adhesion Capacity of hMSCs Cultured in PLA-CH(L), PLA-CH(M), and PLA-CH(H) Scaffolds

Following 28 days in either GM FBS or GM hPL, the viability of cells in the scaffolds was assessed in three replicates using a live/dead assay. Live cells were easily identifiable as they appeared green, a result of the enzymatic conversion of "calcein AM" to calcein (excitation 494 nm, emission 517 nm). Dead cells, on the other hand, emitted red fluorescence (excitation 517 nm, emission 617 nm) due to the binding of ethidium homodimer-1 to the nucleic acids of cells with damaged membranes. Both bright-field and fluorescence images were captured of the same area.

The results from the fluorescence microscopy analysis of PLA-CH(L), PLA-CH(M), and PLA-CH(H) demonstrated the sustained viability of BM-hMSCs in the scaffolds after

28 days of culture, in both GM hPL (Figure 2a) and GM FBS (Supplementary Figure S1). The microscopic data clearly showed visible intact cell nuclei stained with DAPI (in blue), indicating a uniform distribution of cells within the scaffolds (in green), and a minimal presence of dead cells (in red) (Figure 2a, Supplementary Figure S1).

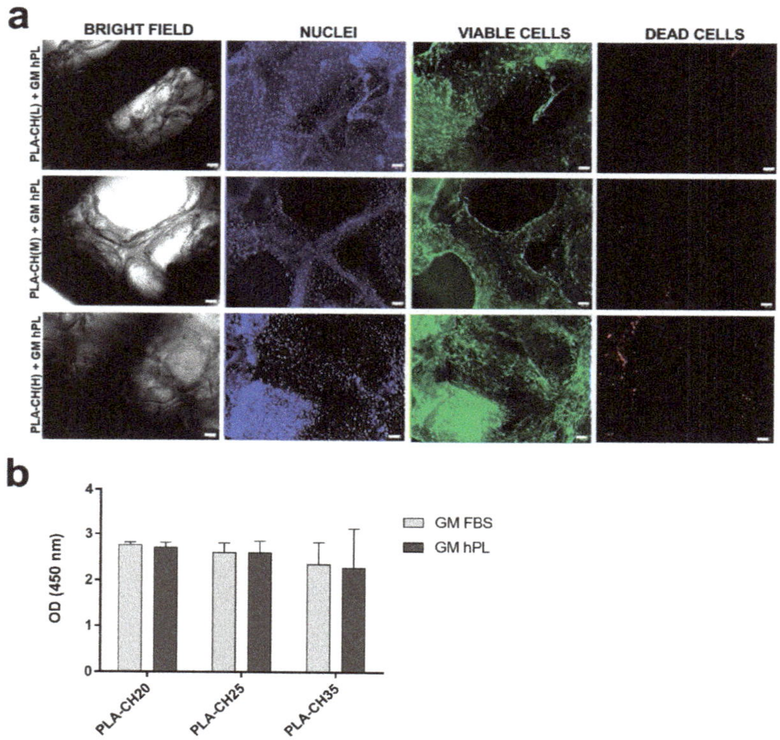

Figure 2. (**a**) Live/dead staining of BM-hMSCs cultivated in PLA-CH (L), PLA-CH(M), and PLA-CH(H) for 28 days in GM hPL. Scale bar: 100 µm. (**b**) Three-dimensional culture proliferation of BM-hMSCs cultivated in PLA-CH (L), PLA-CH (M), and PLA-CH (H) in the GM FBS or GM hPL at 28 days measured by the CCK8 assay.

Cell proliferation was assessed by quantifying the number of viable cells within the scaffolds 28 days after cell seeding using the Cell Counting Kit-8 colorimetric assay (see Figure 2b). The CCK8 assay accurately detects viable cell activity. Previous studies have shown that CCK-8 solution permeates the entire scaffold without causing unspecific staining [47]. The testing showed no significant differences in cell proliferation with different scaffold types and culture media (GM FBS and GM hPL). In particular, comparisons between scaffolds with BM-hMSCs and GM FBS demonstrated no statistically significant differences after 28 days of culture (PLA-CH(L) vs. PLA-CH(M), p-value = 0.7013; PLA-CH(L) vs. PLA-CH(H), p-value = 0.4117; PLA-CH(M) vs. PLA-CH(H), p-value = 0.2487); the comparative analysis of scaffolds with BM-hMSCs and GM hPL did not reveal any statistically significant differences (PLA-CH(L) vs. PLA-CH(M), p-value = 0.1123, PLA-CH(L) vs. PLA-CH(H), p-value = 0.4307, PLA-CH(M) vs. PLA-CH(H), p-value = 0.3183); no significant differences were observed between the same scaffolds in GM FBS or GM hPL (PLA-CH(L) GM FBS vs. PLA-CH(L) GM hPL, p-value = 0.3481; PLA-CH(M) GM FBS vs. PLA-CH(M) GM hPL, p-value = 0.3481; PLA-CH(H) GM FBS vs. PLA-CH(H) GM hPL, p-value = 0.3481). The mean absorbance values, as determined by the CCK-8 assay after a 28-day incubation period, were 2766 ± 0.050 for PLA-CH(L) GM FBS; 2603 ± 0.203 for PLA-CH(M) GM FBS; 2354 ± 0.465 for PLA-CH(H) GM FBS; 2713 ± 0.100 for PLA-CH(L) GM hPL; 2600 ± 0.244 for PLA-CH(M) GM hPL; 2282 ± 0.844 for PLA-CH(H) GM hPL. Therefore, we considered the scaffolds to be equivalent irrespective of the cell proliferation in the different cultures.

To better assess the scaffold morphology and cell colonization, H&E studies were accomplished on PLA-CH(L), PLA-CH(M), and PLA-CH(H) seeded with BM-hMSCs and grown in GM FBS and GM hPL. The scaffolds showed affinity for staining and revealed heterogeneous large pores. The cells appeared well integrated into the scaffold with an elongated morphology creating a 3D network in the pores and settled on the meshes of the hydrogel (Figure 3, Supplementary Figure S2) but not on the PLA surface without a preference for the scaffold type.

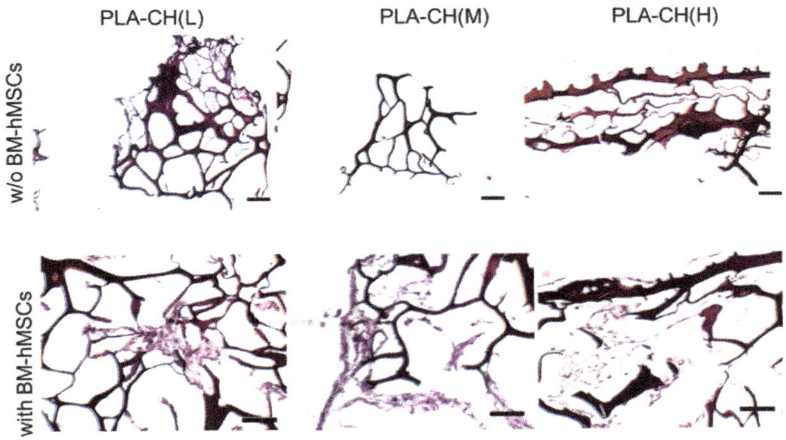

Figure 3. BM-hMSC viable cells cultivated for 28 days in scaffolds in GM hPL, examined by hematoxylin–eosin staining. Scale bar: 100 μm.

In addition, the morphology, adhesion, and dispersion of BM-hMSCs on the PLA-CH(L), PLA-CH(M), and PLA-CH(H) samples were observed with SEM after 28 days of culture in vitro. In fact, the scaffold surface was essential for effective cell adhesion. Figure 4a–c shows the framework of the scaffolds, with detailed information on PLA, hydrogel, and the overall structure. Figure 4 and Supplementary Figure S3 show the interaction of the cells with the scaffolds and their adhesion and spreading preferentially in the pores of the hydrogel shell compared to the PLA core structure, where the cells were mainly localized on the rough surface of the sectioned PLA struts (Figure 4d–f). BM-hMSCs adhered to the hydrogel shell, even bridging the open porosity. The cells extended on a large area, were predominantly elongated, and had a fibroblast-like morphology with evidence of cytoplasmic processes facilitating adhesion and cell communication. These results clearly demonstrate that BM-hMSCs exhibit a remarkable affinity for attachment, proliferation, and migration in PLA-CH(L), PLA-CH(M), and PLA-CH(H) scaffolds, with no significant variations observed between the different scaffolds.

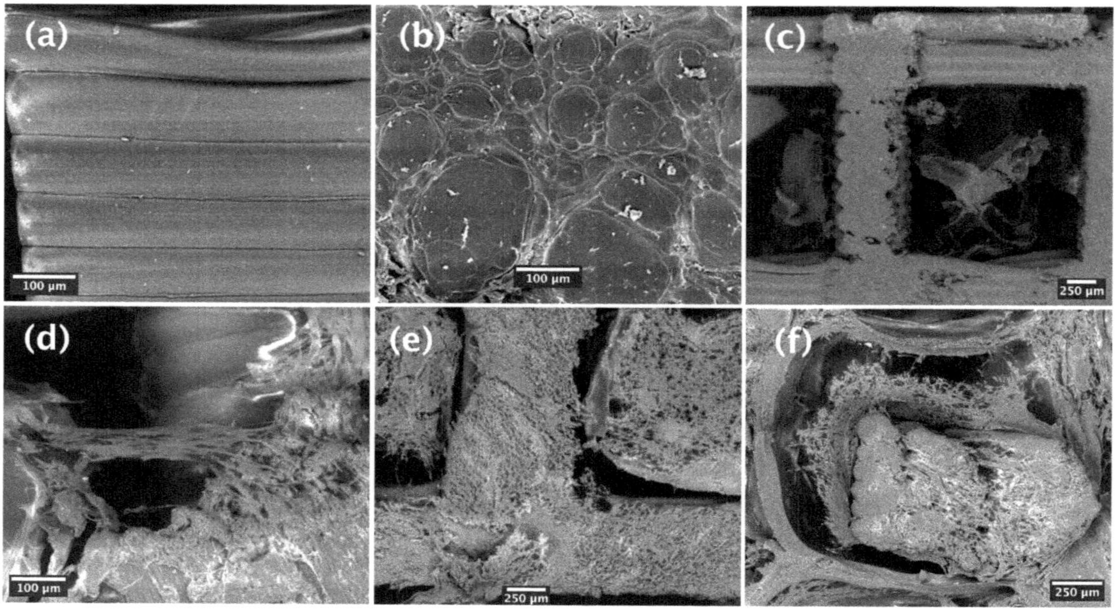

Figure 4. SEM images of PLA-CH(M) scaffolds. (inset (**a**)): the PLA structure; (inset (**b**)), detail of the hydrogel structure; (inset (**c**)), view of the entire structure. SEM images of cellular growth over the scaffolds with BM-hMSCs in GM hPL: PLA-CH(L), PLA-CH(M), and PLA-CH(H) in the inset (**d**–**f**) respectively.

3.3. *Analysis of the Osteogenic Differentiation and Mineralization of the Scaffolds Seeded with Osteo-Differentiated BM-hMSCs*

3.3.1. Histochemical and Immunohistochemical Stainings

The results demonstrated the presence of calcium deposits (von Kossa stain) in the structures of the differentiated scaffolds (Figure 5). In particular, the calcium deposits showed a higher presence in the scaffold cultured with OM hPL with respect to GM ($p < 0.0001$) and OM hPL with respect to OM FBS ($p < 0.0001$ for CH(M) and CH(H); $p = 0.0002$ for CH(L)). In GM, a minimal presence of calcium deposits has been evidenced, especially in GM hPL. The three scaffolds did not show statistical significance among them related to the condition tested. Regarding the immunohistochemical evaluation focusing on scaffold PLA-CH(M), the cells appeared well integrated in the scaffold and produced an extracellular matrix with a positivity for collagen I (Supplementary Figure S4). Figure 6 shows a clear differentiation state as osteoblasts with a positivity for OSC in OM hPL. Moreover, the immunopositivity of the differentiated cells with hPL for OSC, but not for OSP, showed a consolidated differentiation state as osteoblasts in this treatment. On the contrary, the immunopositivity of the differentiated cells with FBS for OSP, which showed a delay in the differentiation process, confirmed a lower presence of calcium deposits. Finally, the immunopositivity for BMP2 in both differentiated groups indicated a remodeling activity of the extracellular matrix.

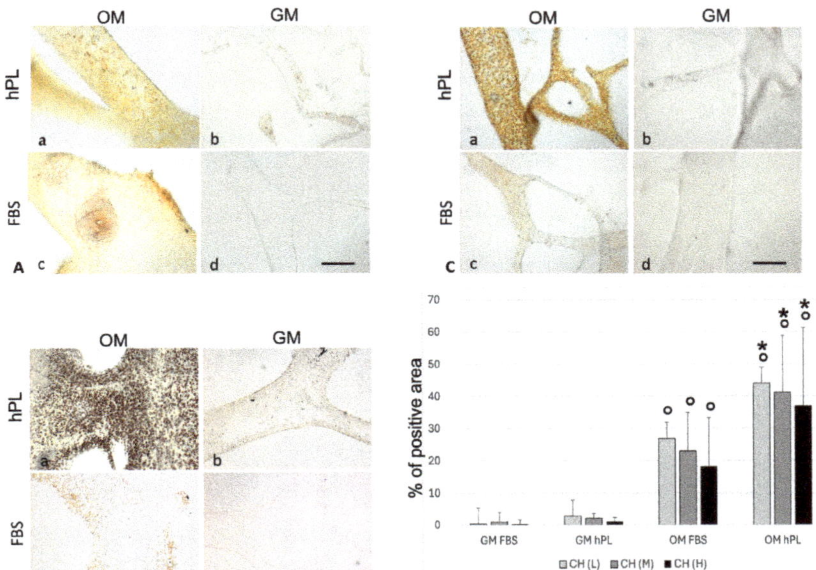

D

	GM FBS		GM hPL		OM FBS		OM hPL	
	MEAN	SD	MEAN	SD	MEAN	SD	MEAN	SD
CH (L)	0.40	1.67	2.80	2.14	26.81	19.65	44.02	20.57
CH (M)	0.98	2.96	2.11	1.52	22.90	12.02	41.15	17.72
CH (H)	0.31	1.30	1.02	1.31	18.12	15.19	36.96	24.30

Figure 5. Calcium deposit distribution in hydrogels PLA-CH cultured with cells using von Kossa staining (brown/black dots). Microphotographs of scaffolds with differentiated (**a**,**c**) and undifferentiated (**b**,**d**) BM-hMSCs (400× magnification, scale bar: 40 µm) in (**A**) PLA-CH(L), (**B**) PLA-CH(M), and (**C**) PLA-CH(H). Calcium deposits appeared as dots in brown/black color. Quantification of percentage of positive area (**D**) within scaffold meshes; ° $p < 0.001$ vs. respective GM; * $p < 0.001$ vs. OM FBS. No statistical differences were observed when comparing the scaffolds with the same treatment.

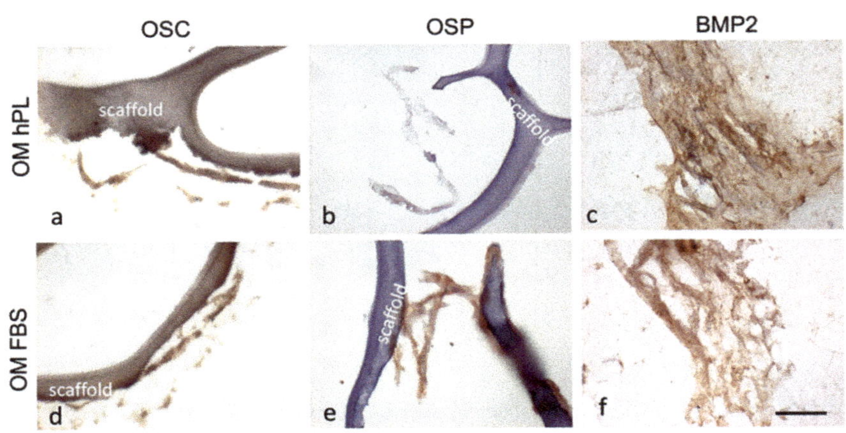

Figure 6. Micrographs of the hydrogels PLA-CH(M) with differentiated (OM) BM-hMSCs in hPL (**a**–**c**) and FBS (**d**–**f**) immunostained for OSC (**a**,**d**), OSP (**b**,**e**), and BMP2 (**c**,**f**) (brown color) with hematoxylin counterstaining (blue/violet) (400× magnification, bar 40 µm).

3.3.2. SEM–EDS Analysis of Scaffold Mineralization

SEM investigation confirmed that the PLA-CH(M) scaffold supported cell differentiation towards the osteogenic lineage by calcium phosphate deposition (Figure 7). BM-hMSCs differentiated into osteoblast cells with evidence of preferential calcium phosphate deposition on the hydrogel shell compared to the PLA core. Calcium phosphate deposition was present using osteogenic induction media, both with hPL and FBS, in the PLA-CH(M) seeded with BM-hMSCs. Significant mineral deposits were visible within the scaffolds, particularly in the porous structure of the hydrogel shell, which was ideal.

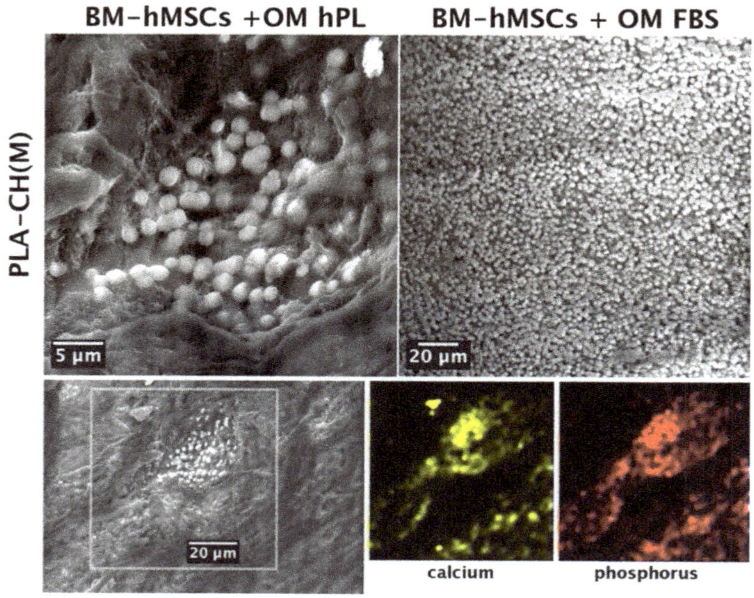

Figure 7. SEM images and EDS investigation of calcium phosphate deposition in PLA-CH(M) with BM-hMSCs in the OM hPL and OM FBS at day 28. Upper part: the bright round particles observed on both samples (upper part of the figure) are ascribed to hydroxyapatite formation. Lower part: evaluation of calcium and phosphorous with SEM–EDS of PLA-CH(M) with BM-hMSCs in OM with FBS at day 28. The mapping of Ca and P in the boxed area ($78 \times 78\ \mu^2$) indicates the extended presence of these elements below the biological film visible at the surface.

To confirm the presence of mineral deposits in the PLA-CH scaffolds bioengineered with BM-hMSCs, SEM–EDS compositional analysis was performed at 28 days. This analysis conclusively demonstrates that the inorganic phase is predominantly composed of calcium and phosphorus, the key elements in hydroxyapatite microparticles formed during osteogenic differentiation. The samples incubated in an osteogenic medium showed the detection of both calcium and phosphorous. The chemical distribution maps collected clearly indicate that these elements co-localize in the same areas of the sections (Figure 7, lower part). In a previous publication [28], the mineral deposits have been deeply investigated through a correlative approach of SEM–EDS and optical microscopy coupled to Raman spectroscopy. It has been demonstrated that similar mineral deposits are indeed Ca-P-O compounds with hydroxyapatite features in the Raman spectra. This correlative technique features more significant spatial resolution than XRD or FTIR and could be successfully implemented for the detection of the early stage of the mineralization process. In the present work, the three-dimensional spatial arrangement of the scaffold and the significant fluorescence from the PLA affected the collection of the Raman signal and the consequent evidence of hydroxyapatite formation. Based on this previous analysis, the bright, round particles visible in Figure 7 could be confidently identified as hydroxyapatite mineral deposits, owing to the combination of SEM image contrast and the simultaneous detection of Ca and P from EDS. On the contrary, the large, irregular detail visible in the SEM image of Figure 7 and in the elemental mapping of Ca only was not considered as a hydroxyapatite.

The evidence of calcium and phosphorous ions was shown in the presence of both FBS and hPL as supplements in the osteogenic culture media. The scaffolds seeded with BM-hMSCs treated with GM FBS or GM hPL, as well as the unseeded scaffold, did not show any significant presence of calcium or phosphorous. This demonstrates that these chemical elements primarily accumulated in the scaffolds seeded with the cells that were grown in OM.

4. Discussion

Stem-cell-based therapy is a crucial topic in regenerative medicine, with hMSCs playing a significant role in combination with three-dimensional biocompatible scaffolds [48]. hMSCs have been demonstrated to be appropriate for the treatment of bone tissue in regenerative medicine [49]. At the same time, there is increasing interest in biocompatible materials that offer an adequate environment for cell growth and differentiation and are, therefore, fundamental in increasing the efficiency in healing injured bone and bone regeneration [27]. While numerous tissue engineering strategies have been developed and researched in detail, only a few approaches have translated into clinical applications [4].

In previous work, the authors have developed 3D gelatin–chitosan hybrid hydrogels in combination with hMSCs to improve osteogenic and chondrogenic differentiation [22,29]. However, the mechanical properties of hydrogels are much lower than those of bone and generally unsuitable for temporary load bearing. Looking for stiffer bioresorbable polymers commercially available in medical grades and offering better control over their mechanical and degradation behaviors, synthetic polymers, such as PCL, PLGA, and PLA, are often used for bone tissue engineering [38,50]. PLA, an FDA-approved biodegradable synthetic polymer, is widely utilized in various biomedical applications due to its biocompatible and biodegradable properties [51]. In bone tissue regeneration systems, PLA is subjected to diverse fabrication methods involving the modification of polymer structures, the development of blended polymer fibers, and the integration of nanoparticles to create nanocomposite fibers [52]. Particularly, given the bioactive role of hydrogels on one side and the mechanical support offered by PLA on the other, combining these two materials improves both the mechanical properties and the bioactivity of the scaffolds [41]. Hybrid combinations of materials have received considerable attention in the inducement of osteogenic and chondrogenic differentiation [52]. Recently, the authors have developed bioresorbable scaffolds with an innovative core–shell architecture. These scaffolds are based on 3D-printed PLA cores with different lattice structures and grafted with CH-hydrogel to meet the diverse requirements for bone tissue engineering applications [44].

The goal was to create a hybrid structure that imitates natural bone tissue, customizing the flexible multi-material design to achieve a good balance between mechanical support and regenerative potential. In order to provide the scaffolds with complementary functions, two components were carefully chosen. The PLA core was designed with lattice structures of varying void volume fractions, allowing for precise adjustment of stiffness and strength. The hydrogel filled the entire void volume of the core lattices with a highly porous shell, which presented highly interconnected pores, smaller than the holes of the core and offering a greater surface area for cell adhesion and homogeneous colonization. The PLA core aimed at transiently providing a tailored mechanical device for bone tissue, while the hydrogel shell served as a microenvironment for cell colonization and mineralization and underwent degradation, leaving space for the formation of new bone tissue [44]. An extensive mechanical characterization of scaffolds with a variety of lattice structures highlighted that their compressive stiffness and strength can be modulated over a wide range of values by acting on the lattice geometrical parameters, aiming at matching the properties of specific target tissues [53].

Here, three distinct PLA-hydrogel core–shell scaffolds with varying hydrogel contents were created and thoroughly examined to determine the conditions for bone tissue regeneration in vitro. The scaffolds had a core–shell structure. The lattice core was realized by additive manufacturing, while the shell was enclosed throughout the core by inserting and crosslinking a hydrogel-forming solution. The hydrogel network created an open porous structure with heterogeneous pore size (around 100 μm on average) that covered and stuck to the struts of the PLA lattice (Figure 4). By changing the size of the lattice unit cells, core void volume fractions between about 69% and 83% were obtained, allowing us to host hydrogel contents between about 12 wt%, for PLA-CH(L), and 22 wt%, for PLA-CH(H) (Table 1). While the hydrogel content increased with the core void volume fraction, the mechanical properties were higher for denser lattices (Table 2). More in detail, the highest stiffness (about 380 MPa) was obtained for the lattices with the smallest holes and lowest void volume fraction (69% ca.), whereas the lowest modulus (about 180 MPa) was achieved for those with the largest holes and highest void volume fraction (83% ca.). In other words, the PLA structure density and the lattice design determined the stiffness values of the three scaffolds. Furthermore, PLA-CH exhibited a compressive stiffness that was significantly higher than that of the hydrogel alone, remaining unaffected by the hydrogel incorporation, and surpassing several composite or core–shell bioresorbable scaffolds that had been suggested in the literature for the objective of bone tissue engineering (Table 3). Therefore, the PLA-CH core–shell scaffolds are a very promising bioresorbable tool for reconstructing hard tissues. While PLA maintained the stiffness stable for several weeks in water at body temperature, the hydrogel gradually lost mass due to hydrolytic degradation, leaving space for new bone tissue formation (Table 2).

Table 3. Compressive stiffness of various hydrogel-based scaffolds.

Material Components *	Void Volume of the Stiff Component [%]	Compressive Modulus [MPa]	Ref.
Gelatin-chitosan (CH) hydrogel	not applicable	0.04 ÷ 0.25	Dey K. et al. [37]
PLA core + CH-hydrogel shell	40 ÷ 90	50 ÷ 550	Pasini C. et al. [54]
Calcium-deficient hydroxyapatite + alginate hydrogel	not available	7	Raja N. et al. [55]
PLA + gelatin methacrylate hydrogel + gold nanoparticles	74 ÷ 87	300 ÷ 700	Heo D.N. et al. [43]
PLA/PVA + chitosan hydrogel + hydroxyapatite	not available (>80)	0.7 ÷ 1.2	Li T.T. et al. [56]
PCL + gelatin-heparin cryogel	80	0.004	Lee S.S. et al. [57]
PCL + nanohydroxyapatite + collagen	50	90 ÷ 110	Cho Y.S. et al. [58,59]
PCL + peptide-based hydrogel	45	11	Wu T. et al. [60]

* Note that particulate composites were excluded from the examples due to the lower mechanical properties typically offered by discontinuous reinforcement.

To analyze the capability of the core–shell scaffolds to simulate the natural microenvironment of the bone, the scaffolds were seeded with BM-hMSCs in different culture conditions. According to previous works, cells were grown in a culture medium containing FBS or hPL as a supplement [22,28]. To induce the osteogenic differentiation, additional stimuli were added as previously described [22]. The analyses were performed after 4 weeks of culture, both in growth medium and differentiation medium, as generally occurred [61].

The viability assay confirmed that all the cells were proliferating and viable as required after 28 days of culture (Figure 2). The cells were anchored to the area of the scaffold, preferentially in the hydrogel pores rather than the PLA structure (Figures 3 and 4). The homogeneous cell distribution of the hydrogel was better appreciated with SEM analysis (Figure 4), compared to the histological approach (Figure 3), probably due to the technical procedure that could lead to the loss of some cells adhered to the structure of scaffolds. However, histological staining allows us to better reveal the capacity of BM-hMSCs to cluster in certain regions of the scaffold, preferentially in the lower size pores and on the surface of the scaffolds, homogeneously covering them preferentially in the outer region compared to the inner region of the scaffolds (Figure 3). No significant difference in cell proliferation was observed in the three types of core–shell scaffolds and different culture media (GM FBS and GM hPL), even if a trend favoring a decrease could be observed when PLA-CH(H) was seeded with BM-hMSCs and analyzed after 28 days of culture (Figure 2). This is probably due to a slight dilution of the test due to the increased hydrogel content in the void volume fraction. Calcium deposition was detected in BM-hMSCs seeded in GM hPL, though having a very low concentration, while GM FBS was inefficient (Figure 2). It is noteworthy that a cell culture procedure using hPL, instead of FBS, stimulated both cell expansion and differentiation [62,63]. Previous works emphasize the role of hPL in increasing the osteogenic and chondrogenic differentiation capabilities of BM-hMSCs without requiring additional differentiation stimuli in hydrogels [22,28,29].

To induce the differentiation of BM-hMSCs toward osteogenic lineage in the core–shell scaffolds, additional differentiation stimuli were added to the culture medium. The calcium deposits were primarily observed using a histochemical method (von Kossa staining), and, subsequently, protein markers for osteogenic differentiation were highlighted using an immunohistochemical technique, both for optical microscopy evaluation. In detail, OPN is an early and pivotal osteogenic marker for the expression of secondary late markers [64], such as OSC [65–67]. The greater presence of deposits was associated with the consolidated, differentiated state of osteoblast (Figure 5), which was confirmed by the OSC immunopositivity and OSP immunonegativity of OM hPL (Figure 6). In addition, the delay in the differentiation state of cells cultured in OM FBS was associated with the presence of immunopositive cells for OSC and OSP. In both cases, BMP-2 was activated for extracellular matrix remodeling (production of collagen I), and bone regeneration itself confirmed its essential role during this process [68]. Finally, these data confirmed the high potentiality of hPL in a regenerative context. The chemical elements analyzed with SEM–EDS confirmed the accumulation of granular deposits only in the scaffolds seeded with cell growth in OM according to the von Kossa staining (Figure 7). No mineral deposits were found when BM-hMSCs grew in the scaffold in basal conditions, without adding differentiation stimuli (Figure 4). After 28 days of osteogenic induction, the formation of granular deposits was found in the core–shell scaffolds, preferentially on the hydrogel surface and between its porosities, while no mineralization was observed on the PLA lattice structure (Figure 7). EDS spectra associated with SEM revealed that the mineral deposits were composed of calcium and phosphorous ions, and these elements co-localized in the same areas of the sections, confirming the presence of hydroxyapatite. The co-localization of calcium and phosphorus suggests an outstanding ability of the core–shell scaffolds to imitate the physiological environment, as already previously demonstrated in gelatin–chitosan hydrogels [22].

5. Conclusions

In this study, bioresorbable scaffolds with innovational core–shell architecture were prepared by grafting a gelatin–chitosan (CH) shell in 3D-printed PLA cores with different lattice structures and investigated in more detail for their biocompatibility and ability to induce bone tissue regeneration in vitro. The objective was to develop a hybrid structure that mimics native bone tissue, creating a versatile multi-material design that is essential for achieving an optimal balance between mechanical support and regenerative potential. Two components were selected to provide the scaffolds with complementary functions: a PLA core, designed with lattice structures and different void volume fractions, and a hydrogel shell that filled in the entire void volume of the core lattices. The PLA core aims to supply tailored mechanical support for bone tissue in the long term, and the hydrogel shell exhibits the favored microenvironment for cell expansion. Overall, our results demonstrated that PLA-CH(L), PLA-CH(M), and PLA-CH(H), differing in the amount of hydrogel content, are biocompatible, and their morphological, mechanical, and degradation properties are appropriate to support cell adhesion, growth, proliferation, and osteogenic differentiation. Additionally, the amount of hydrogel content nor the geometry of the core–shell structure showed influence on the cell behavior. For the cell culture procedure, the cell osteogenic differentiation results were obtained using both hPL and FBS with the addition of differentiation stimuli in the cell culturing. A consolidated differentiation state of BM-hMSCs, as osteoblasts, has been shown to perform better with hPL compared with FBS. Indeed, the cells cultured in differentiation medium with FBS also showed a calcium and phosphorous deposition but with a lower expression. The formation of granular deposits was observed in PLA-CH(M) seeded with osteo-differentiated BM-hMSCs, both in OM FBS and OM hPL. These results suggest that the combination of BM-hMSCs and PLA-CH in the presence of osteogenic differentiation stimuli can provide a construct for guiding bone tissue regeneration. In fact, the scaffolds analyzed presented different mechanical and degradation properties but similar biocompatibility, so they could be used in different clinical settings relating to the anatomical site required for bone regeneration.

Supplementary Materials: The following supporting information can be downloaded at: https://www.mdpi.com/article/10.3390/jfb15080217/s1. Figure S1: Live/dead staining of BM-hMSCs cultivated in PLA-CH (L), PLA-CH(M), and PLA-CH(H) for 28 days in GM FBS. Scale bar: 100 μm. Figure S2: BM-hMSCs viable cells cultivated for 28 days in scaffolds in GM FBS examined by hematoxylin–eosin staining. Scale bar: 100 μm. Figure S3: SEM images of PLA-CH(L), PLA-CH(M), and PLA-CH(H) with BM-hMSCs in GM FBS. Figure S4: Micrographs of the hydrogels PLA-CH(M) with differentiated (OM) BM-hMSCs in hPL (a) and FBS (b) immunostained for collagen I (Coll I) (brown color) with hematoxylin counterstaining (blue/violet) (400× magnification, bar 40 μm).

Author Contributions: Conceptualization, D.R., L.S. and F.R.; methodology, F.R., L.S., C.P., M.F. (Matteo Ferroniand), E.B. and S.P.; validation, F.R., L.S., C.P., M.F. (Matteo Ferroniand) and E.B.; formal analysis, F.R., L.S., C.P., M.F. (Matteo Ferroniand) and E.B.; fluorescence microscope support, R.B.; hPL supply, A.B. and C.A.; technical support, L.G., S.M. and F.T.; investigation, S.B. and M.F. (Mirko Farina); writing—original draft preparation, F.R.; writing—review and editing, F.R., L.S., C.P., M.F. (Matteo Ferroniand) and E.B.; supervision, D.R.; funding acquisition, D.R. All authors have read and agreed to the published version of the manuscript.

Funding: This work was realized with the contributions of University of Brescia "Health & Wealth" Research Program (Brescia, Italy) and Fondazione Comunità Bresciana (Brescia, Italy).

Data Availability Statement: The original contributions presented in the study are included in the article/Supplementary Material, further inquiries can be directed to the corresponding author/s.

Acknowledgments: The authors are highly grateful to Gammatom (Como, Italy) for the gamma sterilization of scaffolds. The authors are highly grateful to the Brescia section of Associazione Donatrici Italiane Sangue Cordone Ombelicale (ADISCO) for supporting the research with hemopoietic and mesenchymal stem.

Conflicts of Interest: The authors declare no conflicts of interest.

References

1. Starr, J.; Tay, Y.K.D.; Shane, E. Current Understanding of Epidemiology, Pathophysiology, and Management of Atypical Femur Fractures. *Curr. Osteoporos. Rep.* **2018**, *16*, 519–529. [CrossRef] [PubMed]
2. Di Pietro, L.; Palmieri, V.; Papi, M.; Lattanzi, W. Translating Material Science into Bone Regenerative Medicine Applications: State-of-The Art Methods and Protocols. *Int. J. Mol. Sci.* **2022**, *23*, 9493. [CrossRef] [PubMed]
3. Flores-Jiménez, M.S.; Garcia-Gonzalez, A.; Fuentes-Aguilar, R.Q. Review on Porous Scaffolds Generation Process: A Tissue Engineering Approach. *ACS Appl. Bio Mater.* **2023**, *6*, 1–23. [CrossRef] [PubMed]

4. Re, F.; Borsani, E.; Rezzani, R.; Sartore, L.; Russo, D. Bone Regeneration Using Mesenchymal Stromal Cells and Biocompatible Scaffolds: A Concise Review of the Current Clinical Trials. *Gels* **2023**, *9*, 389. [CrossRef]
5. Renesme, L.; Pierro, M.; Cobey, K.D.; Mital, R.; Nangle, K.; Shorr, R.; Lalu, M.M.; Thébaud, B. Definition and Characteristics of Mesenchymal Stromal Cells in Preclinical and Clinical Studies: A Scoping Review. *Stem Cells Transl. Med.* **2022**, *11*, 44–54. [CrossRef]
6. Campagnoli, C.; Roberts, I.A.; Kumar, S.; Bennett, P.R.; Bellantuono, I.; Fisk, N.M. Identification of mesenchymal stem/progenitor cells in human first-trimester fetal blood, liver, and bone marrow. *Blood* **2001**, *98*, 2396–2402. [CrossRef]
7. Dominici, M.; Le Blanc, K.; Mueller, I.; Slaper-Cortenbach, I.; Marini, F.; Krause, D.; Deans, R.; Keating, A.; Prockop, D.; Horwitz, E. Minimal criteria for defining multipotent mesenchymal stromal cells. The International Society for Cellular Therapy position statement. *Cytotherapy* **2006**, *8*, 315–317. [CrossRef]
8. Gronthos, S.; Mankani, M.; Brahim, J.; Robey, P.G.; Shi, S. Postnatal human dental pulp stem cells (DPSCs) in vitro and in vivo. *Proc. Natl. Acad. Sci. USA* **2000**, *97*, 13625–13630. [CrossRef]
9. Seo, B.M.; Miura, M.; Gronthos, S.; Bartold, P.M.; Batouli, S.; Brahim, J.; Young, M.; Robey, P.G.; Wang, C.Y.; Shi, S. Investigation of multipotent postnatal stem cells from human periodontal ligament. *Lancet* **2004**, *364*, 149–155. [CrossRef]
10. Ntege, E.H.; Sunami, H.; Shimizu, Y. Advances in regenerative therapy: A review of the literature and future directions. *Regen. Ther.* **2020**, *14*, 136–153. [CrossRef]
11. Gómez-Barrena, E.; Padilla-Eguiluz, N.; Rosset, P.; Gebhard, F.; Hernigou, P.; Baldini, N.; Rouard, H.; Sensebé, L.; Gonzalo-Daganzo, R.M.; Giordano, R.; et al. Early efficacy evaluation of mesenchymal stromal cells (MSC) combined to biomaterials to treat long bone non-unions. *Injury* **2020**, *51* (Suppl. S1), S63–S73. [CrossRef] [PubMed]
12. Machado, A.; Pereira, I.; Costa, F.; Brandão, A.; Pereira, J.E.; Maurício, A.C.; Santos, J.D.; Amaro, I.; Falacho, R.; Coelho, R.; et al. Randomized clinical study of injectable dextrin-based hydrogel as a carrier of a synthetic bone substitute. *Clin. Oral. Investig.* **2023**, *27*, 979–994. [CrossRef] [PubMed]
13. Manzini, B.M.; Machado, L.M.R.; Noritomi, P.Y.; DA Silva, J.V.L. Advances in Bone tissue engineering: A fundamental review. *J. Biosci.* **2021**, *46*, 17. [CrossRef]
14. Lin, H.; Sohn, J.; Shen, H.; Langhans, M.T.; Tuan, R.S. Bone marrow mesenchymal stem cells: Aging and tissue engineering applications to enhance bone healing. *Biomaterials* **2019**, *203*, 96–110. [CrossRef] [PubMed]
15. Bunpetch, V.; Zhang, Z.Y.; Zhang, X.; Han, S.; Zongyou, P.; Wu, H.; Hong-Wei, O. Strategies for MSC expansion and MSC-based microtissue for bone regeneration. *Biomaterials* **2019**, *196*, 67–79. [CrossRef] [PubMed]
16. Harrell, C.R.; Markovic, B.S.; Fellabaum, C.; Arsenijevic, A.; Volarevic, V. Mesenchymal stem cell-based therapy of osteoarthritis: Current knowledge and future perspectives. *Biomed. Pharmacother.* **2019**, *109*, 2318–2326. [CrossRef] [PubMed]
17. Filomeno, P.; Dayan, V.; Touriño, C. Stem cell research and clinical development in tendon repair. *Muscles Ligaments Tendons J.* **2012**, *2*, 204–211.
18. Gao, F.; Chiu, S.M.; Motan, D.A.; Zhang, Z.; Chen, L.; Ji, H.L.; Tse, H.F.; Fu, Q.L.; Lian, Q. Mesenchymal stem cells and immunomodulation: Current status and future prospects. *Cell Death Dis.* **2016**, *7*, e2062. [CrossRef]
19. Re, F.; Gabusi, E.; Manferdini, C.; Russo, D.; Lisignoli, G. Bone Regeneration Improves with Mesenchymal Stem Cell Derived Extracellular Vesicles (EVs) Combined with Scaffolds: A Systematic Review. *Biology* **2021**, *10*, 579. [CrossRef]
20. Han, Y.; Li, X.; Zhang, Y.; Chang, F.; Ding, J. Mesenchymal Stem Cells for Regenerative Medicine. *Cells* **2019**, *8*, 886. [CrossRef]
21. Rauch, C.; Feifel, E.; Amann, E.M.; Spötl, H.P.; Schennach, H.; Pfaller, W.; Gstraunthaler, G. Alternatives to the use of fetal bovine serum: Human platelet lysates as a serum substitute in cell culture media. *ALTEX-Altern. Anim. Exp.* **2011**, *28*, 305–316. [CrossRef]
22. Re, F.; Sartore, L.; Moulisova, V.; Cantini, M.; Almici, C.; Bianchetti, A.; Chinello, C.; Dey, K.; Agnelli, S.; Manferdini, C.; et al. 3D gelatin-chitosan hybrid hydrogels combined with human platelet lysate highly support human mesenchymal stem cell proliferation and osteogenic differentiation. *J. Tissue Eng.* **2019**, *10*, 2041731419845852. [CrossRef]
23. Raghav, P.K.; Mann, Z.; Ahlawat, S.; Mohanty, S. Mesenchymal stem cell-based nanoparticles and scaffolds in regenerative medicine. *Eur. J. Pharmacol.* **2022**, *918*, 174657. [CrossRef] [PubMed]
24. Roseti, L.; Parisi, V.; Petretta, M.; Cavallo, C.; Desando, G.; Bartolotti, I.; Grigolo, B. Scaffolds for Bone Tissue Engineering: State of the art and new perspectives. *Mater. Sci. Eng. C Mater. Biol. Appl.* **2017**, *78*, 1246–1262. [CrossRef] [PubMed]
25. Pereira, H.F.; Cengiz, I.F.; Silva, F.S.; Reis, R.L.; Oliveira, J.M. Scaffolds and coatings for bone regeneration. *J. Mater. Sci. Mater. Med.* **2020**, *31*, 27. [CrossRef] [PubMed]
26. Shuai, C.; Shi, X.; Yang, F.; Tian, H.; Feng, P. Oxygen vacancy boosting Fenton reaction in bone scaffold towards fighting bacterial infection. *Int. J. Extrem. Manuf.* **2024**, *6*, 015101. [CrossRef]
27. Zhang, L.; Yang, G.; Johnson, B.N.; Jia, X. Three-dimensional (3D) printed scaffold and material selection for bone repair. *Acta Biomater.* **2019**, *84*, 16–33. [CrossRef]
28. Re, F.; Sartore, L.; Borsani, E.; Ferroni, M.; Baratto, C.; Mahajneh, A.; Smith, A.; Dey, K.; Almici, C.; Guizzi, P.; et al. Mineralization of 3D Osteogenic Model Based on Gelatin-Dextran Hybrid Hydrogel Scaffold Bioengineered with Mesenchymal Stromal Cells: A Multiparametric Evaluation. *Materials* **2021**, *14*, 3852. [CrossRef]
29. Li, P.; Fu, L.; Liao, Z.; Peng, Y.; Ning, C.; Gao, C.; Zhang, D.; Sui, X.; Lin, Y.; Liu, S.; et al. Chitosan hydrogel/3D-printed poly(ε-caprolactone) hybrid scaffold containing synovial mesenchymal stem cells for cartilage regeneration based on tetrahedral framework nucleic acid recruitment. *Biomaterials* **2021**, *278*, 121131. [CrossRef]

30. Bharadwaz, A.; Jayasuriya, A.C. Recent trends in the application of widely used natural and synthetic polymer nanocomposites in bone tissue regeneration. *Mater. Sci. Eng. C Mater. Biol. Appl.* **2020**, *110*, 110698. [CrossRef]
31. Lei, B.; Guo, B.; Rambhia, K.J.; Ma, P.X. Hybrid polymer biomaterials for bone tissue regeneration. *Front. Med.* **2019**, *13*, 189–201 [CrossRef]
32. Fernandez de Grado, G.; Keller, L.; Idoux-Gillet, Y.; Wagner, Q.; Musset, A.M.; Benkirane-Jessel, N.; Bornert, F.; Offner, D. Bone substitutes: A review of their characteristics, clinical use, and perspectives for large bone defects management. *J. Tissue Eng.* **2018**, *9*, 2041731418776819. [CrossRef] [PubMed]
33. De Leon-Oliva, D.; Boaru, D.L.; Perez-Exposito, R.E.; Fraile-Martinez, O.; García-Montero, C.; Diaz, R.; Bujan, J.; García-Honduvilla, N.; Lopez-Gonzalez, L.; Álvarez-Mon, M.; et al. Advanced Hydrogel-Based Strategies for Enhanced Bone and Cartilage Regeneration: A Comprehensive Review. *Gels* **2023**, *9*, 885. [CrossRef] [PubMed]
34. Zhang, C.; Shi, T.; Wu, D.; Hu, D.; Li, W.; Fei, J.; Liu, W. The application of 3D-printed hydrogels in bone tissue engineering. *Tissue Eng. Part. B Rev.* **2023**. [CrossRef]
35. Manferdini, C.; Gabusi, E.; Sartore, L.; Dey, K.; Agnelli, S.; Almici, C.; Bianchetti, A.; Zini, N.; Russo, D.; Re, F.; et al. Chitosan-based scaffold counteracts hypertrophic and fibrotic markers in chondrogenic differentiated mesenchymal stromal cells. *J. Tissue Eng. Regen. Med.* **2019**, *13*, 1896–1911. [CrossRef]
36. Jayakumar, P.; Di Silvio, L. Osteoblasts in bone tissue engineering. *Proc. Inst. Mech. Eng. H J. Eng. Med.* **2010**, *224*, 1415–1440. [CrossRef] [PubMed]
37. Dey, K.; Agnelli, S.; Re, F.; Russo, D.; Lisignoli, G.; Manferdini, C.; Bernardi, S.; Gabusi, E.; Sartore, L. Rational Design and Development of Anisotropic and Mechanically Strong Gelatin-Based Stress Relaxing Hydrogels for Osteogenic/Chondrogenic Differentiation. *Macromol. Biosci.* **2019**, *19*, e1900099. [CrossRef]
38. Li, C.; Ouyang, L.; Armstrong, J.P.K.; Stevens, M.M. Advances in the Fabrication of Biomaterials for Gradient Tissue Engineering. *Trends Biotechnol.* **2021**, *39*, 150–164. [CrossRef]
39. Taraballi, F.; Sushnitha, M.; Tsao, C.; Bauza, G.; Liverani, C.; Shi, A.; Tasciotti, E. Biomimetic Tissue Engineering: Tuning the Immune and Inflammatory Response to Implantable Biomaterials. *Adv. Healthc. Mater.* **2018**, *7*, e1800490. [CrossRef]
40. Park, J.Y.; Park, S.H.; Kim, M.G.; Yoo, T.H.; Kim, M.S. Biomimetic Scaffolds for Bone Tissue Engineering. *Adv. Exp. Med. Biol.* **2018**, *1064*, 109–121. [CrossRef]
41. Reddy, M.S.B.; Ponnamma, D.; Choudhary, R.; Sadasivuni, K.K. A Comparative Review of Natural and Synthetic Biopolymer Composite Scaffolds. *Polymers* **2021**, *13*, 1105. [CrossRef] [PubMed]
42. Sartore, L.; Pasini, C.; Pandini, S.; Dey, K.; Ferrari, M.; Taboni, S.; Chan, H.H.L.; Townson, J.; Viswanathan, S.; Mathews, S.; et al. Hybrid Core-Shell Polymer Scaffold for Bone Tissue Regeneration. *Int. J. Mol. Sci.* **2022**, *23*, 4533. [CrossRef] [PubMed]
43. Heo, D.N.; Castro, N.J.; Lee, S.J.; Noh, H.; Zhu, W.; Zhang, L.G. Enhanced bone tissue regeneration using a 3D printed microstructure incorporated with a hybrid nano hydrogel. *Nanoscale* **2017**, *9*, 5055–5062. [CrossRef] [PubMed]
44. Pasini, C.; Pandini, S.; Re, F.; Ferroni, M.; Borsani, E.; Russo, D.; Sartore, L. New Poly(lactic acid)-Hydrogel Core-Shell Scaffolds Highly Support MSCs' Viability, Proliferation and Osteogenic Differentiation. *Polymers* **2023**, *15*, 4631. [CrossRef]
45. Raise3D, Raise3D Premium PLA Technical Data Sheet. 2022. Available online: https://s1.raise3d.com/2020/12/PLA-Data-Sheet.zip (accessed on 20 January 2023).
46. Gibson, L.J.; Ashby, M.F. *Cellular Solids: Structure and Properties*, 2nd ed.; Cambridge University Press: Cambridge, UK, 2014. [CrossRef]
47. Tonello, S.; Bianchetti, A.; Braga, S.; Almici, C.; Marini, M.; Piovani, G.; Guindani, M.; Dey, K.; Sartore, L.; Re, F.; et al. Impedance-Based Monitoring of Mesenchymal Stromal Cell Three-Dimensional Proliferation Using Aerosol Jet Printed Sensors: A Tissue Engineering Application. *Materials* **2020**, *13*, 2231. [CrossRef] [PubMed]
48. Funda, G.; Taschieri, S.; Bruno, G.A.; Grecchi, E.; Paolo, S.; Girolamo, D.; Del Fabbro, M. Nanotechnology Scaffolds for Alveolar Bone Regeneration. *Materials* **2020**, *13*, 201. [CrossRef] [PubMed]
49. Mazzoni, E.; Iaquinta, M.R.; Mosaico, M.; De Pace, R.; D'Agostino, A.; Tognon, M.; Martini, F. Human Mesenchymal Stem Cells and Innovative Scaffolds for Bone Tissue Engineering Applications. *Tissue Eng. Part. B Rev.* **2023**, *29*, 514–531. [CrossRef]
50. Shi, M.; Cheng, T.; Zou, H.; Zhang, N.; Huang, J.; Xian, M. The Preparation and Biomedical Application of Biopolyesters. *Mini Rev. Med. Chem.* **2020**, *20*, 331–340. [CrossRef]
51. Hasan, A.; Waibhaw, G.; Saxena, V.; Pandey, L.M. Nano-biocomposite scaffolds of chitosan, carboxymethyl cellulose and silver nanoparticle modified cellulose nanowhiskers for bone tissue engineering applications. *Int. J. Biol. Macromol.* **2018**, *111*, 923–934. [CrossRef]
52. Lee, J.H.; Park, T.G.; Park, H.S.; Lee, D.S.; Lee, Y.K.; Yoon, S.C.; Nam, J.D. Thermal and mechanical characteristics of poly(L-lactic acid) nanocomposite scaffold. *Biomaterials* **2003**, *24*, 2773–2778. [CrossRef] [PubMed]
53. Álvarez-Blanco, M.; Infante-García, D.; Marco, M.; Giner, E.; Miguélez, M.H. Development of bone surrogates by material extrusion-based additive manufacturing to mimic flexural mechanical behaviour and fracture prediction via phase-field approach. *Comput. Methods Programs Biomed.* **2024**, *254*, 108287. [CrossRef] [PubMed]
54. Pasini, C.; Pandini, S.; Ramorino, G.; Sartore, L. Tailoring the properties of composite scaffolds with a 3D-Printed lattice core and a bioactive hydrogel shell for tissue engineering. *J. Mech. Behav. Biomed. Mater.* **2024**, *150*, 106305. [CrossRef] [PubMed]
55. Raja, N.; Yun, H.S. A simultaneous 3D printing process for the fabrication of bioceramic and cell-laden hydrogel core/shell scaffolds with potential application in bone tissue regeneration. *J. Mater. Chem. B* **2016**, *4*, 4707–4716. [CrossRef]

56. Li, T.T.; Zhang, Y.; Ren, H.T.; Peng, H.K.; Lou, C.W.; Lin, J.H. Two-step strategy for constructing hierarchical pore structured chitosan-hydroxyapatite composite scaffolds for bone tissue engineering. *Carbohydr. Polym.* **2021**, *260*, 117765. [CrossRef] [PubMed]
57. Lee, S.S.; Santschi, M.; Ferguson, S.J. Correction to "A Biomimetic Macroporous Hybrid Scaffold with Sustained Drug Delivery for Enhanced Bone Regeneration". *Biomacromolecules* **2022**, *23*, 1474. [CrossRef] [PubMed]
58. Cho, Y.S.; Ghim, M.S.; Hong, M.W.; Kim, Y.Y.; Cho, Y.S. Strategy to improve endogenous bone regeneration of 3D-printed PCL/nano-HA composite scaffold: Collagen designs with BMP-2 and FGF-2. *Mater. Des.* **2023**, *229*, 111913. [CrossRef]
59. Cho, Y.S.; Quan, M.; Kang, N.U.; Jeong, H.J.; Hong, M.W.; Kim, Y.Y.; Cho, Y.S. Strategy for enhancing mechanical properties and bone regeneration of 3D polycaprolactone kagome scaffold: Nano hydroxyapatite composite and its exposure. *Eur. Polym. J.* **2020**, *134*, 109814. [CrossRef]
60. Wu, T.; Wu, Y.; Cao, Z.; Zhao, L.; Lv, J.; Li, J.; Xu, Y.; Zhang, P.; Liu, X.; Sun, Y.; et al. Cell-free and cytokine-free self-assembling peptide hydrogel-polycaprolactone composite scaffolds for segmental bone defects. *Biomater. Sci.* **2023**, *11*, 840–853. [CrossRef] [PubMed]
61. Yurova, K.; Melashchenko, E.S.; Khaziakhmatova, O.G.; Malashchenko, V.V.; Melashchenko, O.B.; Shunkin, E.O.; Norkin, I.K.; Ivanov, P.A.; Khlusov, I.A.; Litvinova, L.S. Osteogenic Differentiation Factors of Multipotent Mesenchymal Stromal Cells in the Current Understanding. *Curr. Pharm. Des.* **2021**, *27*, 3741–3751. [CrossRef]
62. Mareschi, K.; Marini, E.; Niclot, A.G.S.B.; Barone, M.; Pinnetta, G.; Adamini, A.; Spadea, M.; Labanca, L.; Lucania, G.; Ferrero, I.; et al. A New Human Platelet Lysate for Mesenchymal Stem Cell Production Compliant with Good Manufacturing Practice Conditions. *Int. J. Mol. Sci.* **2022**, *23*, 3234. [CrossRef]
63. Oeller, M.; Laner-Plamberger, S.; Krisch, L.; Rohde, E.; Strunk, D.; Schallmoser, K. Human Platelet Lysate for Good Manufacturing Practice-Compliant Cell Production. *Int. J. Mol. Sci.* **2021**, *22*, 5178. [CrossRef] [PubMed]
64. Holm, E.; Gleberzon, J.S.; Liao, Y.; Sørensen, E.S.; Beier, F.; Hunter, G.K.; Goldberg, H.A. Osteopontin mediates mineralization and not osteogenic cell development in vitro. *Biochem. J.* **2014**, *464*, 355–364. [CrossRef] [PubMed]
65. Boskey, A.L.; Wians, F.H.; Hauschka, P.V. The effect of osteocalcin on in vitro lipid-induced hydroxyapatite formation and seeded hydroxyapatite growth. *Calcif. Tissue Int.* **1985**, *37*, 57–62. [CrossRef] [PubMed]
66. Weinreb, M.; Shinar, D.; Rodan, G.A. Different pattern of alkaline phosphatase, osteopontin, and osteocalcin expression in developing rat bone visualized by in situ hybridization. *J. Bone Miner. Res.* **1990**, *5*, 831–842. [CrossRef] [PubMed]
67. Tsao, Y.T.; Huang, Y.J.; Wu, H.H.; Liu, Y.A.; Liu, Y.S.; Lee, O.K. Osteocalcin Mediates Biomineralization during Osteogenic Maturation in Human Mesenchymal Stromal Cells. *Int. J. Mol. Sci.* **2017**, *18*, 159. [CrossRef]
68. Salazar, V.S.; Gamer, L.W.; Rosen, V. BMP signalling in skeletal development, disease and repair. *Nat. Rev. Endocrinol.* **2016**, *12*, 203–221. [CrossRef]

Journal of
Functional
Biomaterials

MDPI

Article

Chemically Pretreated Densification of Juniper Wood for Potential Use in Osteosynthesis Bone Implants

Laura Andze [1,2,*], Vadims Nefjodovs [2,3], Martins Andzs [1], Marite Skute [1], Juris Zoldners [1], Martins Kapickis [3], Arita Dubnika [4,5], Janis Locs [4,5] and Janis Vetra [2]

1 Latvian State Institute of Wood Chemistry, Dzerbenes Street 27, LV-1006 Riga, Latvia; martins.andzs@kki.lv (M.A.); polarlapsa@inbox.lv (M.S.); jzoldn@inbox.lv (J.Z.)
2 Department of Morphology, Institute of Anatomy and Anthropology, Riga Stradins University, Dzirciema Street 16, LV-1007 Riga, Latvia; vadims.nefjodovs@gmail.com (V.N.); janis.vetra@rsu.lv (J.V.)
3 Microsurgery Centre of Latvia, Brivibas Gatve 410, LV-1024 Riga, Latvia; kapickis@sr.com
4 Institute of Biomaterials and Bioengineering, Faculty of Natural Science and Technology, Riga Technical University, Pulka Street 3, LV-1048 Riga, Latvia; arita.dubnika@rtu.lv (A.D.); janis.locs@rtu.lv (J.L.)
5 Baltic Biomaterials Centre of Excellence, Headquarters at Riga Technical University, LV-1048 Riga, Latvia
* Correspondence: laura.andze@kki.lv; Tel.: +371-2244-0088

Abstract: The aim of the study was to perform treatment of juniper wood to obtain wood material with a density and mechanical properties comparable to bone, thus producing a potential material for use in osteosynthesis bone implants. In the first step, partial delignification of wood sample was obtained by Kraft cooking. The second step was extraction with ethanol, ethanol–water mixture, saline, and water to prevent the release of soluble compounds and increase biocompatibility. In the last step, the thermal densification at 100 °C for 24 h was implemented. The results obtained in the dry state are equivalent to the properties of bone. The swelling of chemically pre-treated densified wood was reduced compared to chemically untreated densified wood. Samples showed no cytotoxicity by in vitro cell assays. The results of the study showed that it is possible to obtain noncytotoxic wood samples with mechanical properties equivalent to bones by partial delignification, extraction, and densification. However, further research is needed to ensure the material's shape stability, water resistance, and reduced swelling.

Keywords: juniper wood; chemical pretreatment; kraft cooking; partial delignification; compressed solid wood; wood densification; wood bone implants; in vitro

1. Introduction

There are two forms of surgical procedures used to heal bone disorders and injuries: bone grafting, which replaces missing, worn-out, or irreparably damaged bones, and osteosynthesis, which uses metal plates and screws to fix bone fractures. Material has to fulfill several requirements to be utilized as an osteosynthesis implant, including having mechanical qualities that exceed and are comparable to the host tissue and a good host–implant integration.

The top-performing material in osteosynthesis has been stainless steel, followed by titanium and cobalt-chromium alloys. These materials have become the industry norm for implants, used by millions of patients globally each year [1]. However, despite their success, these alloys are not without limitations. The trend towards non-metallic implants is rising due to various factors. Stainless steel and titanium are materials that are biocompatible, meaning they can coexist safely within the human body. They exhibit isotropic properties, meaning their mechanical characteristics are uniform regardless of the direction of force applied. Additionally, these materials are easily adaptable and can be molded into various configurations for medical applications. However, despite their favorable physical characteristics and ability to maintain their form within the biological environment, neither stainless steel nor titanium possesses the osteogenic properties that support and enhance bone regeneration and healing processes. The density of titanium alloys is typically up to three times higher than that of bones [2], making the metal stronger but potentially leading to increased bone fragility after healing. Hence, there is a risk of aseptic loosening of the metal implants as a potential complication up to 15 years post-surgery [3]. Repeated surgeries may be necessary to remove implants that have been rejected by the bone after healing. Studies have shown a risk of bio-corrosion in stainless steel and titanium alloys [4,5]. Furthermore, metal implants can pose challenges for certain medical tests and procedures [6,7]. Hence, alternative materials such as ceramics [8] like calcium phosphate, calcium carbonate, calcium sulphate, hydroxyapatite [9,10], and various polymers, as well as numerous copolymers, are currently undergoing testing [11,12]. Despite their excellent biocompatibility, these materials are often too brittle and do not possess the required mechanical strength for orthopedic applications.

The growing focus on environmental issues has prompted experts in the field of bone implant technology to delve into biomaterial research [13]. Recent studies have compared the ecological impact of various materials [13,14], explored the use of natural and bio-based polymers in bone regeneration [14–16], and investigated 3D printing as an eco-friendly option for bone implants [17]. Researchers have even looked into using materials sourced directly from nature, such as corals [14]. Despite these advancements, current materials do not meet the necessary mechanical standards for osteosynthesis. As a result, scientists continue to seek a strong and mechanically suitable bone implant made from renewable resources.

To the best of our knowledge, there are currently nine research papers discussing the potential use of solid wood as osteosynthesis of load-bearing bone implants, which have also been examined in a recent review article [18]. Six of these studies focused on untreated wood [19–24], while two looked at thermally treated wood [25,26] and only one study tested densified solid wood [27]. These studies investigated various types of wood such as birch, ash, willow, lime, bamboo, and juniper. Interestingly, only willow and lime showed a strong inflammatory reaction towards untreated wood, while birch, ash, and juniper exhibited good biocompatibility even without sterilization. Among the nine studies, only two looked at wood for osteosynthesis. Despite the positive findings, the inadequate mechanical properties of untreated wood may hinder its use for the osteosynthesis. By densifying the wood, its mechanical properties can be improved beyond those of bone, which is crucial for the osteosynthesis implants. Thus, the issues associated with metals, such as aseptic loosening and biocorrosion, could be avoided by substituting densified solid wood for metal implants while maintaining the biocompatibility of the wood with the bone [27].

While research on mechanical densification of wood dates back to the early 1900s, it was only recently that chemical or enzymatic pretreatment methods were developed [28,29]. In 2018, Song et al. introduced a new approach to wood densification, using a sulfite process to partially remove lignin from wood before hot-pressing [30]. Several studies have since tested this method of densification [31–33]. However, despite these advancements, the potential of densified wood as a biomaterial for osteosynthesis has not yet been explored. Wood shows potential as a material for osteosynthesis implants, but it also presents several challenges. These include anisotropic mechanical properties and dimensional changes in response to moisture, particularly in densified wood. Additionally, the presence of extractive substances in wood may lead to adverse reactions in bone cells. Both the benefits and drawbacks are discussed comprehensively in the review article [18].

The novelty of the research is confirmed by a previously unused type of chemical pretreatment (Kraft cooking) before densification and the evaluation of the obtained materials in the context of osteosynthesis bone implants, with mechanical properties, swelling and cytotoxicity. Even previous studies have not performed extraction after chemical pretreatment to prevent the release of cytotoxic compounds. Furthermore, we conduct mechanical properties testing following swelling in the wet condition to achieve results that are more representative of the human body's environment.

2. Materials and Methods

2.1. Materials

Samples of solid juniper (Juniperus Communis) wood were gathered from a forest in Kegums, Vidzeme region, Latvia for the research. Sodium sulfide hydrate (≥60%, Sigma-Aldrich, Darmstadt, Germany), sodium hydroxide (>97%, Sigma-Aldrich, Darmstadt, Germany), ethanol (>95%) and deionized water were conducted for pretreatment of juniper solid wood sample.

2.2. Sample Preparation

Juniper logs, manually debarked, were left to air dry for 1 month before being cut into specimens measuring 90 mm × 15 mm × 15 mm (longitudinal × tangential × radial). The initial moisture content of the samples was 7.62%. The research methodology for densifying and characterizing juniper wood is depicted in Figure 1.

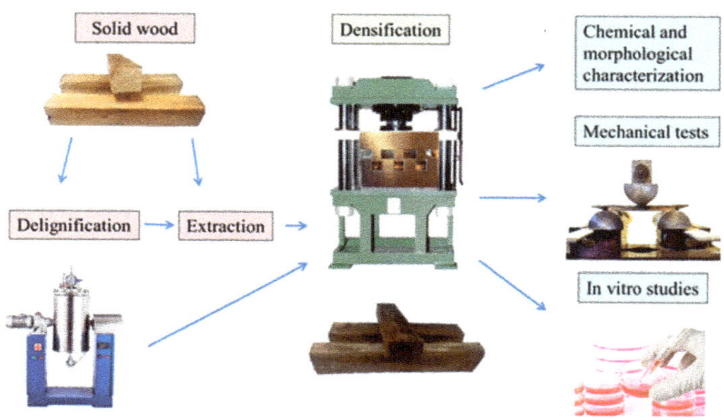

Figure 1. Schematic illustration of the obtained research.

2.3. Chemical Treatment

Before treatment with chemicals, juniper wood specimens were immersed in a typical Kraft cooking solution made from 1.25 M NaOH and 0.25 M Na_2S, commonly used in the pulping industry. The specimens were placed in individual autoclaves filled with the solution for 24 h. Subsequently, the autoclaves were heated to 165 °C in a glycerin bath and cooked at this temperature for various durations—removed immediately after reaching 165 °C (K0), after—10 min (K10), 20 min (K20), and 40 min (K40). The cooked specimens were then thoroughly washed with deionized water until no more coloring was observed, and, finally, stored in water.

The extraction process was as follows. To prevent the potential impact of soluble compounds on bone cells during in vitro assays, sample extraction was performed after cooking/before densification. Alcohol—water/alcohol (1:1)—and alcohol extraction was conducted for 14 h, 36 h, and 7 h, respectively, by simmering the sample in 200 mL of the appropriate solution for 1 h, straining the solution, and continuing the process until the solvent visually no longer changes color. Extraction was performed for only one sample after 10 min of Kraft cooking. The sample was named K10E. Samples' abbreviations and corresponding treatments are summarized in Table 1.

Table 1. Samples' treatments and abbreviations.

Sample Abbreviation	Untreated	DD	WD	K0	K10	K20	K40	K10E
Cooking time (min) after reaching 165 °C	-	-	-	0	10	20	40	10
Treatment before densification	-	Dry	Soaked in water (wet)	Washed and soaked in water				Extracted by ethanol and water

2.4. Densification

After chemical treatment, juniper wood specimens in the radial direction (Figure 2A) underwent hot-pressing in a specially designed mold (Figure 3B) using a single-stage press LAP 40 (Gotfried Joos Maschinefabrik GmbH & Co., Pfalzgrafenweiler, Germany). The pressure applied was 5 MPa at 100 °C for 24 h, followed by an additional 12 h of interrupted heating. Control samples of dry (DD) and wet (WD) juniper wood were also hot pressed using the same method to compare the effects of chemical pretreatment. Immersion in hot (90 °C) water was used to prepare the WD sample before hot pressing. Three specimens of each sample type were produced for evaluation.

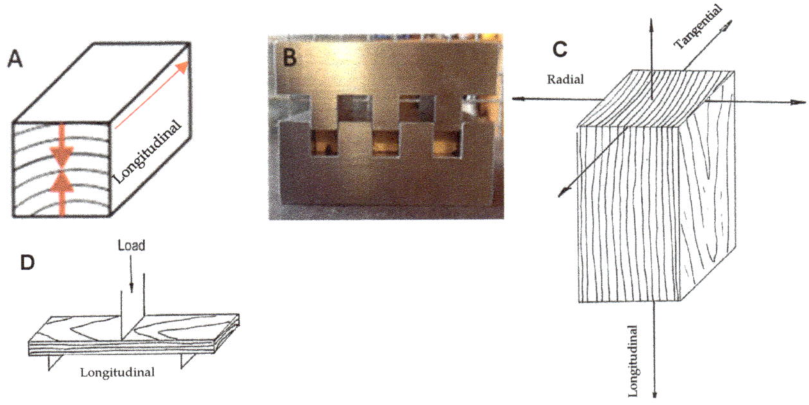

Figure 2. Densification. (**A**)—densification direction (radial). (**B**)—densification mold. (**C**)—anisotropy of the wood sample. (**D**)—direction of 3-point bending measurement.

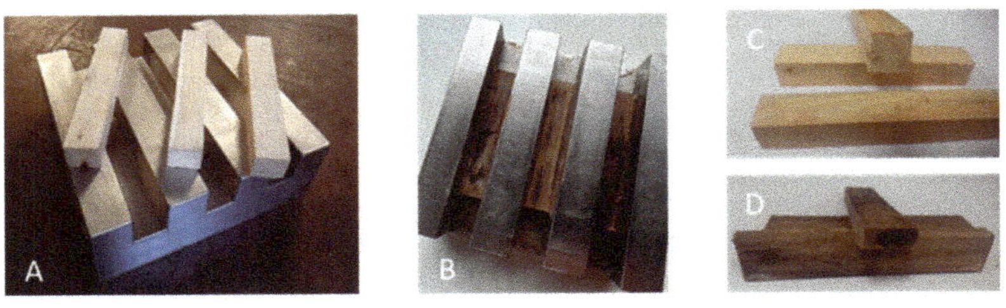

Figure 3. Visual evaluation of untreated sample (**C**) with press mold (**A**) and K10E sample (**D**) with press mold (**B**) after densification.

2.5. SEM

In order to analyze cross-section of the samples using scanning electron microscopy (SEM), a thin layer of gold plasma was applied to the samples using a K550X sputter coater (Emitech, South Petherton, UK). The samples were then examined using a Vega TC microscope (Tescan, Brno, Czech Republic), running software version 2.9.9.21.

2.6. Chemical Characterization

2.6.1. Mass Loss

The method for determining mass loss (ML%) of samples after chemical pretreatment involved calculating the percentage of weight lost using Equation (1):

$$ML_{\%} = \frac{M_1 - M_0}{M_0} \cdot 100\% \quad (1)$$

where M0 and M1 represent the initial and final masses of the completely dried specimen before and after undergoing chemical pretreatment, respectively.

For chemical characterization, wood samples were ground using M20 mill (IKA-WERKE, Breisgau, Germany) before and after chemical pretreatment.

2.6.2. Extractives

The ground samples were Soxhlet-extracted with acetone for 8 h to determine the amount of extractable components. This quantification was done gravimetrically using an ES 225SM-DR scale (Precisa, Zurich, Switzerland) after rotary vacuum evaporation with a PC3001 VARIO equipment (Green Vac, Düsseldorf, Germany). The results were expressed as a percentage of the initial wood sample mass, calculated using Equation (2)

$$Ex_{\%} = \frac{M_2 - M_1}{M_0} \cdot 100\% \quad (2)$$

where, Ex%—the extractives in the wood sample, M—the absolutely dry sample mass, M1—the mass of the absolutely dry round flask, and M2—the mass of the absolutely dry round flask with sample extractives.

2.6.3. Klason Lignin

Klason lignin was conducted under the TAPPI 222om-98 protocol.

2.6.4. Chemical Composition

The HPLC (high-performance liquid chromatography) analysis was conducted by NREL/TP-510-42618 Laboratory Analytical Procedure for the determination of structural carbohydrates and lignin in biomass [34].

Hydrolysates were prepared and analyzed using a Shimadzu LC-20A HPLC system (Shimadzu, Tokyo, Japan) with a refractive index detector. Reference standards with a purity of ≥99.0% (Sigma Aldrich, Steinheim, Germany) were used for the analysis of glucose, cellobiose, arabinose, 2-furaldehyde, acetic acid, 5-HMF, levulinic acid, and formic acid by Shodex Sugar SH1821 column at 60 °C, with eluent 0.008 M H_2SO_4 (Sigma Aldrich, Germany) at a flow rate of 0.6 mL·min^{-1}. Xylose, arabinose, galactose, and mannose were analyzed on Shodex Sugar SP0810 column at 80 °C, with deionized water as the mobile phase under a flow rate of 0.6 mL·min^{-1} using standards with a purity of ≥99.0% (Sigma Aldrich, Germany). Before injection, samples were neutralized to pH 5–7 with $NaHCO_3$ and filtered through a 0.45 μm membrane filter.

2.6.5. FTIR

Fourier transform infrared (FTIR) spectra of wood samples were obtained in KBr (IR grade, Sigma Aldrich, Darmstadt, Germany) pellets using Thermo Fisher Nicolet iS50 spectrometer (Waltham, MA, USA). The spectra were recorded within the range of 4000–450 cm^{-1}, with a spectral resolution of 4 cm^{-1} and 32 scans. A pellet containing 2 mg of a ground sample and 200 mg of KBr was used for the measurements. The resulting spectra were then normalized to the highest absorption maxima.

2.7. Physical-Mechanical Properties

2.7.1. Density

The density of each densified sample was measured after hot pressing and conditioning (25 °C; relative humidity 50%) and determined based on Equation (3):

$$\rho = \frac{M}{t \cdot w \cdot l} \tag{3}$$

where ρ (kg m^{-3}) is the density and M (kg), t (m), w (m), and l (m) are the mass, length, width, and thickness of the conditioned samples, respectively.

2.7.2. Swelling

The swelling test was performed in boiling saline (9% NaCl) at 100 °C for 2 h using samples with initial dimensions of 1.5 cm in length and 1.5 cm in width and corresponding sample thickness. After soaking, the swelling of the sample in volume % was calculated by measuring all dimensions based on Equation (4), and the density of the wood of the sample after swelling was calculated by Equation (4).

$$Sw_{\%} = \frac{V_2 - V_1}{V_1} \cdot 100\% \tag{4}$$

where V_1 (m^3) is the initial volume of the sample and V_2 (m^3) is the final volume of the sample.

2.7.3. Three-Point Bending

The densified juniper samples underwent testing for mechanical properties by the modulus of elasticity, more commonly used as Young's modulus, and modulus of rupture, commonly known as strength, in a three-point bending test conducted on a ZWICK/Z100 (Ulm, Germany) universal testing machine following EN 310 (1993). The distance between the specimens was 70 mm. The three-point bending test was performed in the longitudinal direction as indicated in Figure 2D.

Three specimens of each sample type were analyzed to determine the average property values for strength and resistance. The length and width of all samples retained their original dimensions of 90 mm and 15 mm, respectively. The thickness of the control samples also stayed at the original measurement of 15 mm. In contrast, the thickness of the compacted materials varied according to the results obtained after the densification process (refer to Table 2).

Table 2. Thickness of sample for three-point bending test in dry and wet (after swelling) state.

Sample	Untreated	DD	WD	K0	K10	K20	K40	K10E
thickness, mm (dry)	14.93 ± 0.07	13.89 ± 0.06	8.05 ± 0.03	4.83 ± 0.02	4.81 ± 0.07	4.26 ± 0.04	3.82 ± 0.05	3.99 ± 0.04
thickness, mm (wet)	15.27 ± 0.11	15.20 ± 0.10	15.10 ± 0.11	8.07 ± 0.12	7.5 ± 0.3	8.07 ± 0.08	6.43 ± 0.13	5.70 ± 0.10

2.8. In Vitro Analysis

The cytotoxicity of the sample was assessed using the NIH 3T3 mouse fibroblasts cell line. Cells were pre-seeded in 96-well plates in a density of 5 × 10^3 cells in 100 μL cell medium (89% Dulbecco's Modified Eagle Medium (DMEM), supplemented with 10% (*v*/*v*) calf serum and 1% penicillin/streptomycin (P/S)) and incubated (New Brunswick™ S41i CO_2 Incubator Shaker, Eppendorf, Hamburg, Germany) for 24 h to allow them to attach. Then, the medium was replaced with 100 μL of sample extract dilutions.

To obtain the sample extracts, the sample was placed in 5 mL of fresh cell medium. After 24 and 48 h, all the solution was collected from the samples and replaced with an additional 5 mL of fresh cell medium. The collected solution was then filtered through a 0.2 μm syringe filter and, subsequently, used at 100% for dilution with fresh medium. The extract dilution—1:1—was chosen for this experiment. Untreated cells were used as positive control. The negative control was cell medium with 10% dimethyl sulfoxide (DMSO, Sigma-Aldrich). After incubating the cells with the sample extract and its dilution in a humidified atmosphere with 5% CO_2 and 90% humidity at 37 °C for 24 h and 48 h, all medium in the 96-well plates were replaced with 120 μL CellTiter-Blue (CTB) prepared in cell medium to each well. The plates were put back into the cell incubator for 2 h before optical density was measured by Infinite M Nano microplate reader (Tecan, Männedorf, Switzerland) at 590 nm.

2.9. Statistic

The statistical analysis was carried out using the data analysis tool in SPSS 17.0. Mean values (MV), standard deviation and standard error (SE) were determined from three parallel measurements. The data underwent testing for normality and homogeneity of variances using Levene's test. When the data met the criteria of being normally distributed, a one-way ANOVA test was conducted. For comparisons between two independent samples, Student's *t*-test was employed. All data were reported in the format of MV ± SE. The significance level for all statistical tests was set at $\alpha = 0.05$.

3. Results and Discussion

Some of the research data (density and mechanical properties of the samples—untreated, DD, WD, K0, K20, and K40) have been previously published [27]; however, they have been re-included in this publication to demonstrate the integration of the newest data with the existing dataset and to elaborate on prior results with additional information on chemical analysis, FTIR, SEM, swelling properties, in vitro analysis, and mechanical properties of extracted sample.

3.1. Visual Appearance and SEM

The visual representation of all densified juniper wood samples can be seen in previous publication [27]. The juniper wood K10E sample's appearance has changed after chemical treatment and densification displaying a dark brown color (Figure 3). Changes are linked to the Kraft cooking method, which causes the breakdown of lignin and hemicelluloses in the wood cell walls and the extraction of extractives [35,36]. Chemical changes are described in the Table 1. The change in color in chemically pretreated and densified juniper wood may be attributed to the chemical breakdown of hemicelluloses and lignin, leading to the formation of new chromophoric groups. Additionally, the thermochromatism of the chemicals applied on the wood surface may also play a role [37].

SEM analysis was performed to demonstrated wood structural changes in the cell level after partial delignification and densification. The comparison of microstructures of untreated (A) and densified (B) wood samples in the crossfiber direction at equal magnification is depicted in Figure 4.

An untreated juniper wood sample (Figure 4A) displays cylindrical tracheids with a lumen diameter ranging from 100 to 200 μm. These tracheids are organized in a radial pattern, typical of softwood. The tracheid lumens in the transverse section of the densified wood were nearly imperceptible (Figure 4B). Observing the changes in the microstructure of juniper wood suggests that the wood tracheid underwent sufficient densification and recombination as a result of the chemical treatment and hot pressing. It is likely that the process of partial delignification only softened and plasticized lignin, as well as partially separated readily available lignin in the middle lamella, which weakens the integrity of the cell walls, causing distortions in the lumen structure [37,38].

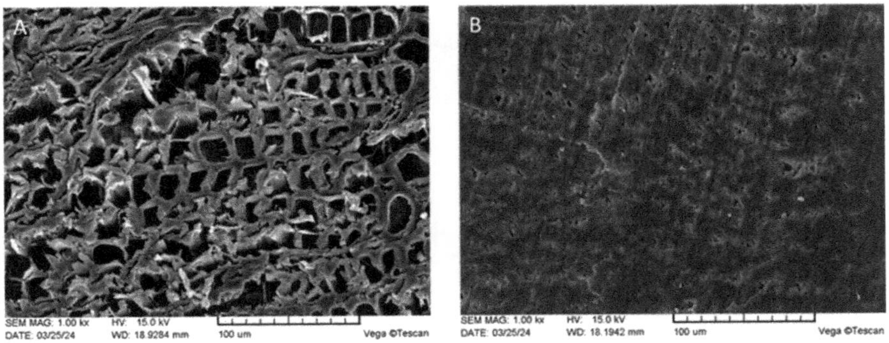

Figure 4. SEM images of untreated (**A**) and densified K10E (**B**) sample at 1000 magnification.

Similar microstructure changes in wood have also been assessed in the works of other authors, using deep eutectic solvents [39], sulfite [30,31,40], and acid or alkyl pretreatments [38], as well as steam or hydrothermal treatment [41].

3.2. Chemical Characterization

The Kraft process, a leading pulping technology globally, involves treating lignocellulosic feedstock in an alkaline solution containing hydroxide, sulfide, and bisulfide ions. This causes the dissolution of lignin and some hemicelluloses [42]. As it can be seen in Table 3, during chemical pretreatment, mass losses of 21–32% occur, primarily due to hemicelluloses dissolution and lignin breakdown. Hemicelluloses content decreases by 7–12% depending on processing severity. Lignin content in juniper wood reduced from 33% in untreated samples to 17% in K20 and K40 samples due to lignin depolymerization during Kraft cooking.

Table 3. Chemical characterization of wood samples before and after treatment.

	Weight Loss, %	Cellulose, %	Hemicelluloses, %	Lignin, %	Extracts, %	Inorganic Compounds, %	Other Components, %
Untreated	0	38.0 ± 2.1	21.1 ± 1.9	33.1 ± 0.5	4.8 ± 0.3	0.060 ± 0.002	2.94
0	21.7 ± 1.8	38.6 ± 0.5	14.4 ± 0.5	20.6 ± 0.1	3.3 ± 0.5	0.050 ± 0.003	1.35
K10	25.9 ± 2.0	38.3 ± 0.9	12.7 ± 1.2	18.2 ± 0.2	3.4 ± 0.2	0.041 ± 0.001	1.459
K20	27.2 ± 0.8	38.1 ± 0.9	12.2 ± 0.5	17.8 ± 0.5	3.2 ± 0.3	0.042 ± 0.005	1.458
K40	32.2 ± 1.2	36.4 ± 1.2	9.1 ± 1.3	17.3 ± 0.7	3.5 ± 0.7	0.040 ± 0.007	1.46

A slight decrease in extractives and inorganics was also observed.

3.3. FTIR

For studying the effects of chemical pretreatment on juniper wood polymers in the cell wall, we conducted an analysis by comparing FTIR spectra of untreated and treated samples (Figure 5). The peaks at 3378 cm^{-1} and 2900 cm^{-1} are associated with the stretching vibrations of O–H and C–H in cellulose, hemicellulose, and lignin [43,44]. These distinctive bands were also observed in both untreated and densified wood samples during analysis. The disappearance of the absorption band at 1729 cm^{-1}, which is associated with the C=O stretching vibration in unconjugated ketones, is a characteristic feature of hemicellulose [31,37,40]. This band was no longer present in the densified wood due to deacetylation in the hemicellulose, demonstrating that the hemicellulose had dissolved in the $NaOH/Na_2S$ solution. Untreated juniper wood showed absorption peaks in the 1588–1560 cm^{-1} region of the spectrum as an absorption shoulder, and similar absorption shoulders were still present after chemical treatment. This indicates that there has been no alteration in the polar functional groups attached to the benzene ring in lignin. In contrast, Shi et al. found changes in these functional groups when using chemical treatments involving NaOH/$NaSO_3$ [31] and H_2SO_4 [37]. Additionally, there was no evidence of a new peak at 1714 cm^{-1} [37], typically indicating pseudo-lignin formation resulting from significant lignin and hemicellulose depolymerization and repolymerization. This demonstrates that the chosen chemical treatment only partially breaks down the lignin and hemicellulose in the cell walls, but rather softens and makes them more plasticize.

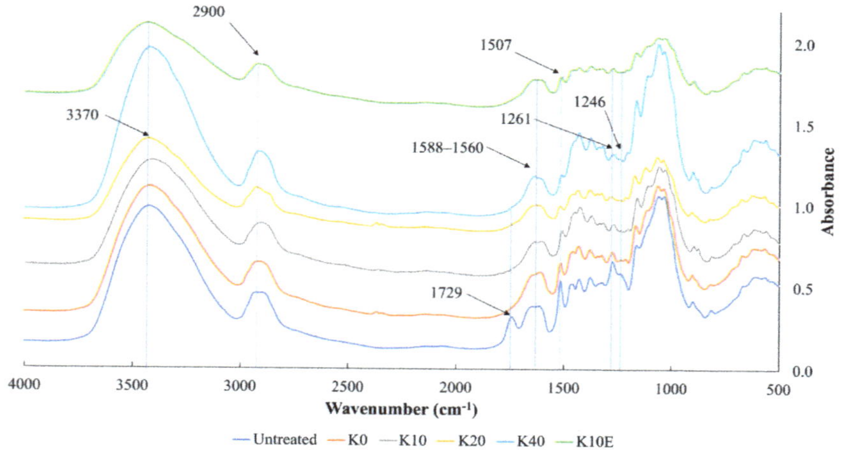

Figure 5. FTIR spectrum of untreated chemically treated sample.

The sharp band at 1507 cm^{-1} at was identified as resulting from the vibrational absorption of the benzene ring structure. In chemically treated wood, the absorption band remains present, but its intensity decreases over time as the treatment progresses. This suggests a reduction in lignin content, aligning with the findings of the chemical analysis detailed in previous section. The decrease of the absorption peaks at the wavelengths of 1261 cm^{-1} and 1246 cm^{-1}, are also related to the decrease of the lignin content, representing guaiacol and syringyl groups, respectively [31,37,44].

3.4. Swelling

Swelling and shrinking are two of the major disadvantages of wood as a material for use in environments with variable moisture content. Considering that muscle tissue consists of 76% water [45], assessing the swelling of wood implants in an aqueous environment is crucial. The most precise method involves measuring the material's swelling in body fluid at 37 °C for an extended period. It takes 2–3 months to reach maximum swelling in water at room temperature [46]. To expedite the process and reach the maximum swelling value, swelling was evaluated by boiling the implant material in saline for two hours, as water is known to be the primary factor influencing swelling [46].

Table 4 shows the swelling results for untreated and densified juniper wood samples. Untreated juniper wood showed a minor swelling of 13%, while chemically untreated dry densified juniper wood (DD) exhibited a 22% of swelling. The most significant swelling was observed in the chemically untreated wet densified wood sample (WD), which fully recovered to its original dimensions and even surpassed them. Similar results were also obtained in the works of other authors [47–49].

Table 4. Untreated and densified sample swelling in boiling saline.

Sample Swelling	Untreated	DD	WD	K0	K10	K20	K40	K10E
Volume, %	13 ± 6	22 ± 6	52 ± 3	34 ± 8	32 ± 4	44 ± 2	38 ± 4	44 ± 1
Longitudial, %	0.08 ± 0.02	0.11 ± 0.03	0.13 ± 0.09	0.09 ± 0.04	0.2 ± 0.1	0.10 ± 0.04	0.17 ± 0.01	0.19 ± 0.11
Tangential (width), %	5.5 ± 0.8	4.0 ± 0.6	4.9 ± 0.3	4.1 ± 0.7	4.8 ± 0.9	3.7 ± 0.1	4.0 ± 0.7	3.8 ± 0.5
Radial (thickness), %	20 ± 3	34 ± 5	67 ± 4	49 ± 5	57 ± 5	52 ± 2	60 ± 4	55 ± 3

Table 4 illustrates the pronounced anisotropy of wood, indicating that swelling primarily occurs in the radial direction—along the growth rings—while the longitudinal dimension remains unaffected. Given the anisotropic characteristics of wood, densification was executed in the radial direction, as this is where the material is most susceptible to physical, mechanical, and chemical influences. Observations reveal that densified wood exhibits the most significant swelling in the same radial direction as the densification process. This suggests that the densification did not establish stable, chemically irreversible bonds sufficient to prevent the material from reverting to its initial form. Future investigations will focus on potential methods such as chemical bonding and/or impregnation to enhance the dimensional stability of the wood during densification and to mitigate additional swelling.

Chemical pretreatment reduced swelling by 15–30% and provided better shape stability. There is no statistically significant variation observed among the chemically treated samples; however, a significant difference is found between the chemically treated compacted samples and the WD sample. To the best of our knowledge, there have been no studies by other authors on the swelling of densified chemically pretreated samples in aqueous media, but there are studies on swelling in different air humidity [50].

Laine et al. and Welzbacher et al. conducted research on the set-recovery in water conditions after densification and following thermal or oil treatment, respectively [47,48]. However, it should be noted that these studies did not perform a chemical pretreatment that reduces the cell density of wood prior to densification. In our study, the calculation of set-recovery may not be accurate due to the variations in wood structure between untreated and chemically treated samples.

3.5. Density

In our research, Kraft cooking pretreatment of wood material led to comparable density results (Figure 6.) to those found in studies using sulphite cooking [30–32,40] or deep eutectic solvent [39] pretreatment conducted by other researchers. Pretreatment with steam [41,51], acid, or alkyl [38] pretreatment in the work of other authors showed less increase in density for wood samples.

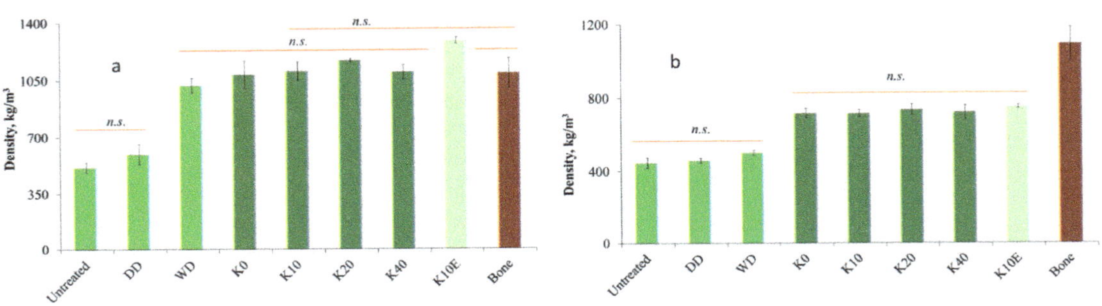

Figure 6. Density of untreated and densified juniper wood vs. cortical bone (**a**) in dry state and (**b**) after swelling in boiling saline [52,53]. n.s.—not statistically significantly different.

As can be seen in Figure 6a, chemical pretreatment of juniper wood and subsequent densification increased the density of the samples by 2.1–2.3 times depending on the treatment time. For sample K10E, the density increased by 2.5 times, reaching 1300 kg m^{-3}, approaching the theoretical maximum density for wood of 1500 kg m^{-3}. This upper limit is due to the density of the wood cell wall, which is typically around 1500 kg m^{-3} and remains relatively constant [54]. The densified juniper wood has a density comparable to that of cortical bone [18,27,52,53], making it a strong candidate for use in bone implant applications. Although the absolute values of the density values of the modified wood samples differ, they do not have a statistically significant difference between each other and also with the K10E sample. Even so, no statistically significant difference was observed between the chemically treated densified samples and the DW sample. In their research, Song et. al. found that reducing the lignin content in wood leads to higher density [30], although our own study did not replicate this result. Despite the fact that lignin has the lowest density (1350 kg m^{-3}) within wood cells, with more delignification potentially boosting wood density. It is important to note that the Kraft process not only removes lignin but also hemicelluloses, which actually have a higher density (1500–1800 kg m^{-3}) [54]. This could explain the lack of significant differences in wood density across various chemical treatment durations.

Although there is a statistically significant difference of swelling in water between the chemically treated and untreated densified juniper wood samples, the alteration in the wood's microstructure due to chemical treatment still remains a factor that cannot be overlooked. In order to compare the swelling levels of untreated and treated samples the results were analyzed in terms of wood density post-swelling (Figure 6b).

Upon swelling, the WD sample returned to the original density of the sample, with no statistically significant difference from the densities of the DD and untreated samples. The density of the WD sample dropped by half while swelling, which is equal to the density increase during densification. The chemically treated densified samples showed a 35–45% decrease in density during swelling, maintaining a density 30–35% higher than the untreated sample before swelling. This indicates microstructural changes in wood during chemical treatment and densification. Wood cell walls have softened, lignin has become more plastic, and new links are formed in the microstructure of wood during thermal densification. There is no statistically significant difference between post-swelling densities of chemically treated densified juniper wood samples.

3.6. Mechanical Properties

The bending properties of juniper wood samples in dry state and after swelling are presented in Figure 7a,b and Figure 8a,b, accordingly. The dry control sample did not show adequate densification, but wet densification led to a notable increase in strength, as shown in Figure 7a. Therefore, the water content in the sample contributes to the formation of hydrogen bonds during the pressing stage, enhancing the densification process and resulting in higher density and improved bending properties.

It is important to note that the hydrogen bonds formed in the WD sample are not stable in a water environment, as illustrated in Figure 7b. The strength of all control samples (Untreated, DD, and WD) after swelling in saline does not show a statistically significant difference among them and is approximately 40% of the strength of untreated wood in a dry state, which aligns with the findings reported in the literature [55,56].

Chemical pretreatment and further densification of juniper wood improved strength in the dry state, with the sample treated for 10 min achieving the highest strength value. The chemical pretreatment of juniper wood has resulted in the partial degradation of lignin and hemicelluloses, leading to the formation of new covalent bonds that further enhance all identified properties. However, no statistically significant difference was observed between the chemically pretreated samples with each other as well as the wet-densified sample (WD). The chemically treated sample extracted pre-densification (K10E) demonstrated a statistically significant increase in strength value compared to the other samples. The combination of chemical pretreatment and densification resulted in a 70–90% increase (160–180 MPa) in strength value reaching a value 40% higher than bone, with the addition of extraction leading to a 2.3 times (215 MPa) greater increase in strength compared to the untreated sample.

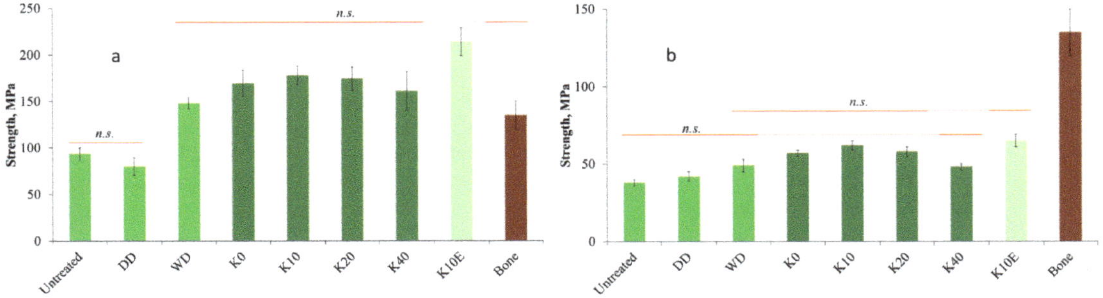

Figure 7. Strength of untreated and densified juniper wood vs. bone (**a**) in wet state and (**b**) after swelling in boiling saline [52]. n.s.—not statistically significantly different.

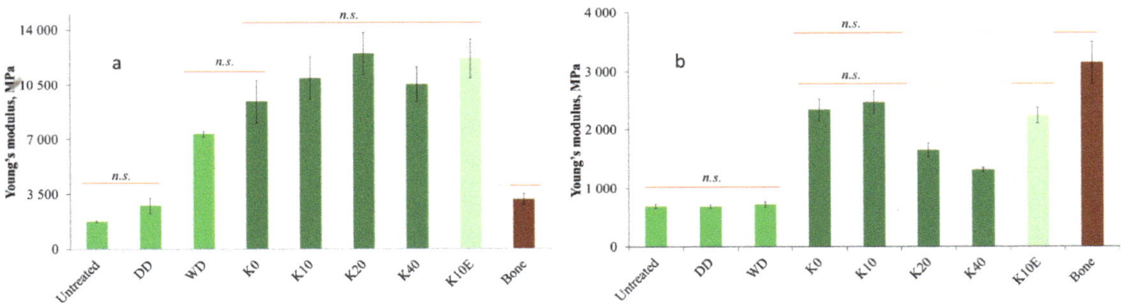

Figure 8. Young's modulus of untreated and densified juniper wood vs. bone (**a**) in wet state and (**b**) after swelling in boiling saline [52]. n.s.—not statistically significantly different.

In the studies of other authors, slightly lower strength absolute values (150 MPa) were achieved on densified poplar by hydrothermal densification [41] or NaOH pretreatment [38], but sulphite cooking pretreatment with subsequent impregnation with sodium silicate increased the poplar strength to 300 MPa [57]. For softwood, the strength value after densification with sulphite cooking pretreatment reached 220 MPa (*Abies*) [31], while deep eutectic solvent pretreatment increased the strength value to 180 MPa (*Pinus*) [39].

Figure 7b illustrates that chemically pretreated densified samples, after being swollen in boiling saline, exhibit a strength increase of 50–60% compared to the control sample, which is statistically significant. However, the chemical bonding formed during the densification process is minimal and does not sufficiently enhance the mechanical properties of the material under wet conditions. This indicates that further investigation is required, potentially involving impregnation or the enhancement of chemical bonds during densification, to minimize sample swelling and improve mechanical strength.

The impact of dry densification (DD) on the Young's modulus value was not found to be statistically significant, whereas wet densification (WD) led to a 4.2-fold increase in the elasticity of untreated juniper (Figure 8a). Chemical pretreatment proved to be effective in improving the flexibility of juniper wood post-densification, resulting in a 6–7.2-fold increase in the Young's modulus value compared to untreated samples. The results obtained are higher than for bone. The chemically treated samples have no statistically significant difference between each other and compared to the extracted sample (K10E).

In other authors' studies on poplar densification by various treatments, higher Young's modulus values (22 GPa) were achieved [38,39,41], but the increase is equivalent, because the Young's modulus of untreated poplar is 4–5 GPa, while the Young's modulus of juniper is 2 GPa. The cross-linked cellulose microfibrils, along with lignin and hemicelluloses, serve as a structural framework in wood cell wall. It has been observed that the elasticity of wood is negatively correlated with the lignin content present [31,32]. The high lignin content (34%) found in juniper wood may explain the lower modulus of elasticity seen in untreated juniper wood samples [58,59].

After boiling in saline, the Young's modulus of the chemically pretreated densified samples decreased for times (Figure 8b). However, this value remains three times higher than that of the control samples (Untreated, DD, and WD). For K0 and K10, the elasticity under wet conditions is similar to the Young's modulus of bone. This indicates that chemical pretreatment followed by densification enhances the mechanical properties of wood materials. Nevertheless, further research is necessary to improve chemical bonding and ensure dimensional stability in moist environments, which is essential for exploring these materials' potential use in osteosynthesis implants.

3.7. In Vitro Analysis

In the conditional medium, untreated juniper wood samples (untreated, DD, and DW) and chemically pretreated densified samples (K0, K10, K20, and K40) both showed pH changes that excluded the indirect in vitro test. Due to this, in vitro tests on NIH 3T3 mouse fibroblasts cells were conducted on the sample K10E that underwent extraction to eliminate water- and alcohol-soluble compounds before densification, in order to avoid potential cytotoxic reactions.

The data from the in vitro cell studies indicated that lower relative metabolic activity is observed after 24 h in the conditioned medium—pure extract from K10E sample (see Figure 9). Nevertheless, the average value of cell viability was over 80%, corresponding to the non-cytotoxic sample according to the ISO 10993-5 standard [60] for medical device evaluation. At 50% or greater dilution, there is no statistical difference in relative metabolic activity between the positive control and evaluated sample. No cytotoxic effect was observed in the condition medium at any dilution after 48 h. The utilization of wood as a bone implant material was mainly explored towards the end of the 20th and beginning of 21st centuries. Publications from that era detail in vivo studies that indicate birch, ash, and juniper exhibit good biocompatibility properties [19–22,24]. On the other hand, lime and willow triggered acute inflammatory responses [23]. Moreover, untreated bamboo displayed cytotoxicity in vitro [61]. The in vitro inflammatory response is influenced by the soluble components of wood (extractives). While Gross et al. conducted in vivo experiments with untreated juniper that showed no toxic reactions [24], our own study found it impractical to conduct in vitro tests on untreated juniper. This suggests that, in certain instances, a material may demonstrate good biocompatibility in vivo analyses despite yielding negative in vitro results.

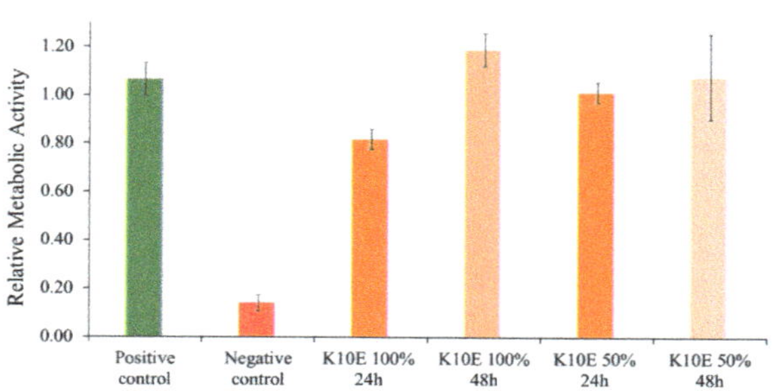

Figure 9. Relative metabolic activity of densified, chemically pretreated, and extracted juniper wood sample (K10E) assessed by CTB assay.

4. Conclusions

The findings of the study indicated that chemical pretreatment has a notable effect on enhancing the mechanical properties of densified wood, suggesting its potential applicability in osteosynthesis implants using densified wood. The results obtained in the study shows equivale mechanical properties as of bone, increasing density by 2.5 times, but strength and Young's modulus by 2.3 and 7.2 times, respectively. Although the swelling is still relatively large, the swelling of chemically pre-treated densified wood was reduced by 15–30% compared to chemically untreated densified wood.

Additionally, it should be noted that the mechanical properties of chemically pretreated densified wood in its wet state do not meet the requirements for use as osteosynthesis implants.

The relative metabolic activity of the NIH 3T3 mouse fibroblasts suggests that the samples exhibit no significant cytotoxic effects at any of the tested dilutions.

The results of the study showed that it is possible to obtain nontoxic wood samples with mechanical properties equivalent to bones by partial delignification, extraction, impregnation and densification in dry state. However, it can be concluded that additional research is needed to enhance the stability of densified wood, the integrity of chemical bonds, and its mechanical strength in wet conditions, particularly in relation to human tissue. This research is essential for evaluating the material's potential for application in osteosynthesis bone implants.

Author Contributions: Conceptualization L.A. and V.N.; methodology, M.A., J.Z., and J.L.; formal analysis, M.A., M.S., J.Z., and A.D.; investigation, L.A., V.N., M.A., M.K., and J.V.; resources, M.A. and L.A.; data curation, L.A., A.D., M.A., and M.S.; writing—original draft preparation, L.A.; writing—review and editing, V.N., M.A., M.K., A.D., J.L., and J.V.; visualization, L.A.; supervision, L.A.; project administration, L.A.; funding acquisition, L.A., V.N., A.D., and J.L. All authors have read and agreed to the published version of the manuscript.

Funding: Funding for this research is provided by the Latvian State Institute of Wood Chemistry through the Bio-economic grants "WoodBone" 2023, grant number 03-23 and "HardBone" 2024, grant number 05-24. The study is implemented with the support of Federation of European Societies for Surgery of the Hand (FESSH Research grant 2024). The authors acknowledge the access to the infrastructure and expertise of the BBCE—Baltic Biomaterials Centre of Excellence (European Union's Horizon 2020 research and innovation program under the grant agreement No. 857287).

Institutional Review Board Statement: Not applicable.

Data Availability Statement: The original contributions presented in the study are included in the article, further inquiries can be directed to the corresponding author.

Acknowledgments: We extend our sincere gratitude to Maris Puke for HPLC analysis and Velta Fridrihsone for SEM analyses.

Conflicts of Interest: The authors declare no conflicts of interest.

References

1. Wu, A.M.; Bisignano, C.; James, S.L.; Abady, G.G.; Abedi, A.; Abu-Gharbieh, E.; Alhassan, R.K.; Alipour, V.; Arabloo, J.; Asaad, M.; et al. Global, Regional, and National Burden of Bone Fractures in 204 Countries and Territories, 1990–2019: A Systematic Analysis from the Global Burden of Disease Study 2019. *Lancet Health Longev.* **2021**, *2*, e580–e592. [CrossRef] [PubMed]
2. Wang, X.; Wang, B.; Hou, X.; Cheng, X.; Zhang, T. Unilateral External Fixator Combined with Lateral Auxiliary Frame for Ultimate Treatment of Tibia and Fibula Shaft Fractures with Poor Soft Tissue Conditions. *Biomed. Res. Int.* **2022**, *2022*, 9990744. [CrossRef] [PubMed]
3. Sansone, V.; Pagani, D.; Melato, M. The Effects on Bone Cells of Metal Ions Released from Orthopaedic Implants. A Review. *Clin. Cases Miner. Bone Metab.* **2013**, *10*, 34–40. [CrossRef]
4. Cadosch, D.; Al-Mushaiqri, M.S.; Gautschi, O.P.; Meagher, J.; Simmen, H.P.; Filgueira, L. Biocorrosion and Uptake of Titanium by Human Osteoclasts. *J. Biomed. Mater. Res.-Part A* **2010**, *95*, 1004–1010. [CrossRef]
5. Cadosch, D.; Chan, E.; Gautschi, O.P.; Simmen, H.P.; Filgueira, L. Bio-Corrosion of Stainless Steel by Osteoclasts-in Vitro Evidence. *J. Orthop. Res.* **2009**, *27*, 841–846. [CrossRef] [PubMed]
6. Hargreaves, B.A.; Worters, P.W.; Pauly, K.B.; Pauly, J.M.; Koch, K.M.; Gold, G.E. Metal-Induced Artifacts in MRI. *Am. J. Roentgenol.* **2011**, *197*, 547–555. [CrossRef]

7. Zhang, X.; Wang, J.; Xing, L. Metal Artifact Reduction in X-Ray Computed Tomography (CT) by Constrained Optimization. *Med Phys.* **2011**, *38*, 701–711. [CrossRef]
8. Sandhu, H.S.; Boden, S.D. Biologic Enhancement of Spinal Fusion. *Orthop. Clin. N. Am.* **1998**, *29*, 621–631. [CrossRef]
9. Lu, J.; Yu, H.; Chen, C. Biological Properties of Calcium Phosphate Biomaterials for Bone Repair: A Review. *RSC Adv.* **2018**, *8*, 2015–2033. [CrossRef] [PubMed]
10. Wang, S.; Liu, J.; Caroprese, M.; Gianfreda, F.; Melloni, F.; de Santis, D. Exploring the Potential of Calcium-Based Biomaterials for Bone Regeneration in Dentistry: A Systematic Review. *Minerva Dent. Oral Sci.* **2023**, *73*, 169–180. [CrossRef]
11. Dinesh, K.R.; Hatti, G. Polymers Used as Implant Biomaterials: A Review. 2018. Available online: https://www.google.com hk/url?sa=t&source=web&rct=j&opi=89978449&url=https://www.researchgate.net/profile/Gururaj-Hatti/publication/330753275_Polymers_used_as_implant_Biomaterials_A_review/links/5c52b272299bf12be3eff424/Polymers-used-as-implant-Biomaterials-A-review.pdf&ved=2ahUKEwj4stSPsdqIAxU2oK8BHWdWOC0QFnoECBYQAQ&usg=AOvVaw3eDFzudZPHd-Nmm849_qN0 (accessed on 18 September 2024).
12. Al-Shalawi, F.D.; Mohamed Ariff, A.H.; Jung, D.W.; Mohd Ariffin, M.K.A.; Seng Kim, C.L.; Brabazon, D.; Al-Osaimi, M.O. Biomaterials as Implants in the Orthopedic Field for Regenerative Medicine: Metal versus Synthetic Polymers. *Polymers* **2023**, *15*, 2601. [CrossRef] [PubMed]
13. Ramos-Rodriguez, D.H.; Pashneh-Tala, S.; Bains, A.K.; Moorehead, R.D.; Kassos, N.; Kelly, A.L.; Paterson, T.E.; Orozco-Diaz, C.A.; Gill, A.A.; Asencio, I.O. Demonstrating the Potential of Using Bio-Based Sustainable Polyester Blends for Bone Tissue Engineering Applications. *Bioengineering* **2022**, *9*, 163. [CrossRef] [PubMed]
14. De Bortoli, L.S.; Schabbach, L.M.; Fredel, M.C.; Hotza, D.; Henriques, B. Ecological Footprint of Biomaterials for Implant Dentistry: Is the Metal-Free Practice an Eco-Friendly Shift? *J. Clean. Prod.* **2019**, *213*, 723–732. [CrossRef]
15. Trimeche, M. Biomaterials for Bone Regeneration: An Overview. *Biomater. Tissue Technol. Rev.* **2017**, *1*, 1–5.
16. Oladapo, B.I.; Zahedi, S.A.; Ismail, S.O.; Olawade, D.B. Recent Advances in Biopolymeric Composite Materials: Future Sustainability of Bone-Implant. *Renew. Sustain. Energy Rev.* **2021**, *150*, 111505. [CrossRef]
17. Yadav, D.; Garg, R.K.; Ahlawat, A.; Chhabra, D. 3D Printable Biomaterials for Orthopedic Implants: Solution for Sustainable and Circular Economy. *Resour. Policy* **2020**, *68*, 101767. [CrossRef]
18. Nefjodovs, V.; Andze, L.; Andzs, M.; Filipova, I.; Tupciauskas, R.; Vecbiskena, L.; Kapickis, M. Wood as Possible Renewable Material for Bone Implants—Literature Review. *J. Funct. Biomater.* **2023**, *14*, 266. [CrossRef]
19. Kristen, H.; Bosch, P.; Bednar, H.; Plenk, H. Biocompatibility of Wood in Bone Tissue. *Arch. Orthop. Trauma Surg.* **1977**, *89*, 1–14.
20. Kristen, H.; Bösch, P.; Bednar, H.; Plenk, H. The Effects of Dynamic Loading on Intracalcaneal Wood Implants and on the Tissues Surrounding Them. *Arch. Orthop. Trauma. Surg.* **1979**, *93*, 287–292. [CrossRef]
21. Bösch, P.; Kristen, H.; Braun, F.; Kovac, W. Reaction of Connective Tissue and Striated Muscle Tissue to Implanted Ashwood. *Wien Med. Wochenschr.* **1979**, *15*, 419–423.
22. Aho, A.J.; Rekola, J.; Matinlinna, J.; Gunn, J.; Tirri, T.; Viitaniemi, P.; Vallittu, P. Natural Composite of Wood as Replacement Material for Ostechondral Bone Defects. *J. Biomed. Mater. Res.-Part B Appl. Biomater.* **2007**, *83*, 64–71. [CrossRef] [PubMed]
23. Horský, I.; Huraj, E.; Paukovic, J. Utilization of Wood in the Manufacture of Orthopedic Implants. *Acta Chir. Orthop. Traumatol. Cech.* **1987**, *1*, 13.
24. Gross, K.A.; Ezerietis, E. Juniper Wood as a Possible Implant Material. *J. Biomed. Mater. Res.-Part A* **2003**, *64*, 672–683. [CrossRef]
25. Rekola, J.; Lassila, L.V.; Hirvonen, J.; Lahdenperä, M.; Grenman, R.; Aho, A.J.; Vallittu, P.K. Effects of Heat Treatment of Wood on Hydroxylapatite Type Mineral Precipitation and Biomechanical Properties in Vitro. *J. Mater. Sci. Mater. Med.* **2010**, *21*, 2345–2354. [CrossRef]
26. Rekola, J.; Lassila, L.V.J.; Nganga, S.; Ylä-Soininmäki, A.; Fleming, G.J.P.; Grenman, R.; Aho, A.J.; Vallittu, P.K. Effect of Heat Treatment of Wood on the Morphology, Surface Roughness and Penetration of Simulated and Human Blood. *Biomed. Mater. Eng.* **2014**, *24*, 1595–1607. [CrossRef]
27. Andze, L.; Andzs, M.; Skute, M.; Nefjodovs, V.; Kapickis, M.; Tupciauskas, R. Preliminary Study of Chemically Pretreated Densification of Juniper Wood for Use in Bone Implants. *Mater. Sci. Forum* **2022**, *1071*, 101–108. [CrossRef]
28. Kutnar, A.; Šernek, M. Densification of Wood. *Zb. Gozdarstva Lesar.* **2007**, *82*, 53–62.
29. Cabral, J.P.; Kafle, B.; Subhani, M.; Reiner, J.; Ashraf, M. Densification of Timber: A Review on the Process, Material Properties, and Application. *J. Wood Sci.* **2022**, *68*, 20. [CrossRef]
30. Song, J.; Chen, C.; Zhu, S.; Zhu, M.; Dai, J.; Ray, U.; Li, Y.; Kuang, Y.; Li, Y.; Quispe, N.; et al. Processing Bulk Natural Wood into a High-Performance Structural Material. *Nature* **2018**, *554*, 224–228. [CrossRef]
31. Shi, J.; Peng, J.; Huang, Q.; Cai, L.; Shi, S.Q. Fabrication of Densified Wood via Synergy of Chemical Pretreatment, Hot-Pressing and Post Mechanical Fixation. *J. Wood Sci.* **2020**, *66*, 5. [CrossRef]
32. Mania, P.; Wróblewski, M.; Wójciak, A.; Roszyk, E.; Moliński, W. Hardness of Densified Wood in Relation to Changed Chemical Composition. *Forests* **2020**, *11*, 506. [CrossRef]
33. Raman, V.; Liew, K.C. Density of Densified Paraserianthes Falcataria Wood Pre-Treated with Alkali. *Proc. IOP Conf. Ser. Earth Environ. Sci.* **2020**, *549*, 012030. [CrossRef]
34. Sluiter, A.; Hames, B.; Ruiz, R.; Scarlata, C.; Sluiter, J.; Templeton, D.; Crocker, D. *Determination of Structural Carbohydrates and Lignin in Biomass—NREL/TP-510-42618*; National Renewable Energy Laboratory: Golden, CO, USA, 2008.

35. Sable, I.; Grinfelds, U.; Vikele, L.; Rozenberga, L.; Lazdina, D.; Zeps, M.; Jansons, A. Chemical Composition and Fi Ber Properties of Fast-Growing Species in Latvia and Its Potential for Forest Bioindustry. *For. Stud.* **2017**, *66*, 27–32. [CrossRef]
36. Sable, I.; Grinfelds, U.; Vikele, L.; Rozenberga, L.; Zeps, M.; Neimane, U.; Jansons, A. Effect of Refining on the Properties of Fibres from Young Scots (Pinus Sylvestris) and Lodgepole Pines (Pinus Contorta). *Balt. For.* **2017**, *23*, 529–533.
37. Shi, J.; Lu, Y.; Zhang, Y.; Cai, L.; Shi, S.Q. Effect of Thermal Treatment with Water, H_2SO_4 and NaOH Aqueous Solution on Color, Cell Wall and Chemical Structure of Poplar Wood. *Sci. Rep.* **2018**, *8*, 17735. [CrossRef] [PubMed]
38. Wang, J.; Fishwild, S.J.; Begel, M.; Zhu, J.Y. Properties of Densified Poplar Wood through Partial Delignification with Alkali and Acid Pretreatment. *J. Mater. Sci.* **2020**, *55*, 14664–14676. [CrossRef]
39. Ran, Y.; Lu, D.; Jiang, J.; Huang, Y.; Wang, W.; Jinzhen, C. Deep Eutectic Solvents-Assisted Wood Densification: A Promising Strategy for Shape-Fixation. *Chem. Eng. J.* **2023**, *471*, 144476. [CrossRef]
40. Maturana, J.C.; Guindos, P.; Lagos, J.; Arroyave, C.; Echeverría, F.; Correa, E. Two-Step Hot Isostatic Pressing Densification Achieved Non-Porous Fully-Densified Wood with Enhanced Physical and Mechanical Properties. *Sci. Rep.* **2023**, *13*, 14324. [CrossRef]
41. Bao, M.; Huang, X.; Jiang, M.; Yu, W.; Yu, Y. Effect of Thermo-Hydro-Mechanical Densification on Microstructure and Properties of Poplar Wood (*Populus tomentosa*). *J. Wood Sci.* **2017**, *63*, 591–605. [CrossRef]
42. Vila, C.; Romero, J.; Francisco, J.L.; Santos, V.; Parají, J.C. On the Recovery of Hemicellulose before Kraft Pulping. *BioResources* **2012**, *7*, 4179–4189. [CrossRef]
43. Lipeh, S.; Schimleck, L.R.; Mankowski, M.E.; McDonald, A.G.; Morrell, J.J. Relationship between Attenuated Total Reflectance Fourier Transform Infrared Spectroscopy of Western Juniper and Natural Resistance to Fungal and Termite Attack. *Holzforschung* **2020**, *74*, 246–259. [CrossRef]
44. Emmanuel, V.; Odile, B.; Céline, R. FTIR Spectroscopy of Woods: A New Approach to Study the Weathering of the Carving Face of a Sculpture. *Spectrochim. Acta-Part A Mol. Biomol. Spectrosc.* **2015**, *136*, 1255–1259. [CrossRef] [PubMed]
45. Lorenzo, I.; Serra-Prat, M.; Carlos Yébenes, J. The Role of Water Homeostasis in Muscle Function and Frailty: A Review. *Nutrients* **2019**, *11*, 1857. [CrossRef] [PubMed]
46. Mantanis, G.I.; Young, R.A.; Rowell, R.M. Swelling of Wood Part III. Effect of Temperature and Extractives on Rate and Maximum Swelling. *Holzforschung* **1995**, *49*, 239–248. [CrossRef]
47. Welzbacher, C.R.; Wehsener, J.; Rapp, A.O.; Haller, P. Thermo-Mechanische Verdichtung Und Thermische Modifikation von Fichtenholz (Picea Abies Karst) Im Industriellen Maßstab—Betrachtung Der Dimensionsstabilität Und Dauerhaftigkeit. *Holz Als Roh-Und Werkst.* **2008**, *66*, 39–49. [CrossRef]
48. Laine, K.; Belt, T.; Rautkari, L.; Ramsay, J.; Hill, C.A.S.; Hughes, M. Measuring the Thickness Swelling and Set-Recovery of Densified and Thermally Modified Scots Pine Solid Wood. *J. Mater. Sci.* **2013**, *48*, 8530–8538. [CrossRef]
49. Xu, B.H.; Wang, B.L.; Yu, K.B.; Bouchaïr, A. An Optional Connection Material in Timber Structures: Densified Poplar. *J. Mater. Sci.* **2021**, *56*, 14114–14125. [CrossRef]
50. Grönquist, P.; Schnider, T.; Thoma, A.; Gramazio, F.; Kohler, M.; Burgert, I.; Rüggeberg, M. Investigations on Densified Beech Wood for Application as a Swelling Dowel in Timber Joints. *Holzforschung* **2019**, *73*, 559–568. [CrossRef]
51. Fang, C.H.; Mariotti, N.; Cloutier, A.; Koubaa, A.; Blanchet, P. Densification of Wood Veneers by Compression Combined with Heat and Steam. *Eur. J. Wood Wood Prod.* **2012**, *70*, 155–163. [CrossRef]
52. Singh, D.; Rana, A.; Jhajhria, S.K.; Garg, B.; Pandey, P.M.; Kalyanasundaram, D. Experimental Assessment of Biomechanical Properties in Human Male Elbow Bone Subjected to Bending and Compression Loads. *J. Appl. Biomater. Funct. Mater.* **2019**, *17*, 2280800018793816. [CrossRef]
53. Meema, H.E.; Meema, S. Compact Bone Mineral Density of the Normal Human Radius. *Acta Oncol.* **1978**, *17*, 342–352. [CrossRef] [PubMed]
54. Björklund, J.; Fonti, M.V.; Fonti, P.; Van den Bulcke, J.; von Arx, G. Cell Wall Dimensions Reign Supreme: Cell Wall Composition Is Irrelevant for the Temperature Signal of Latewood Density/Blue Intensity in Scots Pine. *Dendrochronologia* **2021**, *65*, 125785. [CrossRef]
55. Bockel, S.; Harling, S.; Grönquist, P.; Niemz, P.; Pichelin, F.; Weiland, G.; Konnerth, J. Characterization of Wood-Adhesive Bonds in Wet Conditions by Means of Nanoindentation and Tensile Shear Strength. *Eur. J. Wood Wood Prod.* **2020**, *78*, 449–459. [CrossRef]
56. Bader, M.; Nemeth, R. Moisture-dependent Mechanical Properties of Longitudinally Compressed Wood. *Eur. J. Wood Wood Prod.* **2019**, *77*, 1009–1019. [CrossRef]
57. Kuai, B.; Wang, Z.; Gao, J.; Tong, J.; Zhan, T.; Zhang, Y.; Lu, J.; Cai, L. Development of Densified Wood with High Strength and Excellent Dimensional Stability by Impregnating Delignified Poplar by Sodium Silicate. *Constr. Build. Mater.* **2022**, *344*, 128282. [CrossRef]
58. Bogolitsyn, K.G.; Gusakova, M.A.; Khviyuzov, S.S.; Zubov, I.N. Physicochemical Properties of Conifer Lignins Using Juniperus Communis as an Example. *Chem. Nat. Compd.* **2014**, *50*, 337–341. [CrossRef]
59. Hänninen, T.; Tukiainen, P.; Svedström, K.; Serimaa, R.; Saranpää, P.; Kontturi, E.; Hughes, M.; Vuorinen, T. Ultrastructural Evaluation of Compression Wood-like Properties of Common Juniper (*Juniperus communis* L.). *Holzforschung* **2012**, *66*, 389–395. [CrossRef]

60. *ISO 10993-5*; Biological Evaluation of Medical Devices. Part 5: Tests for In Vitro Cytotoxicity. ISO: Geneva, Switzerland, 2009.
61. Li, S.H.; Liu, Q.; De Wijn, J.R.; Zhou, B.L.; De Groot, K. In Vitro Calcium Phosphate Formation on a Natural Composite Material Bamboo. *Biomaterials* **1997**, *18*, 389–395. [CrossRef]

MDPI AG
Grosspeteranlage 5
4052 Basel
Switzerland
Tel.: +41 61 683 77 34

Journal of Functional Biomaterials Editorial Office
E-mail: jfb@mdpi.com
www.mdpi.com/journal/jfb

www.ingramcontent.com/pod-product-compliance
Lightning Source LLC
Chambersburg PA
CBHW061508210926
51850CB00008B/1856